Glencoe McGraw-Hill

# Math Triumphs

## Foundations for Algebra 2

McGraw-Hill Glencoe

## Photo Credits

**iv v** Getty Images; **vi** CORBIS; **vii** Comstock Images; **viii** Steve Hamblin/CORBIS; **1** PunchStock; **2–3** Stan Liu/Getty Images; **7** Getty Images; **14** PunchStock; **22** Getty Images; **28** PunchStock; **30** CORBIS; **34** Alamy; **42** Brand X Pictures; **45** Jill Braaten/The McGraw-Hill Companies; **46–47** CORBIS; **51** Getty Images; **56** C. Sherburne/Getty Images; **62** Todd Gipstein/Getty Images; **66** John Kelly/ Getty Images; **80–81** E. Audras/Getty Images; **91** Ulrich Niehoff/Alamy; **116** Norman Pogson/Alamy; **126–127** Ed Darack/Getty Images; **138** Getty Images; **162** Lars A. Niki/The McGraw-Hill Companies; **175** Ryan McVay/Getty Images; **176–177** Jupiterimages; **181** SuperStock; **182** Mike Rinnan/Alamy; **185** Gerald Wofford/The McGraw-Hill Companies; **186** PunchStock; **187** Steve Cole/Getty Images; **190** Alamy; **191** PunchStock; **192** Lawrence M. Sawyer/Getty Images; **197 198 203** PunchStock; **207** Getty Images; **208–209** Jochen Sand/ Getty Images; **213** Getty Images; **218** PunchStock; **229** CORBIS; **230** William Leaman/Alamy; **236** CORBIS; **239** PunchStock.

*The McGraw·Hill Companies*

**Glencoe**



Send all inquiries to:
Glencoe/McGraw-Hill
8787 Orion Place
Columbus, OH 43240-4027

ISBN: 978-0-07-891634-2
MHID: 0-07-891634-8

*Math Triumphs: Foundations for Algebra 2*
*Student Edition*

Printed in the United States of America.

8 9 10 HSO 17 16 15 14 13 12

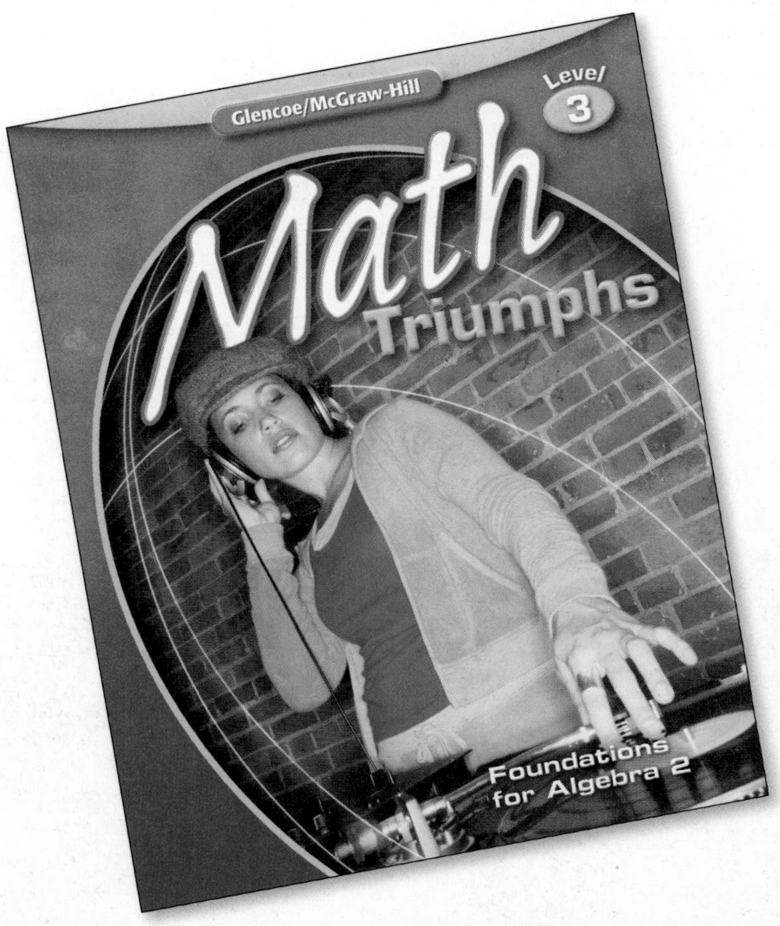

Glencoe/McGraw-Hill
Level
3
Math
Triumphs
Foundations for Algebra 2

# Contents

## Chapter 1 Expressions, Equations, and Functions



*Death Valley National Park, California*

# Equations and Inequalities



*Autumn in New England*

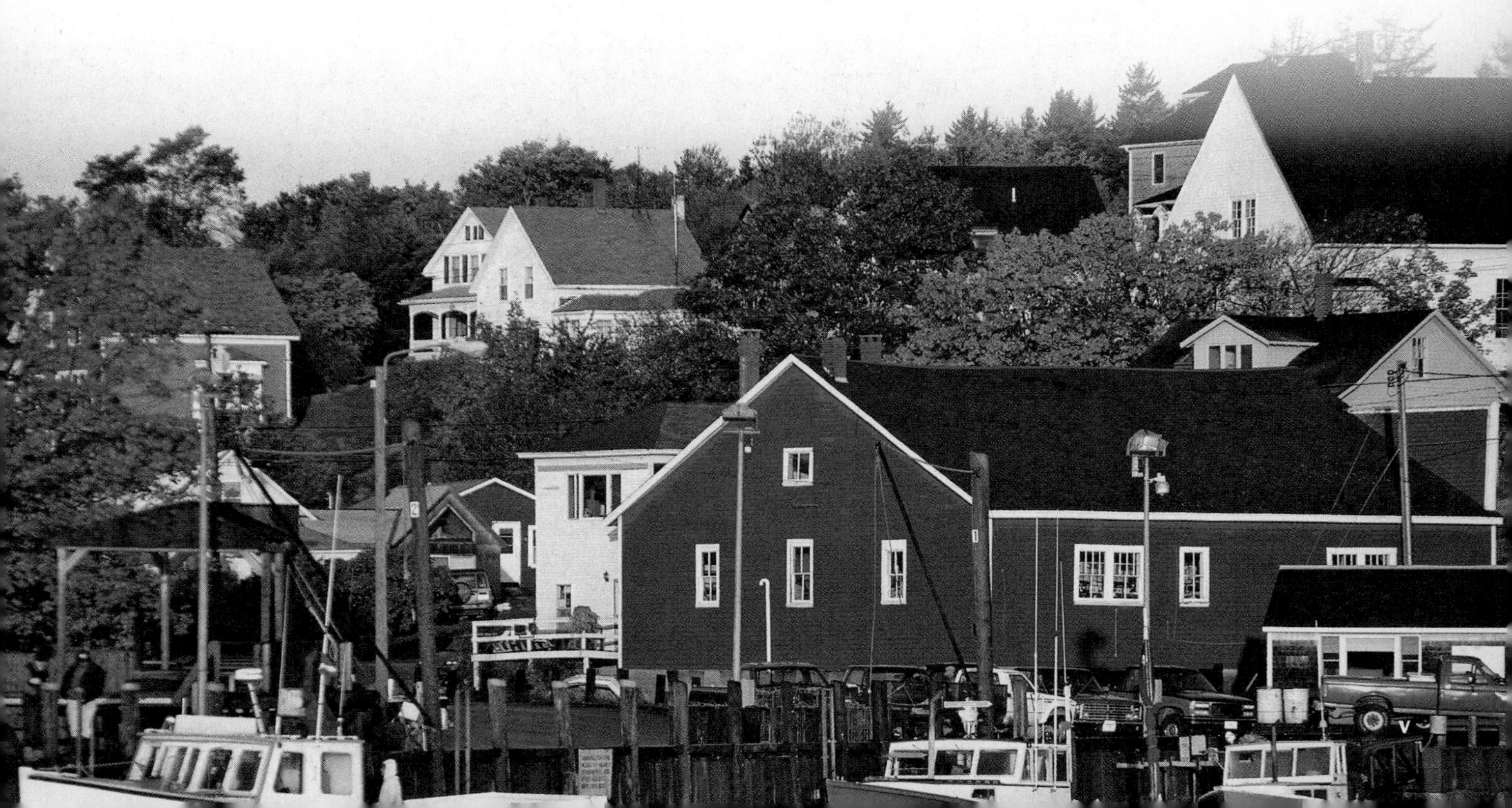

# Contents

## Chapter 3 Graphs

*Memphis, Tennessee*

# Chapter 4 Polynomials

*Ellis Island, New York*

# Contents

## Chapter 5 Probability



*Carlsbad Caverns National Park, New Mexico*

# Chapter 6 Right Triangles and Trigonometry



*Grand Tetons, Wyoming*

Chapter 1

# Expressions, Equations, and Functions

## *How much money was spent?*

A skateboarder visited a skate park for a day. Admission is \$5 and drinks cost \$2 each. You can use the function $y = 2x + 5$ to model how much a skateboarder will spend for a day at the skatepark.

## STEP 1 Chapter Pretest

Are you ready for Chapter 1? Take the Chapter 1 Pretest to find out.

## STEP 2 Preview

Get ready for Chapter 1. Review these skills and compare them with what you will learn in this chapter.

| What You Know | What You Will Learn |
|---|---|
| You know how to evaluate expressions.<br><br>**Example:** $4(3 + 5) = 4(8)$<br>$= 32$<br><br>**TRY IT!**<br><br>1 $6(2 + 8) =$ _______<br><br>2 $9(5 + 7) =$ _______<br><br>3 $7(3 + 11) =$ _______ | *Lesson 1-1*<br><br>The **Distributive Property** states that to multiply a sum or difference by a number, multiply each term inside the parentheses by the number outside the parentheses.<br><br>**Example:** $4(3 + 5) = 4(3) + 4(5)$<br>$= 12 + 20$<br>$= 32$ |
| You know how to multiply a number by itself.<br><br>**Example:** $3 \cdot 3 = 9$<br><br>**TRY IT!**<br><br>4 $4 \cdot 4 =$ _______<br><br>5 $6 \cdot 6 =$ _______<br><br>6 $14 \cdot 14 =$ _______ | *Lesson 1-2*<br><br>A **power** is an expression that has a base and an exponent. The **exponent** indicates how many times to use the **base** as a factor.<br><br>$2^3$ means use 2 as a factor 3 times.<br><br>$2^3 = 2 \cdot 2 \cdot 2$<br><br>$= 8$ |
| You know what a vertical line is.<br><br>**Example:**<br><br>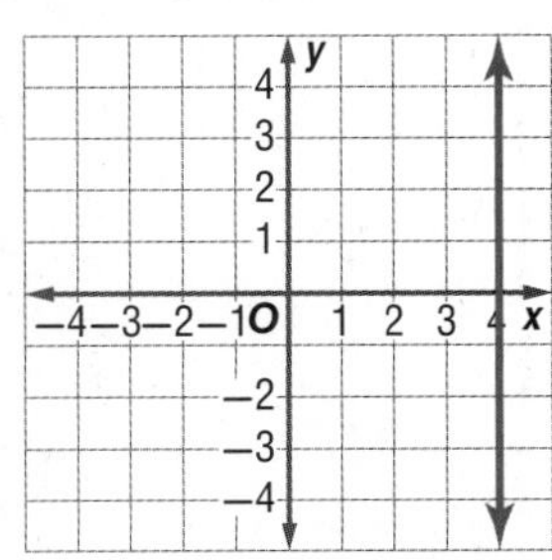<br> | *Lesson 1-6*<br><br>You can use the **vertical line test** to determine if a relation is a function. If a vertical line can be placed anywhere on a graph and it passes through only one point, then the graph represents a function. |

Lesson 1-1

# Number Properties

## KEY Concept

### Commutative Properties

**Addition**

$$12 + 13 = 13 + 12$$
$$25 = 25$$

The order of the addends changes, but the sum does not change.
$a + b = b + a$

**Multiplication**

$$6 \cdot 12 = 12 \cdot 6$$
$$72 = 72$$

The order of the factors changes, but the product does not change.
$c \cdot d = d \cdot c$

### Associative Properties

**Addition**

$$(3 + 6) + 7 = 3 + (6 + 7)$$
$$9 + 7 = 3 + 13$$
$$16 = 16$$

The order of the addends does not change, but the grouping changes.
$(a + b) + c = a + (b + c)$

**Multiplication**

$$(5 \cdot 9) \cdot 3 = 5 \cdot (9 \cdot 3)$$
$$45 \cdot 3 = 5 \cdot 27$$
$$135 = 135$$

The order of the factors does not change, but the grouping changes.
$(c \cdot d) \cdot f = c \cdot (d \cdot f)$

### Identity Properties

**Addition**

$$3 + 0 = 3$$

Any number plus 0 equals that number.
$a + 0 = a$

**Multiplication**

$$4 \cdot 1 = 4$$

Any number times 1 equals that number.
$b \cdot 1 = b$

### Distributive Property

$$2(6 + 8) = (2 \cdot 6) + (2 \cdot 8)$$
$$2(14) = 12 + 16$$
$$28 = 28$$

Multiply the number outside the parentheses by each number inside the parentheses.
$a(b + c) = (a \cdot b) + (a \cdot c)$

## VOCABULARY

**Associative Property of Addition**
the grouping of the addends does not change the sum

**Associative Property of Multiplication**
the grouping of the factors does not change the product

**Commutative Property of Addition**
the order in which two numbers are added does not change the sum

**Commutative Property of Multiplication**
the order in which two numbers are multiplied does not change the product

**Distributive Property**
to multiply a sum by a number, you can multiply each addend by the number and add the products

Check your solution to an equation by substituting your answer into the original equation for the variable.

## Example 1

**Name the property illustrated.**

$(3 \cdot 2) \cdot 5 = 3 \cdot (2 \cdot 5)$

1. Verify that the expressions on each side of the equation are equal.

$(3 \cdot 2) \cdot 5 = 3 \cdot (2 \cdot 5)$

$6 \cdot 5 = 3 \cdot 10$

$30 = 30$

Notice the three factors are listed in the same order on each side of the equation.

2. The factors are grouped differently, but the products are the same.

3. The Associative Property of Multiplication is illustrated.

### YOUR TURN!

**Name the property illustrated.**

$12 + 17 = 17 + 12$

1. Verify that the expressions on each side of the equation are equal.

$12 + 17 = 17 + 12$

____ = ____

2. The __________ are in a different order, but the sums are __________.

3. The ____________________________ __________________ is illustrated.

## Example 2

**Use the Distributive Property to simplify 3(11 + 4).**

1. Multiply the 3 times each number inside the parentheses.

$3(11 + 4) = (3 \cdot 11) + (3 \cdot 4)$

2. Verify that the expressions on each side of the equation are equal.

$3(11 + 4) \stackrel{?}{=} (3 \cdot 11) + (3 \cdot 4)$

$3 \cdot 15 \stackrel{?}{=} 33 + 12$

$45 = 45$

### YOUR TURN!

**Use the Distributive Property to simplify 5(12 + 3).**

1. Multiply the ____ times each number inside the parentheses.

5(12 + 3) = (____ • ____) + (____ • ____)

2. Verify that the expressions on each side of the equation are equal.

5(12 + 3) ≟ (____ • ____) + (____ • ____)

____ • ____ ≟ ____ + ____

____ = ____

GO ON

## Guided Practice

**Fill in the blanks to make a true sentence. Name the property illustrated.**

1. $13 \cdot 12 = 12 \cdot 13$

   $156 =$ ______

   ______________________

2. $19 = 19 \cdot 1$

   $19 =$ ______

   ______________________

### Step by Step Practice

3. Name the property illustrated. $14 + (16 + 11) = (16 + 11) + 14$

**Step 1** Verify that the expression on each side of the equation is equal.

$14 + (16 + 11) = (16 + 11) + 14$

______ + ______ = ______ + ______

______ = ______

Notice the order of the addends is different on each side of the equation.

**Step 2** The ______________ are grouped in the same way, but the order has changed. The ______________________ is illustrated.

**Name the property illustrated.**

4. $14 + 8 = 8 + 14$

   Has the order of the numbers changed? ______________

   The ______________________ is illustrated.

5. $(8 + 1) + 4 = 8 + (1 + 4)$

   Has the grouping of the numbers changed? ______________

   The ______________________ is illustrated.

**Use the Distributive Property to simplify.**

6. $2(12 + 4)$

   ______ + ______ = ______

7. $6(1 + 10)$

   ______ + ______ = ______

## Step by Step Problem-Solving Practice

**Solve.**

**8** Brittany sold 4 bags of marbles with 12 marbles in each bag. Dawn sold 12 bags of marbles with 4 marbles in each bag. Compare the number of marbles Brittany sold to the number of marbles Dawn sold.

Brittany: ____ • ____ = ____ Dawn: ____ • ____ = ____
bags marbles bags marbles

__________

Check off each step.

_____ **Understand: I underlined key words.**

_____ **Plan: To solve the problem, I will** ______________________________

______________________________.

_____ **Solve: The answer is** ______________________________.

_____ **Check: I checked my answer by** ______________________________

______________________________.

## Skills, Concepts, and Problem Solving

**Fill in the blanks to illustrate each property.**

**9** Multiplication Identity Property

16 • ____ = ____

**10** Associative Property of Multiplication

(____ • 4) • 5 = 12 • (____ • ____)

**11** Commutative Property of Addition

____ + 8 = ____ + 3

**12** Distributive Property

____(8 + 1) = (____ • ____) + (____ • ____)

**Use the Distributive Property to simplify.**

**13** 4(25 + 12)

____ + ____ = ____

**14** 6(60 + 2)

____ + ____ = ____

**15** $x(5 + 3)$

____ + ____ = ____

GO ON

**Name the property illustrated.**

16 the total number of segments in 2 squares and 2 triangles

$2(4 + 3) = 2 \cdot 4 + 2 \cdot 3$ ______________________________

17 $(8 + 2) + 7 = 8 + (2 + 7)$ ______________________________

18 $12 + 9 = 9 + 12$ ______________________________

19 $0 + 9 = 9$ ______________________________

20 $6(10 \cdot 2) = (6 \cdot 10)2$ ______________________________

21 $7 \cdot 3 = 3 \cdot 7$ ______________________________

22 $(17 + 3) + 5 = 17 + (3 + 5)$ ______________________________

**Solve.**

23 MONEY The wages for several summer jobs are shown in the table. On Monday, Sharon mowed 2 lawns and spent 3 hours gardening. On Wednesday, she spent 2 hours babysitting and mowed one lawn. On Friday, she spent 2 hours gardening. How much did Sharon earn for all her work?

| Job | Wage |
|---|---|
| Babysitting | $8/h |
| Gardening | $6/h |
| Mowing Lawn | $20/yard |

______________________________

Vocabulary Check **Write the vocabulary word that completes each sentence.**

24 The property that states the order in which two numbers are multiplied does not change the product, is the

______________________________.

25 The ______________________________ shows that grouping addends differently does not change their sum.

26 **Reflect** Suppose you need to clean your room and clean the kitchen. Does the order in which you perform these tasks matter? Relate your answer to one of the number properties.

______________________________

______________________________

______________________________

Lesson 1-2

# Multiplication: Properties of Exponents

## KEY Concept

In the following properties, $m$ and $n$ are positive integers.

A **power** is an expression that has a base and an exponent. The **exponent** indicates how many times to use the **base** as a factor.

$3^5$ means use 3 as a factor 5 times.

$3^5 = 3 \cdot 3 \cdot 3 \cdot 3 \cdot 3$

$3^5 = 243$

**Zero Property of Exponents**

$a^0 = 1$, where $a \neq 0$

$5^0 = 1$

**Product of Powers Property**

$a^m \cdot a^n = a^{m+n}$

$2^3 \cdot 2^4 = 2^{3+4}$

$= 2^7$

$= 128$

The bases are the same. Add the exponents.

**Power of a Product Property**

$(ab)^n = a^n b^n$

$(6 \cdot 2)^3 = 6^3 \cdot 2^3$

$= 216 \cdot 8$

$= 1{,}728$

Raise each base to the exponent.

**Power of a Power Property**

$(a^m)^n = a^{m \cdot n}$

$(4^2)^4 = 4^{2 \cdot 4}$

$= 4^8$

$= 65{,}536$

Multiply the exponents.

### VOCABULARY

**base**
the number used as a factor in an expression involving exponents

**exponent**
the number of times a base is multiplied by itself

**power**
an expression in the form $a^n$, where $a$ is the base, and $n$ is the exponent



Use these properties to simplify expressions with exponents.

GO ON

## Example 1

**Name the property used to simplify the power $12^0$.**

1. The base is raised to an exponent of 0.
2. The base is not 0.
3. Use the Zero Property of Exponents.

$12^0 = 1$

**YOUR TURN!**

**Name the property used to simplify the power $14^0$.**

1. The base is raised to an exponent of ____.
2. The base is not ____.
3. Use the ________________________.

$\square^0 =$ ____

## Example 2

**Simplify the expression. Name the property used.**

$4^2 \cdot 4^3$

1. The base of each power is the same. Add the exponents. Use the Product of Powers Property.
2. $4^2 \cdot 4^3 = (4 \cdot 4) \cdot (4 \cdot 4 \cdot 4)$

$= 4^{2+3} = 4^5$

$= 1{,}024$

**YOUR TURN!**

**Simplify the expression. Name the property used.**

$(3 \cdot 5)^2$

1. Two different bases are raised to the ________ exponent. Use the ________________________.
2. $(3 \cdot 5)^2 =$ ____ $\cdot$ ____

= (____ $\cdot$ ____) $\cdot$ (____ $\cdot$ ____)

= ____ $\cdot$ ____

= ____

## Example 3

**Simplify the expression. Name the property used.**

$(3^3)^2$

1. A power is raised to an exponent. Use the Power of a Power Property.
2. $(3^3)^2 = 3^{3 \cdot 2}$

$= 3^6$

$= 729$

**YOUR TURN!**

**Simplify the expression. Name the property used.**

$(2^4)^2$

1. A ________ is raised to an ________. Use the ________________________.
2. $(2^4)^2 =$ ______

= ______

= ______

## Guided Practice

**Name the property used to simplify the power.**

1 $(b^3c^5)^2 = b^6c^{10}$

______________________

2 $254^0 = 1$

______________________

### Step by Step Practice

3 Simplify the expression $5t^4 \cdot 2t^2$. Name the property used.

**Step 1** Use the Commutative and Associative Properties to group the factors.

$5t^4 \cdot 2t^2 = (\_\_\_ \cdot \_\_\_) \cdot (\_\_\_ \cdot \_\_\_)$

**Step 2** Multiply the integers. $(\_\_\_ \cdot \_\_\_) = \_\_\_$

**Step 3** Multiply the powers. The base of each power is the

______________. ______________ the exponents. Use the

______________________________________.

$t^4 \cdot t^2 = t^{\square} = t^{\square}$

**Step 4** Simplify.

$5t^4 \cdot 2t^2 = \_\_\_\_$

**Simplify each expression. Name each property used.**

4 $15^0 = \_\_\_\_\_\_$

______________________

5 $2^5 \cdot 2^2 = 2^{\square}$

$= 2^{\square}$

$= \_\_\_\_\_\_$

______________________

6 $(x^4)^3 = x^{\square}$

$= \_\_\_\_\_\_$

______________________

7 $(a^9b^7c^8)^0 = \_\_\_\_\_\_$

______________________

GO ON

**Simplify each expression.**

8 $a^3 \cdot a^6 = a^{\square} = a^{\square}$

9 $12b^0 = \_\_\_\_ \cdot \_\_\_\_ = \_\_\_\_$

10 $(3x^2)(4x^3) = (3 \cdot 4) \cdot x^{\square} = \_\_\_\_$

11 $(-2x^7)(3x^2) = (\_\_\_\_ \cdot 3) \cdot x^{\square} = \_\_\_\_$

12 $(4h^3)^2 = 4^{\square} \cdot h^{\square} = \_\_\_\_$

13 $(15d^5f^7)^0 = \_\_\_\_$

## Step by Step Problem-Solving Practice

**Solve.**

14 The amount of power, in watts, dispersed in an electric outlet can be found using the formula $P = I^2R$, where $I$ is the current in amps, and $R$ is the resistance in ohms. A current of $5d^3$ amps is flowing through an electric outlet with a resistance of $0.5d^2$ ohms. Find an expression that represents the power dispersed by the outlet.

$P = I^2R$ $\quad I = 5d^3$ $\quad R = 0.5d^2$

$P = \_\_\_\_\_\_ \cdot \_\_\_\_\_\_$ Substitute expressions into the formula.

$P = \_\_\_\_\_\_ \cdot \_\_\_\_\_\_$ Use the Power of a Power Property.

$P = \_\_\_\_\_\_ \cdot \_\_\_\_\_\_$ Simplify exponents.

$P = \_\_\_\_\_\_\_\_\_\_\_\_$ Use the Commutative Property.

$P = \_\_\_\_\_\_\_\_\_\_\_\_$ Multiply. Use the Product of Powers Property.

$P = \_\_\_\_\_\_\_\_\_\_\_\_$ Simplify.

Check off each step.

______ **Understand: I underlined key words.**

______ **Plan: To solve the problem, I will** ________________________

________________________________.

______ **Solve: The answer is** ________________________.

______ **Check: I checked my answer by** ________________________

________________________________.

## Skills, Concepts, and Problem Solving

**Simplify each expression.**

15 $(2^3)^2$ ______

16 $(3^2 \cdot 1^5)^2$ ______

17 $x^7 \cdot 2x^8$ ______

18 $8(3b^0)$ ______

19 $(2a^6)(ab)^3$ ______

20 $(2g^6)^4(3g)^3$ ______

21 $(5^3 \cdot 4^5)^0$ ______

22 $7xy^2 \cdot 3x$ ______

**Solve.**

23 **BANKING** Interest compounded annually can be calculated using the formula $A = P(1 + r)^t$, where the principle $P$ is the original amount, $A$ is the amount at the end of the period, $r$ is the interest rate, and $t$ is the number of years.

Lei deposits $1,200 into an account that compounds interest annually at a rate of 5%. At the end of 3 years, Lei withdraws all of the money from the account. What is the amount of Lei's withdrawal?

______

24 **AREA** Find the area of the rectangle in terms of $x$ and $y$.

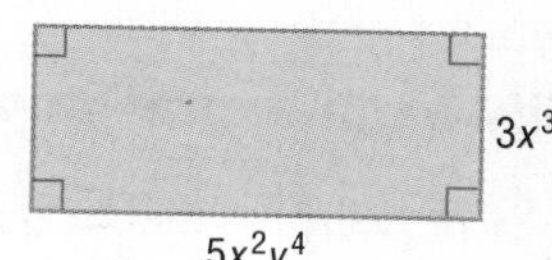


______

25 **SCIENCE** The number of a certain bacteria in a colony after $n$ days is represented by the expression $(3 \cdot 2n)^n$. What is the bacteria population in that colony after 4 days? ______

**Vocabulary Check** **Write the vocabulary word that completes each sentence.**

26 In an expression involving exponents, the number used as the factor is the ______.

27 An expression that has a base raised to an exponent is a(n) ______.

28 **Reflect** Can you list all of the different ways to write $x^{10}$ as a product of powers? Explain.

______

______

______

STOP

Chapter 1

# Progress Check 1 (Lessons 1-1 and 1-2)

**Name the property illustrated.**

1 $6 + 0 = 6$ ______

2 $(4 \cdot 5) \cdot 10 = 4 \cdot (5 \cdot 10)$ ______

3 $0 \cdot 3 = 0$ ______

**Fill in the blanks to illustrate each named property.**

4 Commutative Property of Multiplication

$7 \cdot$ ____ $= 9 \cdot$ ____

5 Associative Property of Addition

(____ + 1) + 12 = 6 + (____ + ____)

6 Distributive Property

8(4 + ____) = ____ + 72

7 Multiplicative Identity

____ $\cdot$ ____ = 2

**Use the Distributive Property to simplify.**

8 $3(11 + 6) =$ ____ + ____ = ____

9 $15(10 + 4) =$ ____ + ____ = ____

10 $a(3 + 2) =$ ____ + ____ = ____

11 $s(1 + 9) =$ ____ + ____ = ____

**Simplify each expression.**

12 $d^5 \cdot d^6 = d^{\square} = d^{\square}$

13 $(5b)^2 =$ ____ $\cdot$ ____ = ____

14 $(h^6g^6)^0 =$ ____

15 $(3p^8z^4)^2 =$ ____

16 $(4k^7)(8k^9) = (4 \cdot 8) \cdot k^{\square} =$ ____

17 $(2^3 \cdot 2^2) =$ ____ = ____

**Solve.**

18 **AREA** The height of the rectangle is represented by $13y$. Find the expression for the area of the rectangle.

______

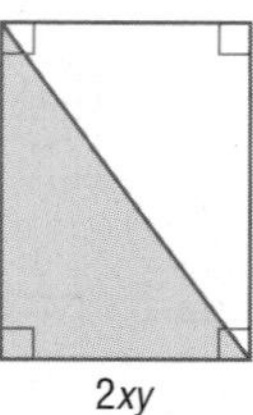


19 **PAINTING** It takes Ken 4 hours to paint a room. It takes Brenda 5 hours to paint a room. Write an equation that shows two ways to calculate how long it takes for each of them to paint two rooms.

______

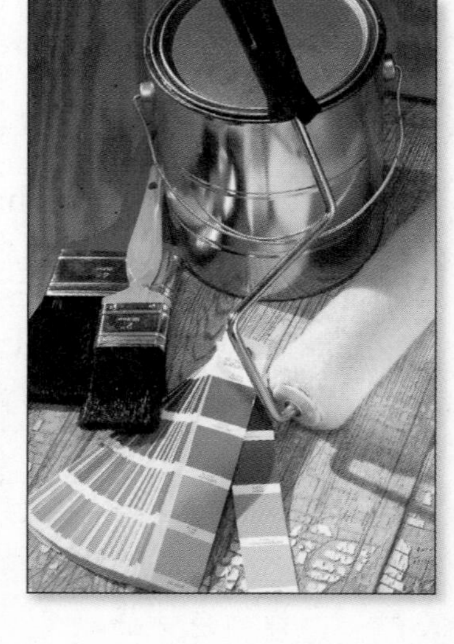

Lesson 1-3

# Division: Properties of Exponents

## KEY Concept

The following properties can be used to simplify expressions with exponents, where the denominator does not equal zero and $m$ and $n$ are integers.

### Property of Negative Exponents

$$a^{-n} = \frac{1}{a^n}$$

$$5^{-2} = \frac{1}{5^2}$$

$$= \frac{1}{5 \cdot 5} = \frac{1}{25}$$

Move the base with the exponent to the denominator. Then change the negative exponent to a positive exponent.

### Quotient of Powers Property

$$\frac{a^m}{a^n} = a^{m-n}$$

$$\frac{3^8}{3^6} = \frac{3 \cdot 3 \cdot 3 \cdot 3 \cdot 3 \cdot 3 \cdot 3 \cdot 3}{3 \cdot 3 \cdot 3 \cdot 3 \cdot 3 \cdot 3}$$

$$= 3^{8-6}$$

$$= 3^2 = 9$$

The bases are the same, so subtract the exponents.

### Power of a Quotient Property

$$\left(\frac{a}{b}\right)^n = \frac{a^n}{b^n}$$

$$\left(\frac{4}{3}\right)^4 = \frac{4 \cdot 4 \cdot 4 \cdot 4}{3 \cdot 3 \cdot 3 \cdot 3}$$

$$= \frac{4^4}{3^4}$$

$$= \frac{256}{81}$$

Raise each base to the exponent. Then divide.

### VOCABULARY

**base**
the number used as a factor in an expression involving exponents

**exponent**
the number of times a base is multiplied by itself

**power**
an expression in the form $a^n$, where $a$ is the base, and $n$ is the exponent

## Example 1

**Simplify the expression $5^{-4}$. Name the property used.**

1. The exponent is negative.
   Use the Property of Negative Exponents to simplify.
2. Move the base of 5 to the denominator with an exponent of 4.

$$5^{-4} = \frac{1}{5^4} = \frac{1}{5 \cdot 5 \cdot 5 \cdot 5} = \frac{1}{625}$$

**YOUR TURN!**

**Simplify the expression $2^{-5}$. Name the property used.**

1. The exponent is ____________.
   Use the ______________________________ to simplify.
2. Move the base of 2 to the denominator with an exponent of ____.

$$2^{-5} = \frac{1}{\square} = \frac{1}{\square \cdot \square \cdot \square \cdot \square \cdot \square} = \_\_\_\_$$

## Example 2

**Simplify the expression $\frac{5^7}{5^4}$. Name the property used.**

1. The base of each power is the same. Use the Quotient of Powers Property to simplify.
2. Subtract the exponents.

$$\frac{5^7}{5^4} = 5^{7-4} = 5^3 = 125$$

**YOUR TURN!**

**Simplify the expression $\left(\frac{3}{2}\right)^3$. Name the property used.**

1. Two different bases are raised to the ____________ exponent.

   Use the ______________________________ to simplify.
2. Raise each base to the exponent, and then ____________.

$$\left(\frac{3}{2}\right)^3 = \frac{\square}{\square} = \frac{\square}{\square}$$

## Guided Practice

**Fill in the blanks to simplify each expression.**

1. $\frac{9^{12}}{9^{10}} = 9^{\square} = 9^{\square} = \_\_\_\_$

2. $2^{-8} = \frac{1}{\square} = \frac{1}{\square}$

## Step by Step Practice

3. Simplify the expression $\left(\frac{x^4}{x^2}\right)^3$. Name the property used.

   **Step 1** Two bases are raised to the ____________ exponent.

   **Step 2** Raise each base to the exponent, and then ____________.

$$\frac{(x^{\square})^{\square}}{(x^{\square})^{\square}} = \frac{x^{\square}}{x^{\square}} = x^{\square} = \_\_\_\_$$

**Simplify each expression. Name the property used.**

4 $\left(\frac{10}{5}\right)^2 = \frac{10^{\square}}{5^{\square}} = \frac{\square}{\square} = \_\_\_\_$

______________________________

______________________________

5 $\frac{c^4}{c^5} = c^{\square} = c^{\square} = \frac{1}{\square}$

______________________________

______________________________

**Simplify each expression.**

6 $\frac{12b^8}{6b^2} = \left(\frac{12}{6}\right)b^{\square} = \_\_\_\_$

7 $\left(\frac{y^2}{2}\right)^2 = \frac{(y^2)^{\square}}{2^{\square}} = \frac{\square}{\square}$

8 $\left(\frac{a^4b^3}{a^2b^6}\right)^2 = \frac{(a^{\square}b^{\square})}{(a^{\square}b^{\square})} = a^{\square}b^{\square} = \frac{a^{\square}}{b^{\square}}$

9 $(7s^4t^7)^{-2} = \frac{1}{\square} \cdot \frac{1}{\square} \cdot \frac{1}{\square} = \frac{1}{\square}$

## Step by Step Problem-Solving Practice

**Solve.**

10 The volume of a rectangular prism is modeled by the expression $48a^{20}b^6c^2$ cubic centimeters. The area of the base of the prism is modeled by the expression $8a^5b^3c^7$ square centimeters. What is the expression for the height of the prism?

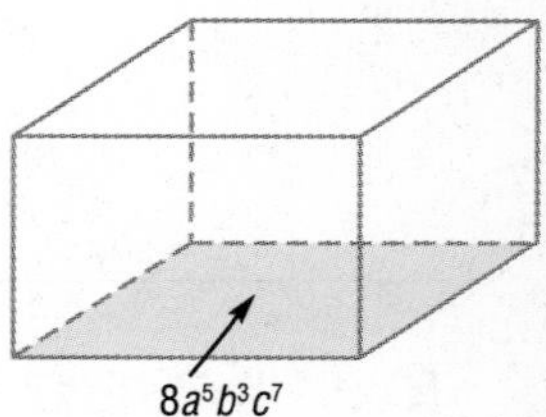


$V \quad = \quad B \quad \cdot \quad h$

________ = ________ $\cdot$ ________ Substitute.

________ $= h$ Divide.

________ $= h$ Group like terms.

________ = ________ $= h$ Simplify.

Check off each step.

______ **Understand: I underlined key words.**

______ **Plan: To solve the problem, I will** ____________________.

______ **Solve: The answer is** ____________________.

______ **Check: I checked my answer by** ____________________.



GO ON

## Skills, Concepts, and Problem Solving

**Simplify each expression.**

11 $2(6)^{-2} =$ ______

12 $\left(\frac{2^4 \cdot 2^{-2}}{3^2}\right) =$ ______

13 $\frac{5^8}{5^7} =$ ______

14 $\frac{14x^9}{2x^5} =$ ______

15 $(3a^3b^5)^{-2} =$ ______

16 $\frac{2x^3y^4}{x^8y^3} =$ ______

17 $\left(\frac{3}{b^4}\right)^2 =$ ______

18 $\left(\frac{x^3y}{x^5y^6}\right)^3 =$ ______

19 $(9a^4)^{-2} =$ ______

**Solve.**

20 **AREA** The area of the triangle below is modeled by the expression $c^5d^7e$. What is the expression that models the height of the triangle? Use the formula for area, $A = \frac{1}{2}bh$.

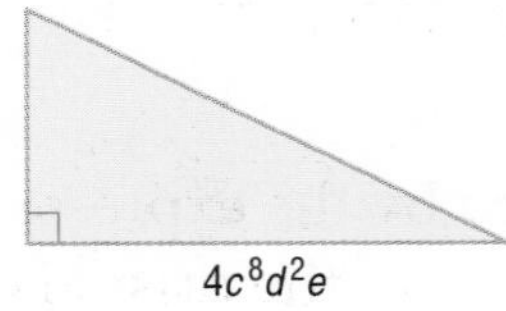


______

21 **MEASUREMENT** The length of an iron beam used by a construction team is modeled by the expression $125x^3y^4$ feet. Using a cutting torch, one of the workers divides the beam into equal pieces measuring $5x^2y^4$ feet. What is the length of each piece of the beam after it is cut?

______

**Vocabulary Check** **Write the vocabulary word that completes each sentence.**

22 In the expression $x^{-4}$, $x$ is the ______.

23 When a base has a negative ______, move the base with the negative ______ to the denominator and change the sign to positive.

24 **Reflect** What number will correctly complete the equation $b^? \cdot b^7 = 1$? Explain.

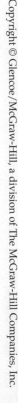

______

______

STOP

# Lesson 1-4 Simplify Square Roots

## KEY Concept

**Square numbers** are named by geometric figures.

The area of the square shown is
$4^2 = 4 \cdot 4 = 16$.

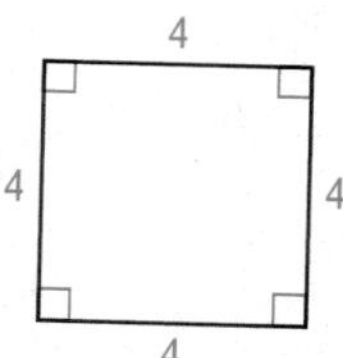


The number 16 is a **perfect square** because 16 is the result of multiplying the factor 4 two times.

The inverse of squaring a number is finding the **square root** of a number. The symbol $\sqrt{\ }$ denotes taking the square root.

Because $4^2 = 16$, $\sqrt{16} = 4$.

$\sqrt{16}$ represents the positive square root of 16.

$-\sqrt{16}$ represents the negative square root of 16.

$\pm\sqrt{16}$ represents both the positive and negative square root of 16.

### VOCABULARY

**perfect square**
a number with a square root that is a rational number

**radical sign**
the sign used to indicate a nonnegative square root, $\sqrt{\ }$

**square number**
the product of a number multiplied by itself

**square root**
one of two equal factors of a number

When a number is not a perfect square, the square root of the number is an irrational number and can only be estimated.

## Example 1

**Simplify the square root.**

$\sqrt{25}$

What number times itself equals 25?

Because $5^2 = 25$, $\sqrt{25} = 5$.

### YOUR TURN!

**Simplify the square root.**

$\sqrt{81}$

Because $\square^2 = 81$, $\sqrt{81} = \square$.

GO ON

## Example 2

**Simplify the square root.**

$-\sqrt{100x^6}$

What expression times itself equals $x^6$?

1. Because $10^2 = 100$, $\sqrt{100} = 10$.
2. Because $(x^3)^2 = x^6$, $\sqrt{x^6} = x^3$.
3. The negative sign before the radical means the negative square root.

   $-\sqrt{100x^6} = -10x^3$

**YOUR TURN!**

**Simplify the square root.**

$\pm\sqrt{49m^{18}}$

1. Because $\square^2 = 49$, $\sqrt{49} = \square$.
2. Because $(\square)^2 = m^{18}$, $\sqrt{m^{18}} = \square$.
3. The plus/minus sign before the radical means the ________ and ________ square root.

   $\pm\sqrt{49m^{18}} =$ ________

## Example 3

**Estimate the square root.**

$\sqrt{40}$

1. 40 is between the perfect square numbers 36 and 49.
2. $\sqrt{40}$ is between $\sqrt{36}$ and $\sqrt{49}$.
3. $\sqrt{40}$ is between 6 and 7.
4. Because 40 is closer to 36 than 49, $\sqrt{40}$ is closer to 6.

**YOUR TURN!**

**Estimate the square root.**

$\sqrt{104}$

1. 104 is between the perfect square numbers ____ and ____.
2. $\sqrt{104}$ is between $\sqrt{\square}$ and $\sqrt{\square}$.
3. $\sqrt{104}$ is between ____ and ____.
4. Because 104 is closer to ____ than 121, $\sqrt{104}$ is closer to ____.

## Guided Practice

**Simplify each square root.**

1. $\sqrt{36}$

   $\square^2 = 36$

   $\sqrt{36} =$ ________

2. $\sqrt{144}$

   $\square^2 = 144$

   $\sqrt{144} =$ ________

3. $-\sqrt{9}$

   $\square^2 = 9$

   $-\sqrt{9} =$ ________

4. $\pm\sqrt{196} =$ ________

5. $-\sqrt{81} =$ ________

6. $\pm\sqrt{256} =$ ________

**Estimate each square root.**

7 $\sqrt{20}$

$\sqrt{20}$ is between $\sqrt{\square}$ and $\sqrt{\square}$.

$\sqrt{20}$ is between ____ and ____,

but closer to ____.

8 $\sqrt{75}$

$\sqrt{75}$ is between $\sqrt{\square}$ and $\sqrt{\square}$.

$\sqrt{75}$ is between ____ and ____,

but closer to ____.

9 $\sqrt{47}$

$\sqrt{47}$ is between $\sqrt{\square}$ and $\sqrt{\square}$.

$\sqrt{47}$ is between ____ and ____,

but closer to ____.

10 $\sqrt{13}$

$\sqrt{13}$ is between $\sqrt{\square}$ and $\sqrt{\square}$.

$\sqrt{13}$ is between ____ and ____,

but closer to ____.

## Step by Step Practice

11 Simplify $\sqrt{169x^8}$.

**Step 1** Because $\square^2 = 169$, $\sqrt{169} =$ ______.

**Step 2** Because $(\square)^2 = x^8$, $\sqrt{x^8} =$ ______.

**Step 3** $\sqrt{169x^8} =$ ______

**Simplify each square root.**

12 $\sqrt{16v^6}$

$\sqrt{\square} \cdot \sqrt{\square}$

$\sqrt{16v^6} =$ ______

13 $\sqrt{144x^2}$

$\sqrt{\square} \cdot \sqrt{\square}$

$\sqrt{144x^2} =$ ______

14 $\sqrt{4k^{16}}$

$\sqrt{\square} \cdot \sqrt{\square}$

$\sqrt{4k^{16}} =$ ______

15 $\sqrt{400z^4}$

$\sqrt{\square} \cdot \sqrt{\square}$

$\sqrt{400z^4} =$ ______



GO ON

**Simplify each square root.**

16 $\pm\sqrt{225} =$ ______

17 $\sqrt{x^6} =$ ______

18 $\sqrt{121x^{10}} =$ ______

19 $-\sqrt{256b^{24}} =$ ______

20 $\sqrt{144m^2n^{14}} =$ ______

21 $\sqrt{64h^4j^{16}k^{12}} =$ ______

## Step by Step Problem-Solving Practice

**Solve.**

22 Ashley is hanging a square picture in her living room. The area of the picture is 400 square inches. What is the length of each side of the picture?

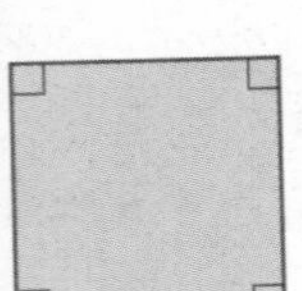

The area of the picture is ______.

A square has all four sides of equal length.
Think: What number times itself equals 400?

Because $\square^2 = 400$, $\sqrt{400} = \square$.

Check off each step.

______ **Understand: I underlined key words.**

______ **Plan: To solve the problem, I will** ______.

______ **Solve: The answer is** ______.

______ **Check: I checked my answer by** ______.

## Skills, Concepts, and Problem Solving

**Estimate each square root.**

23 $\sqrt{87}$

between ____ and ____,

but closer to ____

24 $\sqrt{138}$

between ____ and ____,

but closer to ____

25 $\sqrt{18}$

between ____ and ____,

but closer to ____

26 $\sqrt{63}$

between ____ and ____,

but closer to ____

**Simplify each square root.**

27 $\sqrt{81} =$ ________

28 $\sqrt{25c^{16}} =$ ________

29 $-\sqrt{100m^{10}n^{18}} =$ ________

30 $\pm\sqrt{169x^4} =$ ________

31 $\sqrt{a^2b^{16}c^2} =$ ________

32 $-\sqrt{49x^{22}} =$ ________

33 $-\sqrt{196} =$ ________

34 $\pm\sqrt{9p^6q^{12}} =$ ________

**Solve.**

35 **AREA** The area of a circle $A$ is calculated using the formula $A = \pi r^2$, where $r$ is the radius of the circle. What is the radius of a circle with an area of $121\pi$ square inches?

________________

36 **AREA** If the area of the inner rectangle in the figure below is $\sqrt{225}$ square centimeters, what is the area of the shaded region?

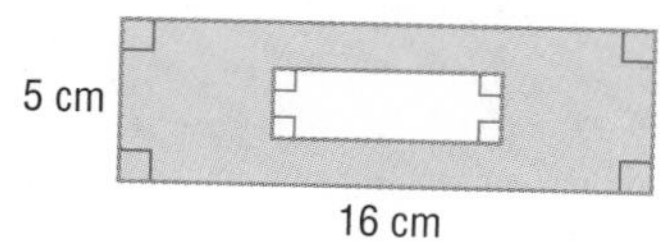


________________

37 **PHYSICS** In a controlled setting, if you drop an object, the time $t$ that it takes the object to fall a distance $d$ is calculated using the formula $t = \sqrt{\frac{2d}{9.8}}$, where time is measured in seconds and distance is measured in meters. Find the approximate time it takes an object to fall 500 meters.

________________

**Vocabulary Check** **Write the vocabulary word that completes each sentence.**

38 100 is an example of a(n) __________ because it is the square of 10.

39 If a number is one of the two equal factors of another number, it is the __________.

40 **Reflect** Explain how to know if a variable term is a perfect square.

________________

________________

STOP

Chapter 1

# Progress Check 2 (Lessons 1-3 and 1-4)

**Simplify each expression.**

1 $4(2)^{-5} =$ ______

2 $\left(\frac{3^2 \cdot 4^{-2}}{3^5}\right) =$ ______

3 $\frac{d^3}{d^{-3}} =$ ______

4 $\frac{7x^4}{56x^{-6}} =$ ______

5 $(5c^2g^4)^2 =$ ______

6 $\frac{6x^3(2y)^{-2}}{x^{-4}y^{-4}} =$ ______

7 $\left(\frac{6t}{t^3}\right)^2 =$ ______

8 $\frac{42z^{17}}{6z^9} =$ ______

9 $\frac{9^{-2} \cdot 9^2 \cdot 3^5}{3^5} =$ ______

10 $\left(\frac{2}{3}hm^3\right)^3 =$ ______

11 $(-n^7)^3 =$ ______

12 $p \cdot p^7 \cdot b^{-4} =$ ______

**Estimate each square root.**

13 $\sqrt{13}$ is between ____ and ____, but closer to ____

14 $\sqrt{71}$ is between ____ and ____, but closer to ____

15 $\sqrt{24}$ is between ____ and ____, but closer to ____

16 $\sqrt{107}$ is between ____ and ____, but closer to ____

**Simplify each square root.**

17 $\sqrt{49} =$ ______

18 $\sqrt{16m^8} =$ ______

19 $-\sqrt{81a^6b^2} =$ ______

20 $\pm\sqrt{144x^{18}} =$ ______

21 $\sqrt{p^8n^4} =$ ______

22 $-\sqrt{9x^2} =$ ______

23 $-\sqrt{225} =$ ______

24 $\pm\sqrt{121p^8q^8} =$ ______

**Solve.**

25 **AREA** The area of a square is calculated using the formula $A = s^2$, where $s$ is the length of the side of the square. What is the length of a side of the square with an area of 64 square feet?

______

Lesson 1-5

# Ordered Pairs and Relations

## KEY Concept

**Ordered Pair**

$$(-5, 4)$$

$x$-coordinate → $-5$; $y$-coordinate → $4$

A **relation** is a set of ordered pairs.

{(1, 6), (7, −3), (−2, 3), (0, 9), (−5, 0)}

The **domain** of a relation is the set of $x$-coordinates. {−5, −2, 0, 1, 7}

The **range** of a relation is the set of $y$-coordinates. {−3, 0, 3, 6, 9}

It is customary to list these values from least to greatest.

A **mapping** shows how the domain elements are paired with the range elements.

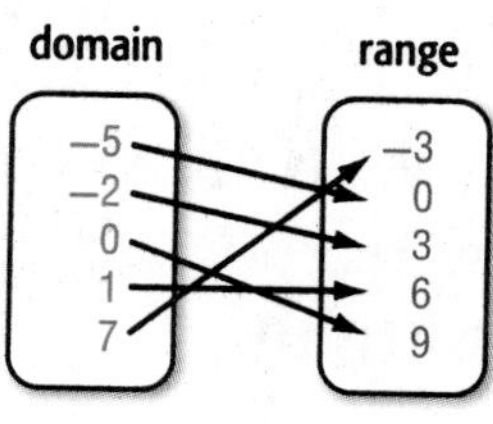


### VOCABULARY

**domain**
the set of the first numbers of the ordered pairs in a relation

**mapping**
a diagram that shows how each domain element is paired with an element in the range

**ordered pair**
a pair of numbers, written in an order, $(x, y)$, that identifies a point in a coordinate grid

**range**
the set of the second numbers of the ordered pairs in a relation

**relation**
a set of ordered pairs

A relation can also be shown in a table or as a graph.

## Example 1

**Name the domain and range.**

| $x$ | −1 | 3 | 4 | 0 |
|---|---|---|---|---|
| $y$ | 8 | 7 | −3 | 2 |

1. The domain is the set of $x$-values.
   domain = {−1, 0, 3, 4}

2. The range is the set of $y$-values.
   range = {−3, 2, 7, 8.}

### YOUR TURN!

**Name the domain and range.**

| $x$ | 2 | 0 | 10 | −2 |
|---|---|---|---|---|
| $y$ | 5 | −4 | 6 | 5 |

Notice there are two 5s. You only need to list one in the range list.

1. The domain is the set of $x$-values.
   domain = {____________}

2. The range is the set of $y$-values.
   range = {____________}

GO ON

## Example 2

**Name the domain and range.**

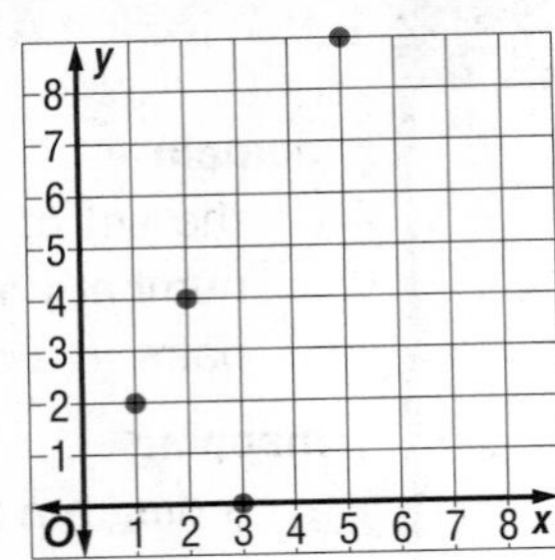

1. The points graphed are (1, 2), (2, 4), (3, 0), and (5, 9).
2. The domain is the set of $x$-values. domain = {1, 2, 3, 5}
3. The range is the set of $y$-values. range = {0, 2, 4, 9}

**YOUR TURN!**

**Name the domain and range.**

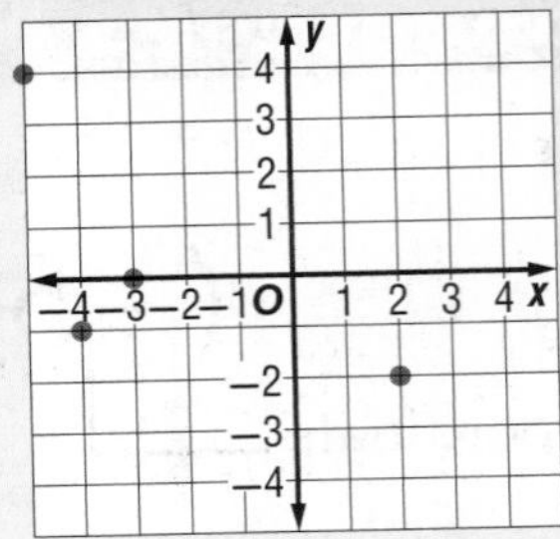

1. The points graphed are (____, ____), (____, ____), (____, ____), and (____, ____).
2. The domain is the set of $x$-values.

   domain = {____, ____, ____, ____}
3. The range is the set of $y$-values.

   range = {____, ____, ____, ____}

## Example 3

**Draw a mapping diagram for the relation.**

**{(−8, 4), (0, −1), (−2, 2), (−2, 7)}**

1. List the $x$-values as the domain in the left column of the mapping.
2. List the $y$-values as the range in the right column of the mapping.
3. Draw arrows to show pairs.

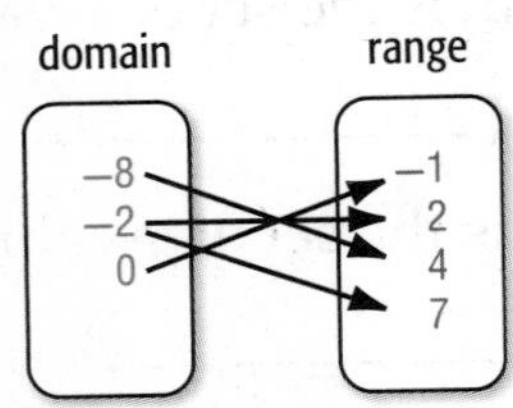


**YOUR TURN!**

**Draw a mapping diagram for the relation.**

**{(6, 0), (−4, −8), (5, −1), (9, 0)}**

1. List the $x$-values as the ________ in the ________ column of the mapping.
2. List the $y$-values as the ________ in the ________ column of the mapping.
3. Draw ________ to show pairs.

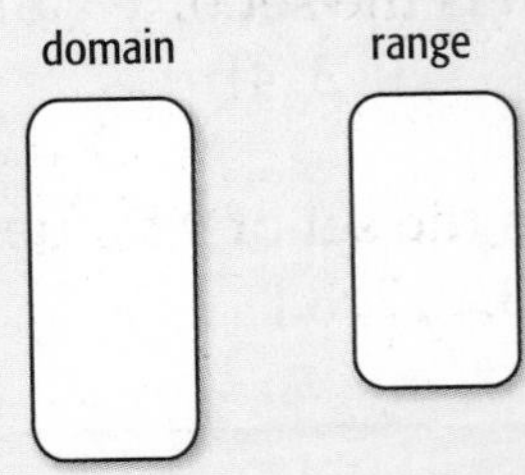

## Guided Practice

**Name the domain and range for each relation.**

1 $\{(2, 0), (1, -3), (4, 4), (-7, 0)\}$

The domain is the set of ________________.

The domain = {________________}.

The range is the set of ________________.

The range = {________________}.

2 $\{(2, 2), (-5, 0), (-6, 3), (2, -1)\}$

domain = {________________}

range = {________________}

**Draw a mapping diagram for each relation.**

3 $\{(-4, -4), (7, 5), (-2, 3), (4, 1)\}$

domain range

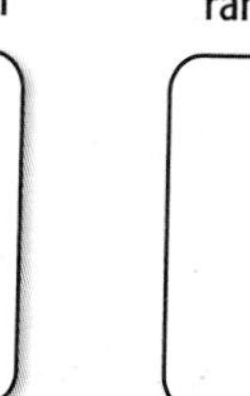

4 $\{(3, 3), (9, 0), (-2, 3), (1, 0)\}$

domain range

### Step by Step Practice

5 Name the domain and range.

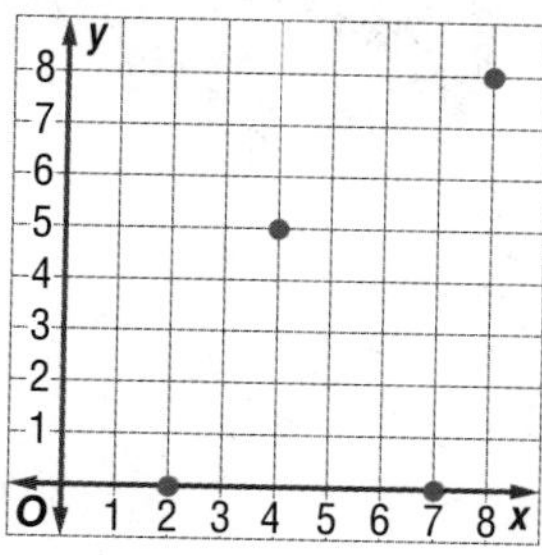


**Step 1** The points graphed are (____, ____), (____, ____),

(____, ____), and (____, ____).

**Step 2** The domain is the set of ________________.

The domain = {________________}.

**Step 3** The range is the set of ________________.

The range = {________________}.

GO ON

**Name the domain and range for each relation.**

6

| x | 2 | −12 | 15 | 3 |
|---|---|---|---|---|
| y | 4 | −10 | 4 | 2 |

domain = ____________

range = ____________

7 $\{(8, 11), (5, -6), (0, 0), (0, 7)\}$

domain = ____________

range = ____________

8 domain range

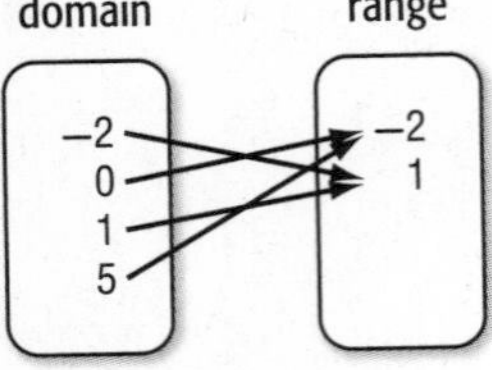


domain = ____________

range = ____________

9

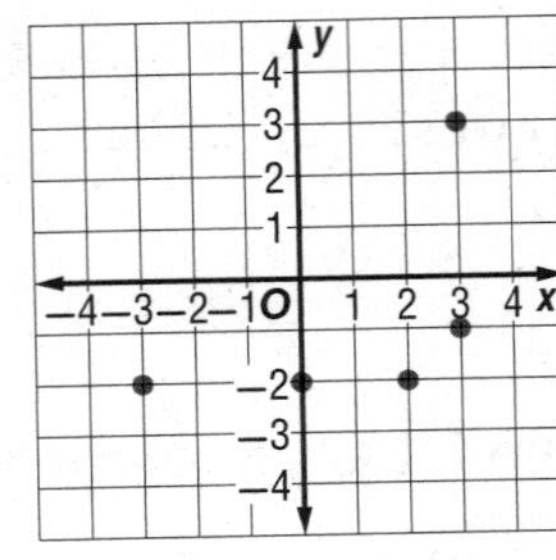


domain = ____________

range = ____________

## Step by Step Problem-Solving Practice

**Solve.**

10 Claudia is purchasing party favors. A set of 8 favors costs $12.99. A set of 12 favors costs $17.99, and a set of 18 favors costs $25.99. Identify the domain of quantities and range of prices from which Claudia has to choose.

| Quantity | | | |
|---|---|---|---|
| Price | | | |

Check off each step.

_____ **Understand: I underlined key words.**

_____ **Plan: To solve the problem, I will** ____________.

_____ **Solve: The answer is** ____________

____________.

_____ **Check: I checked my answer by** ____________

____________.

## Skills, Concepts, and Problem Solving

**Name each domain and range.**

**11** $\{(-1,-1), (2, -2), (4, 3), (7, -5)\}$

domain = ______________________

range = ______________________

**12** $\{(6, 0), (10, -7), (-3, 0), (2, 12)\}$

domain = ______________________

range = ______________________

**13**

| x | 21 | 11 | −13 | 5 |
|---|---|---|---|---|
| y | 15 | −10 | −5 | −5 |

domain = ______________________

range = ______________________

**14**

| x | 7 | 0 | 7 | 3 |
|---|---|---|---|---|
| y | 4 | 2 | 0 | 4 |

domain = ______________________

range = ______________________

**15**

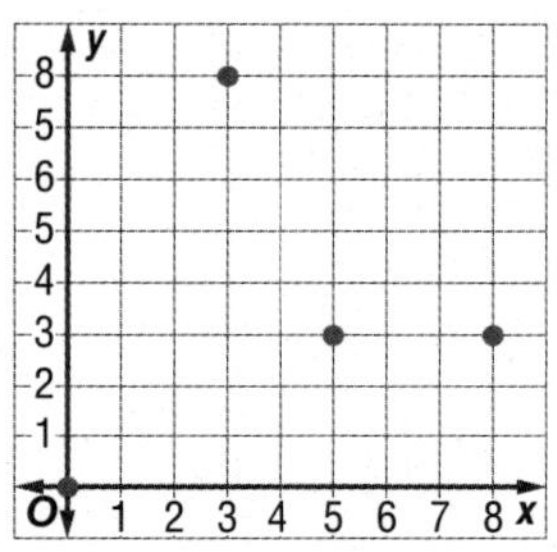


domain = ______________________

range = ______________________

**16**

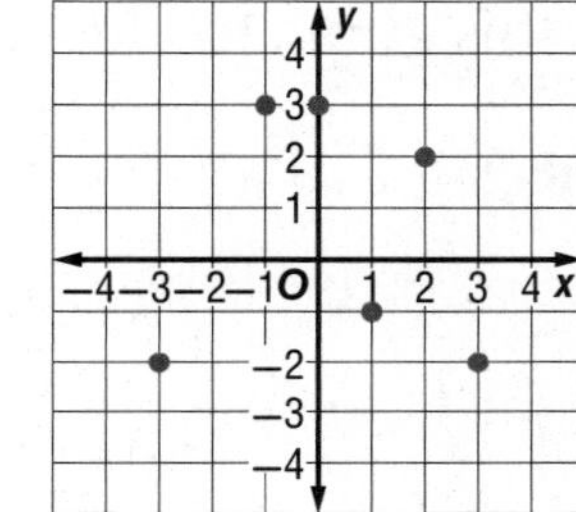


domain = ______________________

range = ______________________

**Draw a mapping diagram for each relation.**

**17** $\{(-1, -3), (6, 5), (0, -3), (12, 0)\}$

domain range

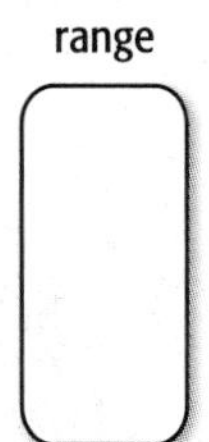

**18** $\{(7, 0), (7, 1), (7, 2), (7, 3)\}$

domain range

**19** $\{(6, 2), (8, -3), (-6, 2)\}$

domain range

**20** $\{(1, 1), (4, 1), (9, 1), (1, 1)\}$

domain range

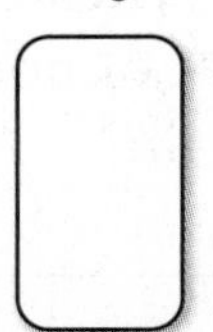

GO ON

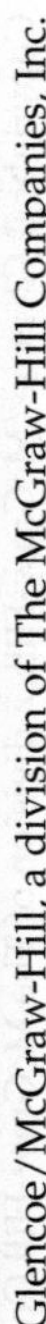

**Solve.**

21 **COOKING** A box of pancake mix shows a table with the amount of water that should be used with each amount of mix.

| Cups of pancake mix | 2 | 3 | 4 | 5 |
|---|---|---|---|---|
| Cups of water | $1\frac{1}{2}$ | $2\frac{1}{4}$ | 3 | $3\frac{3}{4}$ |

Let the domain of this relation be the cups of pancake mix and the range be the cups of water. List the domain and the range.

_______________________________________________

22 **RUNNING** As part of his conditioning before the cross country season begins, Isaac keeps a graph of the number of miles he runs each day. The graph below shows the first 5 days of his conditioning. Use this graph to identify the domain and range.

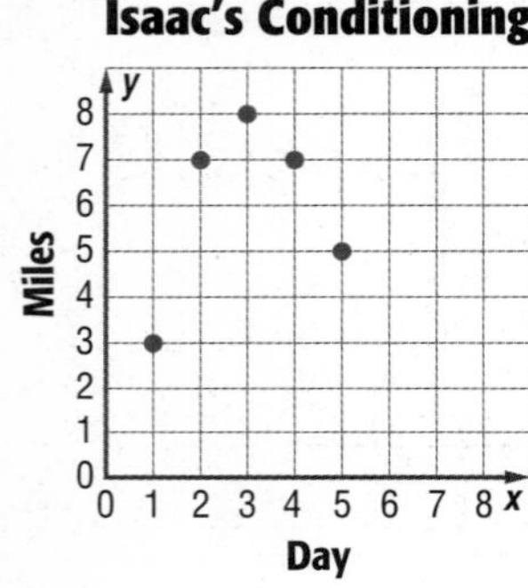


_______________________________________________

**Vocabulary Check** **Write the vocabulary word that completes each sentence.**

23 An example of a(n) ______________ is (0, 4).

24 In a relation, the set of $y$-coordinates is the ______________.

25 A set of ordered pairs is a(n) ______________.

26 **Reflect** A pet shop creates a relation between the number of dogs in a household and the number of pounds of dog food needed per month. The domain is the number of dogs, and the range is the number of pounds of food. Could a non-integer number be part of the domain?

_______________________________________________

_______________________________________________

Lesson 1-6

# Functions

## KEY Concept

A **function** is a relation in which each unique $x$-value is paired with exactly one $y$-value.

**Function**

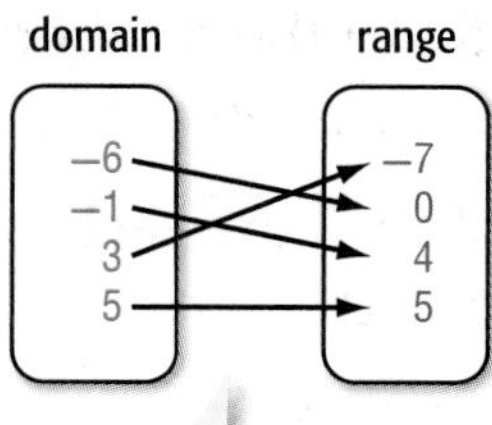


Each domain value is paired with exactly one range value. So the relation is a function.

**Not a Function**

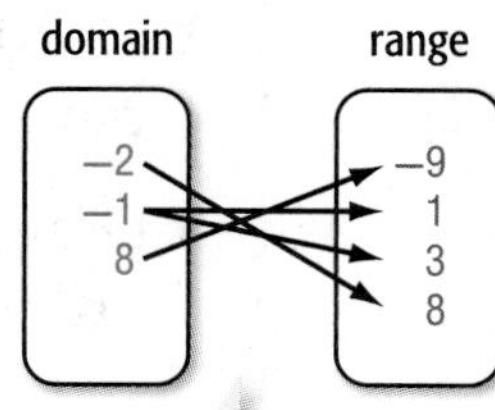


The domain value −1 is paired with two range values, 1 and 3. So the relation is not a function.

A vertical line test is another way to determine if a relation is a function. If a vertical line can be placed anywhere on a graph and it passes through **only one point,** then the graph represents a function.

**Function**

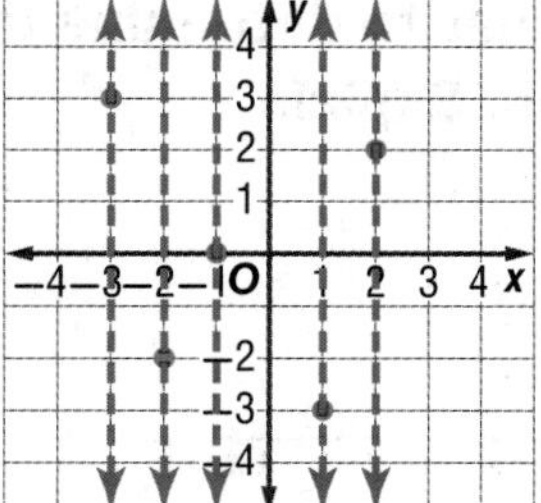

Each vertical line passes through only one point of the graph.

**Not a Function**

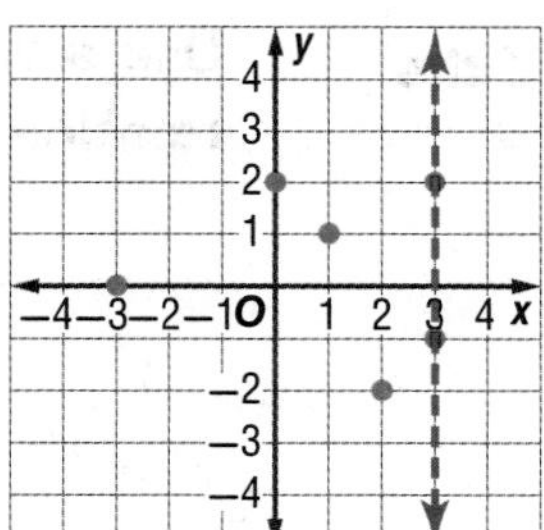

The vertical line that passes through the point (3, 2) also passes through the point (3, −1).

### VOCABULARY

**domain**
the set of the first numbers of the ordered pairs in a relation

**function**
a relation in which each element of the domain is paired with exactly one element of the range

**mapping**
a diagram that shows how each domain element is paired with an element in the range

**range**
the set of the second numbers of the ordered pairs in a relation

**relation**
a set of ordered pairs

**vertical line test**
a test used to determine if a relation is a function

Mappings and vertical line tests are good ways to visually determine if a relation is a function.

## Example 1

**Draw a mapping diagram for the relation. Is the relation a function? Explain.**

**{(4, 6), (−2, 7), (−3, 4), (0, 7)}**

1. List the $x$-coordinates of the ordered pairs in the domain box.
2. List the $y$-coordinates of the ordered pairs in the range box.
3. Draw an arrow to show how each ordered pair is matched.

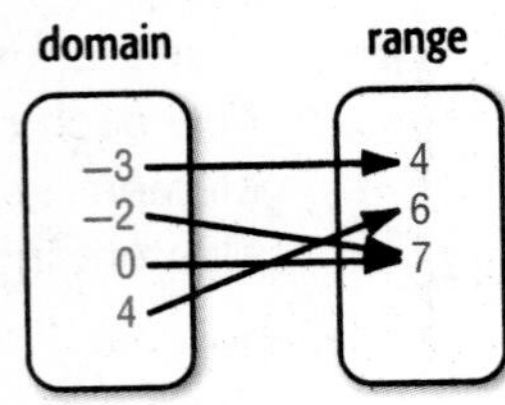


Each domain value is paired with one range value. The relation is a function.

**YOUR TURN!**

**Draw a mapping diagram for the relation. Is the relation a function? Explain.**

**{(−1, 5), (−2, 9), (1, 4), (−1, 3)}**

1. List the $x$-coordinates of the ordered pairs in the ____________ box.
2. List the $y$-coordinates of the ordered pairs in the ____________ box.
3. Draw an arrow to show how each ordered pair is matched.

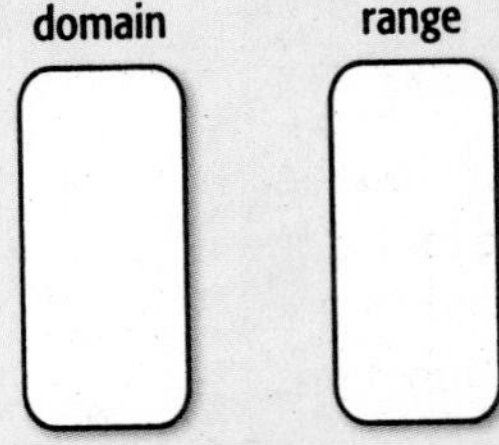


The domain value ______ is paired with two range values, ______ and ______. The relation is ____________.

## Example 2

**Use a vertical line test to determine if the relation is a function. Explain.**

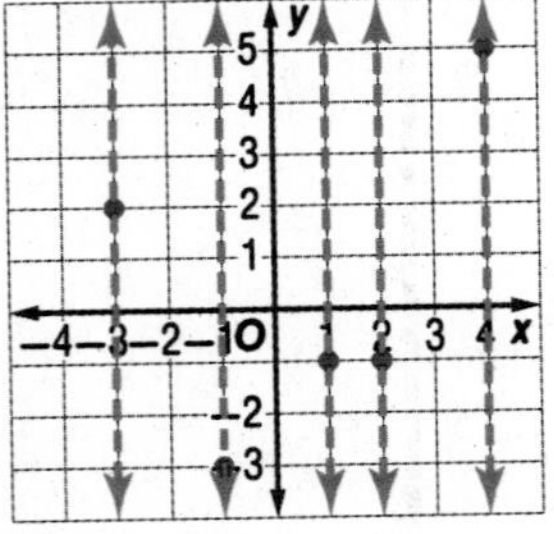

1. Imagine drawing a vertical line through each of the points on the graph. Will any line pass through more than one point? no
2. There is not a vertical line that passes through more than one point. The relation is a function.

**YOUR TURN!**

**Use a vertical line test to determine if the relation is a function. Explain.**

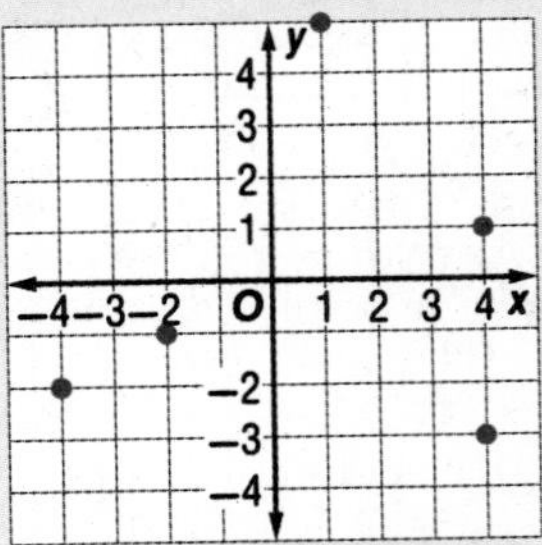

1. Imagine drawing a vertical line through each of the points on the graph. Will any line pass through more than one point?

   ________

2. A vertical line passes through ________ and ________. The relation is ________ ________.

## Guided Practice

**Determine if each relation is a function. Explain.**

1. {(0, 5), (−1, 3), (−2, 0), (−4, 5)}

domain range

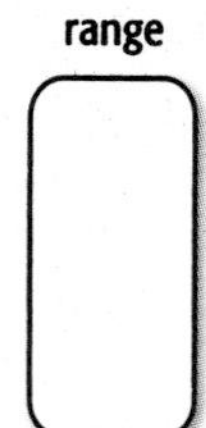

Each ______ value is paired with one ______ value.

So the relation is ______.

2. {(6, −1), (1, 9), (6, 5), (2, 5)}

domain range

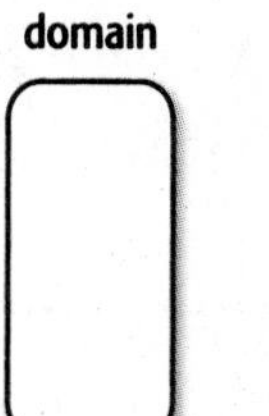

The domain value ______ is paired with two range values, ______ and ______.

So the relation is ______.

## Step by Step Practice

3. A relation is given below. Determine if it is a function.
{(−2, 9), (10, 9), (0, 3), (10, 15), (11, 15)}

**Step 1** List the $x$-coordinates of the ordered pairs.

______

**Step 2** List the $y$-coordinates of the ordered pairs.

______

**Step 3** Draw a mapping diagram of the relation.

**Step 4** Does the mapping show a function? Explain your answer.

The relation is ______.

______

GO ON

**Determine if each relation is a function. Explain.**

4 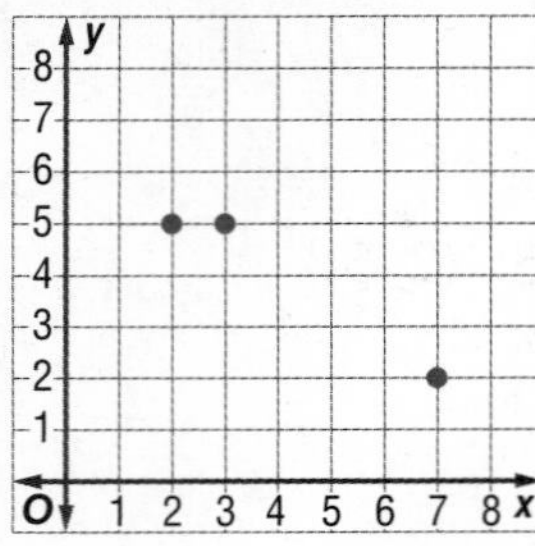

The relation is ______________________

______________________

______________________.

5 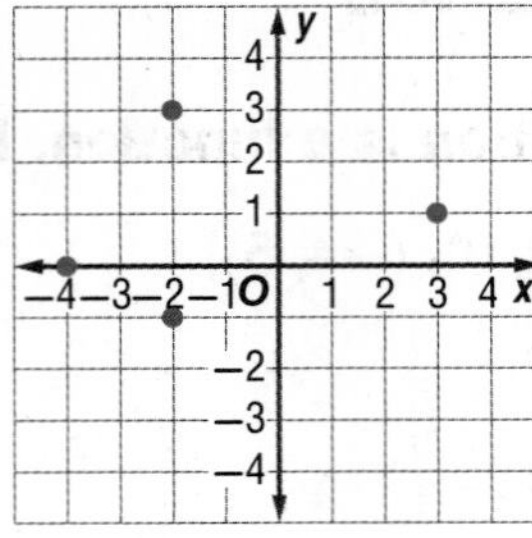

The relation is ______________________

______________________

______________________.

6 {(−8, 6), (3, 2), (−8, 0), (12,0)}

The relation is ______________________

______________________

______________________.

7 {(4, 22), (0, 1), (15,1), (3, −8)}

The relation is ______________________

______________________

______________________.

## *Step by Step* ***Problem-Solving Practice***

**Solve.**

8 The Children's Development Center opened for business three years ago. In the first year, 65 students were enrolled. The enrollment was 82 students for both the second and third years. Let the year of operation be the domain and the enrollment be the range. Is the relation (operating year, enrollment) a function?

Check off each step.

______ **Understand: I underlined key words.**

______ **Plan: To solve the problem, I will** ______________________

______________________.

______ **Solve: The answer is** ______________________.

______ **Check: I checked my answer by** ______________________

______________________.

## Skills, Concepts, and Problem Solving

**Determine if each relation is a function. Explain.**

**9**

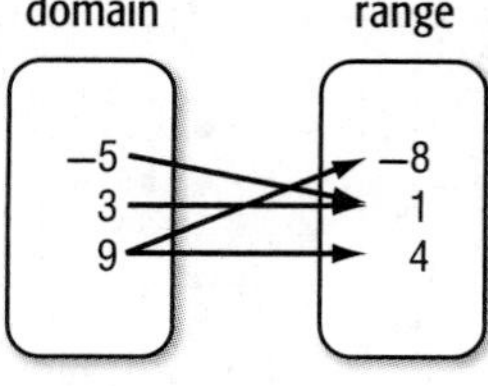


**10**

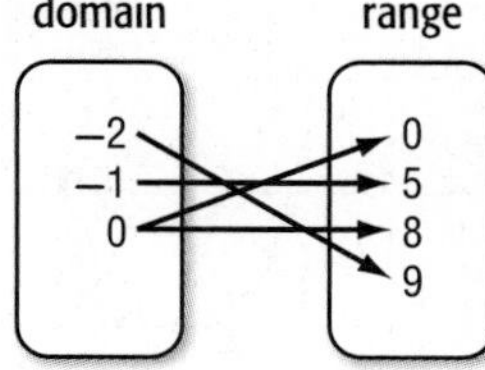


**11** $\{(-13, 10), (-5, 4), (-7, 0), (8, 0)\}$

**12** $\{(2, -2), (-9, 5), (17, 6), (-9, 2)\}$

**13**

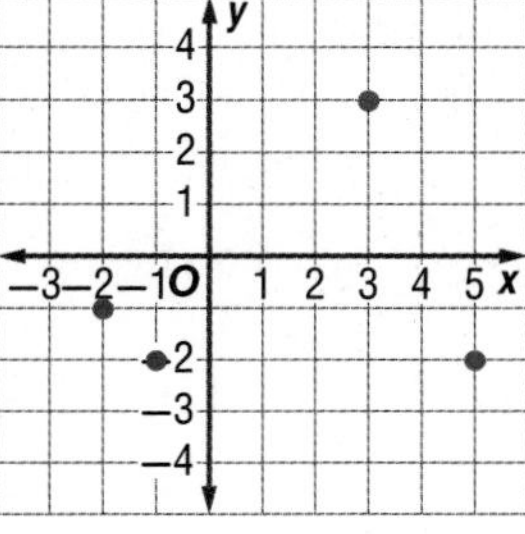


**14**

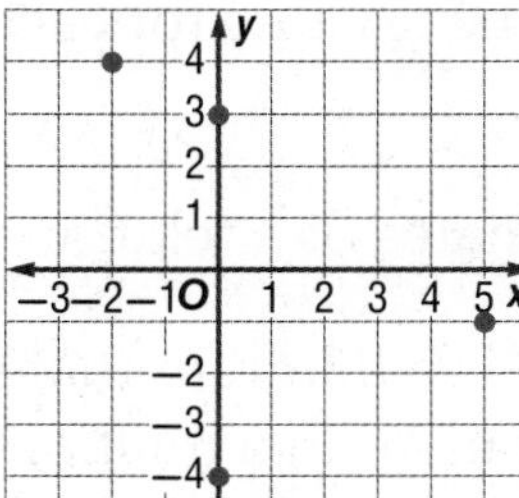




GO ON

**Solve.**

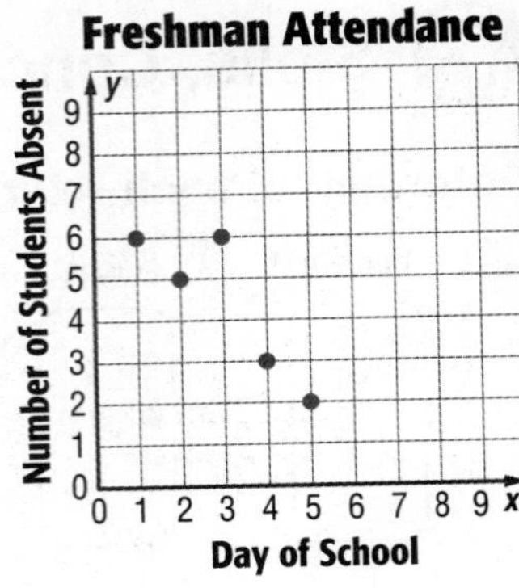


**15** **ATTENDANCE** The graph shows the number of students absent from the freshman class at Big Oak High School for the first five days of the school year. Is the relation shown in the graph a function? Explain.

______________________________

______________________________

**16** **NUTRITION** For her health class, Vi is collecting nutrition information from various drink labels. She makes the following table with the serving size and the number of calories.

| Serving Size (in ounces) | 8 | 10 | 8 | 12 |
|---|---|---|---|---|
| Number of Calories | 50 | 160 | 80 | 170 |

Vi creates a relation in her table, where the serving size is the domain, and the calorie count is the range. Is this relation a function? Why or why not?

______________________________

______________________________

**Vocabulary Check** **Write the vocabulary word that completes each sentence.**

**17** If there is exactly one $y$-value for each $x$-value in a relation, then the relation is a(n) ____________.

**18** The set of $y$-coordinates of a relation is called the ____________.

**19** A(n) ____________ is a diagram that shows how each $x$ element relates to each $y$ element in a relation.

**20** **Reflect** A mapping of a relation has five different number values in the domain and six different number values in the range. Can you determine if this relation is a function? Why or why not?

______________________________

______________________________

______________________________

Lesson 1-7

# Functions and Equations

## KEY Concept

An **equation** is a mathematical sentence that contains an equal sign. Some equations model functions. These equations are often written in function notation, where $f(x)$ replaces $y$ and is read, "$f$ of $x$." $y = x + 1$ and $f(x) = x + 1$.

The function $y = \frac{1}{2}x + 3$ is graphed using a table of $x$- and $y$-values generated from the equation.

| x | y |
|---|---|
| −4 | 1 |
| −2 | 2 |
| 0 | 3 |
| 2 | 4 |
| 4 | 5 |

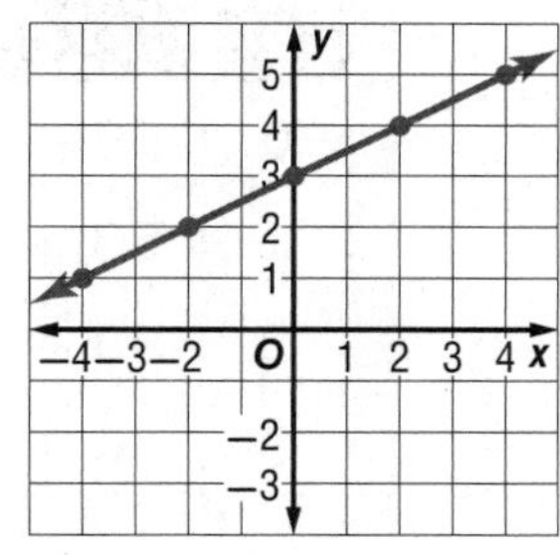


### VOCABULARY

**equation**
a mathematical sentence that contains an equal sign, =, indicating that an expression on the left side of the equal sign has the same value as an expression on the right side

**function**
a relation in which each element of the domain is paired with exactly one element of the range

To make a table, select values for $x$. Substitute them into the equation and solve for $y$.

## Example 1

**Graph the function $y = -3x + 5$.**

1. Plot the points in the table on the graph.

| x | y |
|---|---|
| −1 | 8 |
| 0 | 5 |
| 1 | 2 |
| 2 | −1 |
| 3 | −4 |

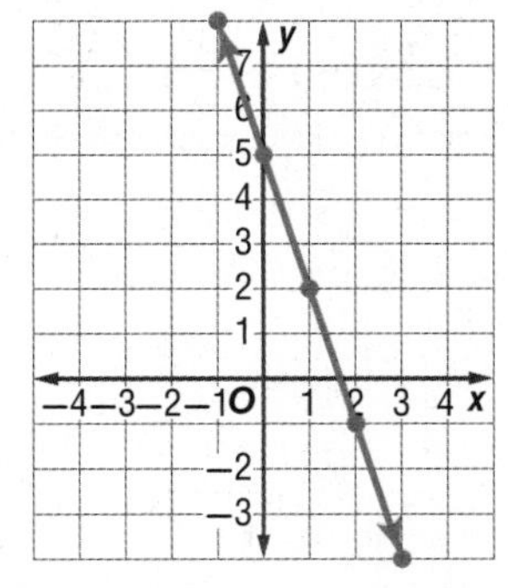


2. Connect the points.

## YOUR TURN!

**Graph the function $y = -\frac{1}{4}x + 1$.**

1. Plot the points in the table on the graph.

| x | y |
|---|---|
| −8 | 3 |
| −4 | 2 |
| 0 | 1 |
| 4 | 0 |
| 8 | −1 |

2. Connect the points.

GO ON

## Example 2

A sporting goods store sells baseballs for \$5 each. The function $C(x) = 5x$ represents the cost $C(x)$ of $x$ baseballs. Make a table and graph the function. How much would 4 baseballs cost?

1. Select values to use for $x$.
2. Substitute each value into the equation and solve for $C(x)$.

| x | C(x) = 5x | (x, C(x)) |
|---|---|---|
| 0 | 5(0) = 0 | (0, 0) |
| 1 | 5(1) = 5 | (1, 5) |
| 2 | 5(2) = 10 | (2, 10) |
| 3 | 5(3) = 15 | (3, 15) |

3. Plot the points. Connect the points.

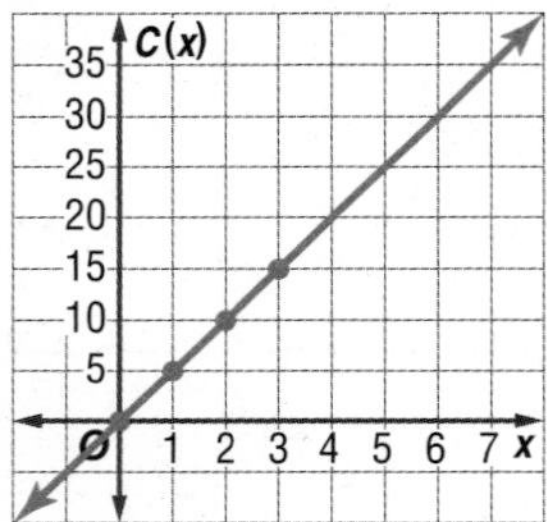


4. Find the value of $C(x)$ when $x = 4$.

**(4, 20) Four baseballs would cost \$20.**

### YOUR TURN!

A boat rental store allows customers to rent a boat for \$50 plus \$25 per hour. The function $C(x) = 50 + 25x$ represents the cost $C(x)$ of renting a boat for $x$ hours. Make a table and graph the function. How much does it cost to rent a boat for 5 hours?

1. Select values to use for $x$.
2. Substitute each value into the equation and solve for $C(x)$.

| x | C(x) = 50x + 25 | (x, C(x)) |
|---|---|---|
| 0 | 50 + 25(____) = ____ | (0, ____) |
| 1 | 50 + 25(____) = ____ | (1, ____) |
| 2 | 50 + 25(____) = ____ | (2, ____) |
| 3 | 50 + 25(____) = ____ | (3, ____) |

3. Plot the points. Connect the points.

4. Find the value of $C(x)$ when $x = 5$.

(5, ____)

The cost is \$____ for 5 hours.

## Guided Practice

**Graph the function.**

1 $y = 2x + 5$

| x | −4 | −2 | 0 | 1 | 2 |
|---|---|---|---|---|---|
| y | −3 | 1 | 5 | 7 | 9 |

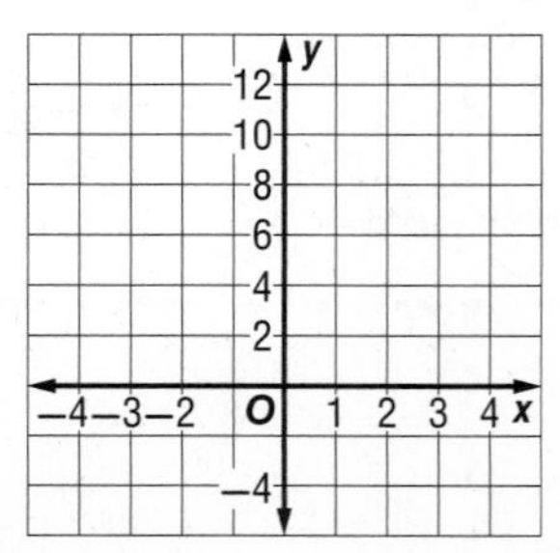

## Step by Step Practice

2 Complete a table of $x$- and $y$-values for the function $y = -2x + 6$. Graph the function.

**Step 1** Select values to use for $x$.

**Step 2** Substitute each value of $x$ into the equation and solve for $y$.

| $x$ | $-2x + 6$ | $y$ |
|---|---|---|
| −1 | −2(_____) + 6 | _____ |
| 0 | −2(_____) + 6 | _____ |
| 1 | −2(_____) + 6 | _____ |
| 2 | −2(_____) + 6 | _____ |
| 3 | −2(_____) + 6 | _____ |
| 4 | −2(_____) + 6 | _____ |

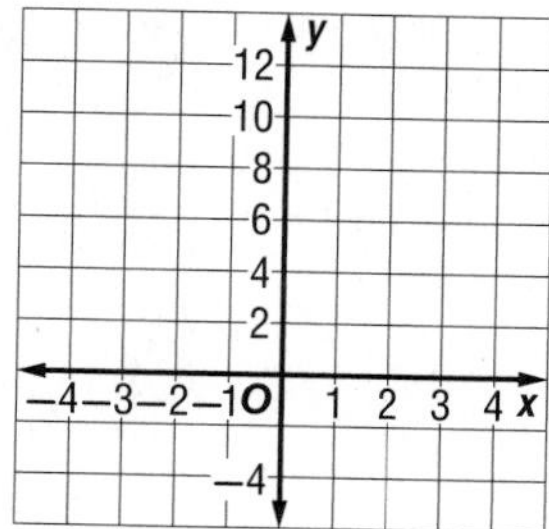

**Step 3** Plot the points. Draw the line.

**Complete a table of *x*- and *y*-values for each function. Graph each function.**

3 $y = \frac{3}{4}x - 2$

| $x$ | $\frac{3}{4}x - 2$ | $y$ |
|---|---|---|
| −8 | $\frac{3}{4}(-8) - 2$ | _____ |
| −4 | $\frac{3}{4}(-4) - 2$ | _____ |
| 0 | $\frac{3}{4}(0) - 2$ | _____ |
| 4 | $\frac{3}{4}(4) - 2$ | _____ |
| 8 | $\frac{3}{4}(8) - 2$ | _____ |

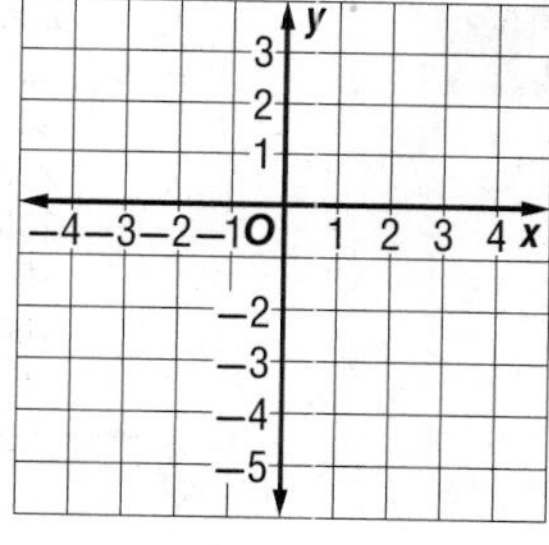

4 $y = -\frac{x}{3}$

| $x$ | −6 | −3 | 0 | 3 | 6 |
|---|---|---|---|---|---|
| $y$ | _____ | _____ | _____ | _____ | _____ |

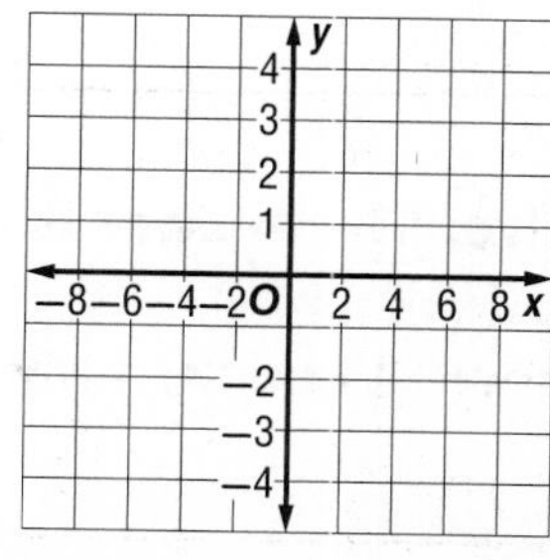

GO ON

**Complete a table of *x*- and *y*-values for each function. Graph each function.**

5 $y = \frac{x}{2} + 4$

| x | | | | | |
|---|---|---|---|---|---|
| y | | | | | |

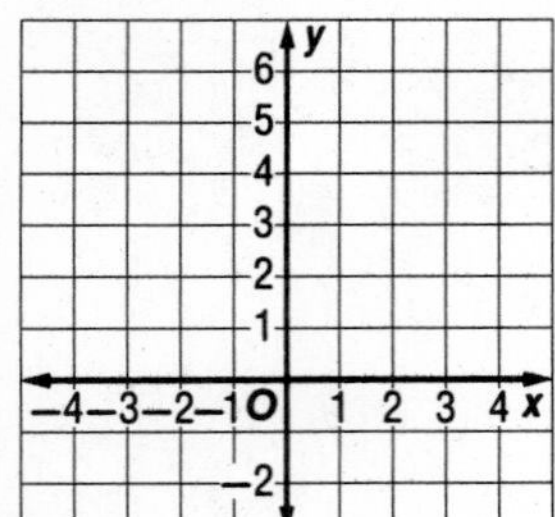


6 $y = 6x - 7$

| x | | | | | |
|---|---|---|---|---|---|
| y | | | | | |

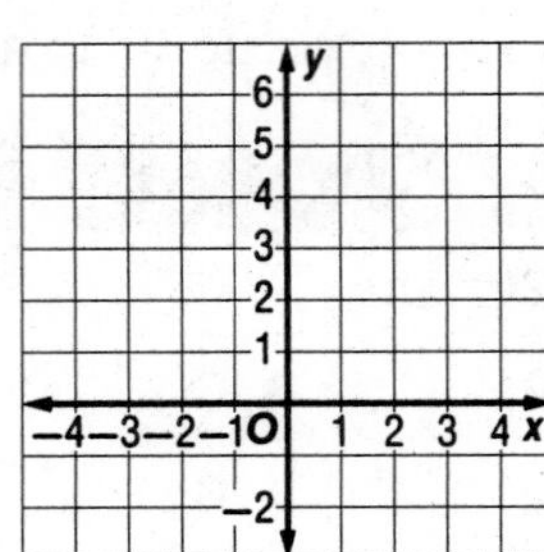


## Step by Step Problem-Solving Practice

**Solve.**

7 For each service call, a plumbing company charges a flat fee of \$75 plus \$100 per hour. The function $C(x) = 75 + 100x$ represents the cost of a plumbing job that takes $x$ hours. Make a table and graph the function. How much would you pay for a service call that took 4 hours?

| x | y |
|---|---|
| | |
| | |
| | |
| | |

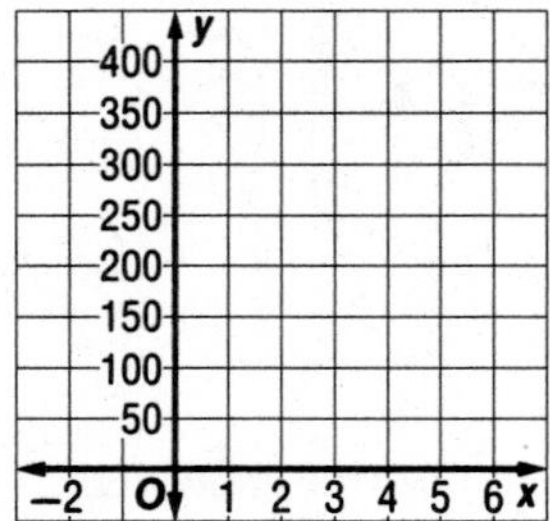


Check off each step.

______ **Understand: I underlined key words.**

______ **Plan: To solve the problem, I will** ______________________________

______________________________.

______ **Solve: The answer is** ______________________________.

______ **Check: I checked my answer by** ______________________________

______________________________.

## Skills, Concepts, and Problem Solving

**Use a table of *x*- and *y*-values to graph each function.**

**8** $y = -3x$

| x | y |
|---|---|
| | |
| | |
| | |
| | |
| | |

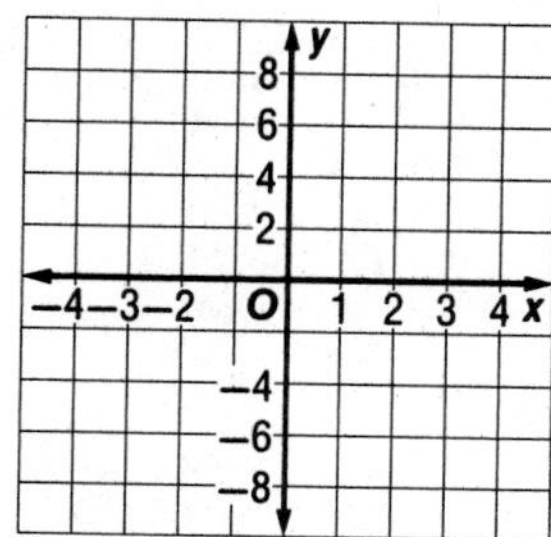


**9** $y = \frac{2}{3}x - 2$

| x | y |
|---|---|
| | |
| | |
| | |
| | |
| | |

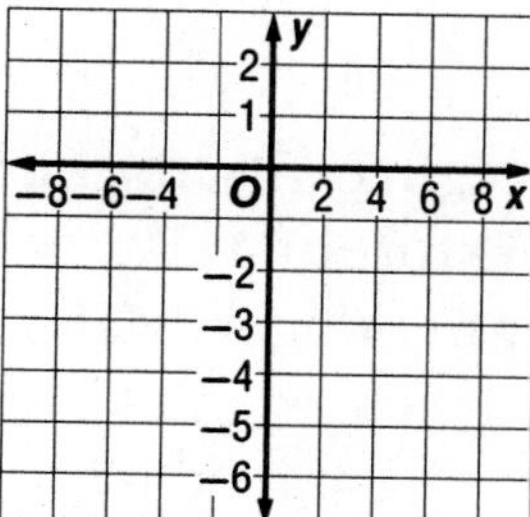


**10** $y = 8 - 2x$

| x | y |
|---|---|
| | |
| | |
| | |
| | |
| | |

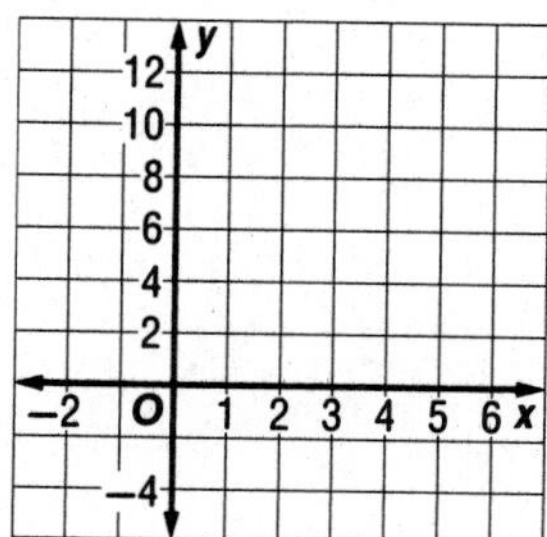


**11** $y = \frac{x}{6} + 5$

| x | y |
|---|---|
| | |
| | |
| | |
| | |
| | |

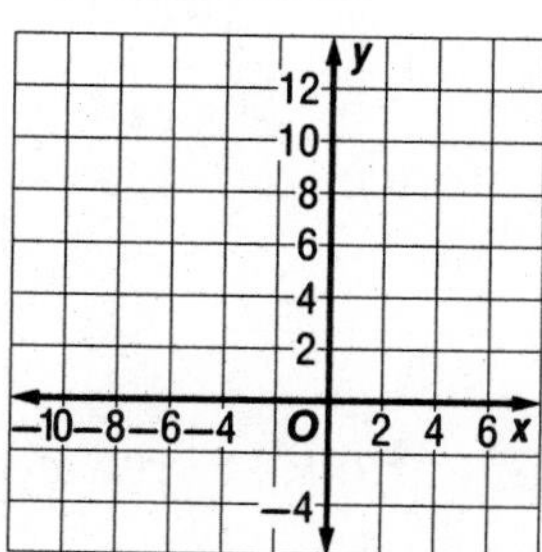


GO ON

**Solve.**

**12** **RENTAL CHARGES** The Big Rig Truck Company charges a flat rate of \$100 plus an additional \$2 per mile for rental of its moving trucks. The function $C(x) = 100 + 2x$ represents the cost of renting a truck for $x$ miles. Make a table and graph the function. How much will it cost to rent a truck that you plan on driving 50 miles?

| $x$ | | | | | |
|---|---|---|---|---|---|
| $C(x)$ | | | | | |

______________________________

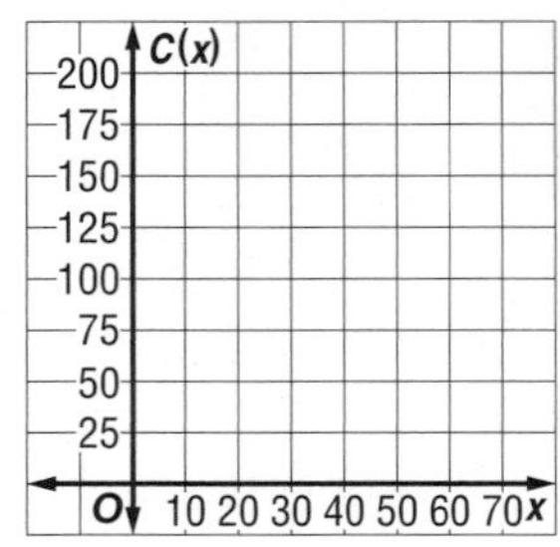


**13** **POSTAGE** The cost, in cents, for mailing a domestic letter is calculated using the formula $C(x) = 41 + 17(x-1)$, where $x$ is the weight of the letter in ounces. Make a table and graph the function. How much would it cost to mail a letter weighing 5 ounces?

| $x$ | $C(x)$ |
|---|---|
| | |
| | |
| | |
| | |
| | |

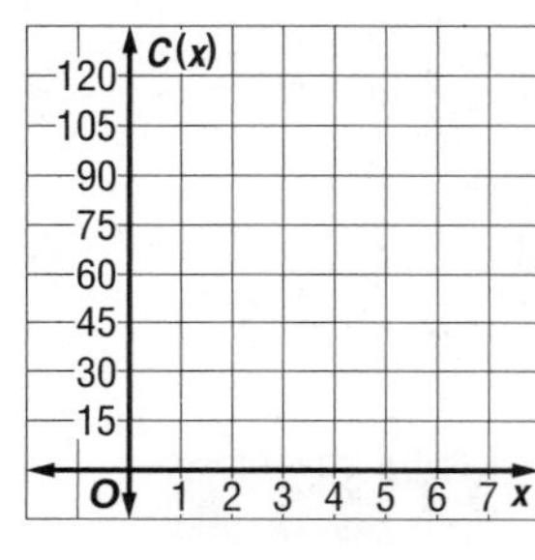


______________________________

**Vocabulary Check** **Write the vocabulary word that completes each sentence.**

**14** A(n) ____________ is a relation that assigns exactly one $y$-value to each $x$-value.

**15** A mathematical sentence that sets two expressions equal to one another is a(n) ____________.

**16** **Reflect** Can an equation be created to model every relation?

______________________________

______________________________

STOP

# Progress Check 3 (Lessons 1-5, 1-6 and 1-7)

**Name each domain and range. Determine if each relation is a function.**

**1** $\{(-8, -1), (3, -2), (7, 9), (0, -5)\}$

domain = ____________________

range = ____________________

The relation is ____________________.

**2** $\{(0, 0), (10, -2), (-3, 5), (6, 11)\}$

domain = ____________________

range = ____________________

The relation is ____________________.

**3**

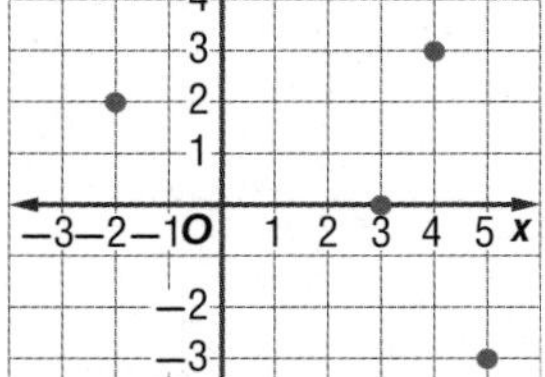

domain = ____________________

range = ____________________

The relation is ____________________.

**4**

| x | −5 | −3 | −1 | 2 | 3 |
|---|---|---|---|---|---|
| y | 0 | 4 | −5 | 1 | −2 |

domain = ____________________

range = ____________________

The relation is ____________________.

**Graph each function.**

**5** $y = 2x + 1$

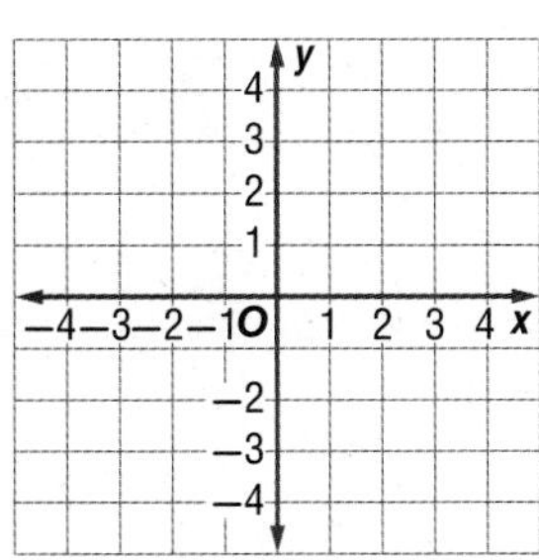

**6** $y = -\frac{1}{3}x$

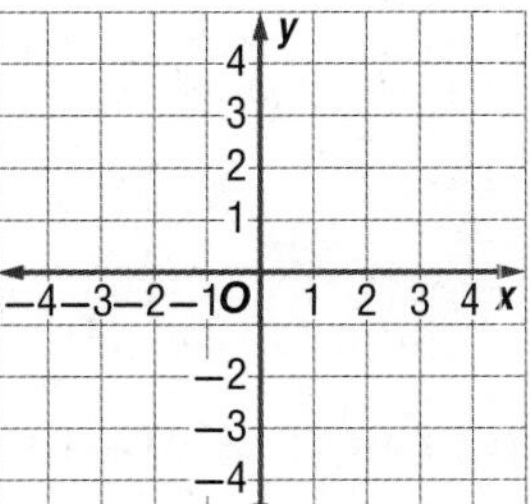

**Solve.**

**7** **GEOMETRY** Jared plotted the points (0, 5), (5, 5), (−2, 0), and (3, 0) on a coordinate graph. After plotting the points, Jared used line segments to form a quadrilateral with the points. Create Jared's relation on a coordinate graph. Then use the vertical line test to determine if it is a function.

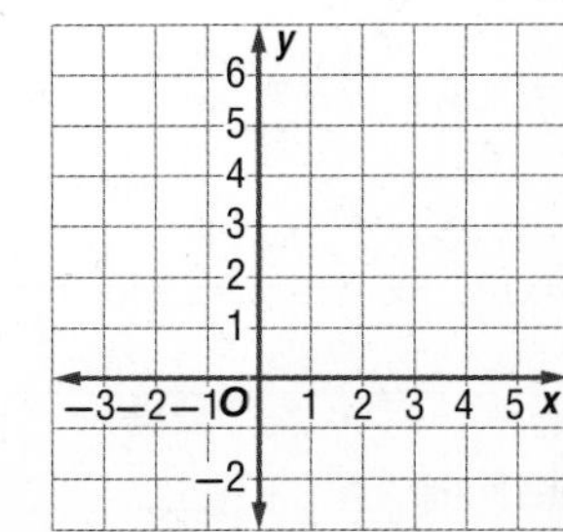

____________________________________

Chapter 1

# Chapter Test

**Name the property illustrated.**

1 $(45 + 8) \cdot 0 = 0$

______________________

2 $4(3 + 17) = 12 + 68$

______________________

**Use the Distributive Property to simplify.**

3 $2(15 + 7)$

_____ + _____ = _____

4 $12(11 + 6)$

_____ + _____ = _____

**Simplify each expression.**

5 $8(4)^{-2} =$

_____

6 $(3w)^3 =$

_____ • _____ = _____

7 $(2p^6)(5p^3) =$

(_____ • _____) • $p^{\square}$ = _____

**Estimate each square root.**

8 $\sqrt{38}$

between _____ and _____,

but closer to _____

9 $\sqrt{130}$

between _____ and _____,

but closer to _____

**Simplify each square root.**

10 $\sqrt{64} =$

_____

11 $\sqrt{9g^4} =$

_____

12 $\sqrt{c^{12}d^2} =$

_____

13 $\pm\sqrt{2500x^{10}} =$

_____

**Name each domain and range. Determine if each relation is a function.**

14 $\{(-1,-1), (0, -2), (5, 6), (0, -5)\}$

domain = ______________________

range = ______________________

The relation is ______________________.

15

x | y
---|---
−3 → −1 |
0 → 0 |
8 → 0 |

domain = ______________________

range = ______________________

The relation is ______________________.

**Use a table of *x*- and *y*-values to graph each function.**

**16** $y = -3 - \frac{1}{2}x$

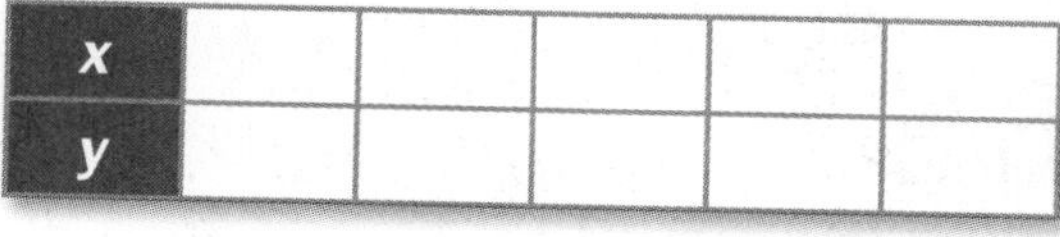

| x | | | | | |
|---|---|---|---|---|---|
| y | | | | | |

**17** $y = -2x - 2$

| x | | | | | |
|---|---|---|---|---|---|
| y | | | | | |

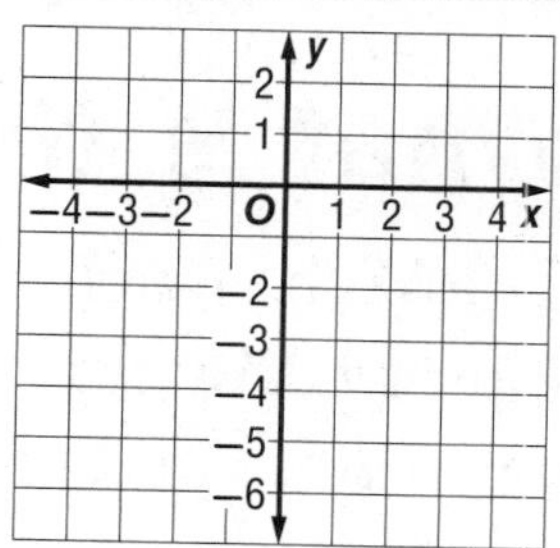


**Solve.**

**18** **GAME RENTALS** The Big Gamer Store charges a flat rate of \$10 to become a member and \$1.50 for each game rental. The function $G(x) = 10 + 1.5x$ represents the yearly cost of renting $x$ games. Make a table and graph the function. How much will Kareem pay if he rents 20 games in one year?

______________________________

| x | | | | | |
|---|---|---|---|---|---|
| y | | | | | |

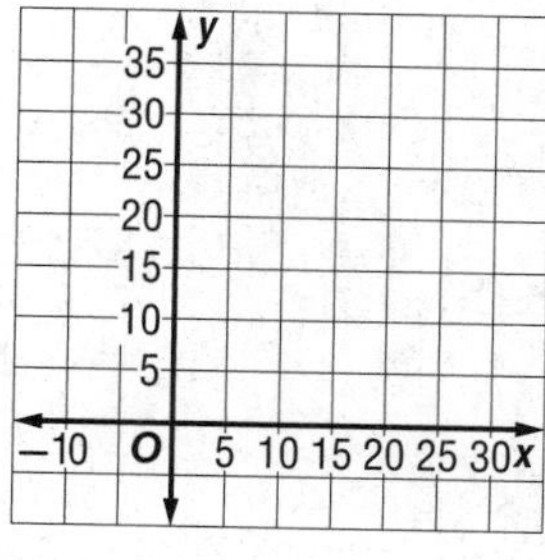

**19** **AREA** The area of the shaded region is represented by $32y^4$. Find the expression for the length of the side of the square.

______________________________

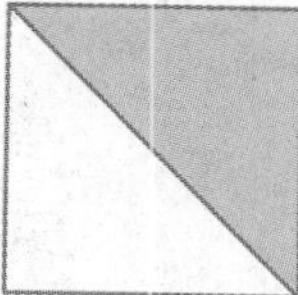

**Correct the Mistake**

**20** Yancy used a property of exponents to simply $(2x^3)^4$. His result was $8x^{12}$. Which properties did he use and did he simplify correctly? If not, what is the correct answer?

______________________________

______________________________

STOP

Chapter 2

# Equations and Inequalities

## *What is the better deal?*

One Web site charges a \$10 membership fee and \$0.50 per song to download music. Another Web site charges \$1.50 per song. You can use the equation $10 + 0.5x = 1.5x$ to find how many songs you would need to buy in order to spend the same amount of money at each site.

**STEP 1 Chapter Pretest** Are you ready for Chapter 2? Take the Chapter 2 Pretest to find out.

**STEP 2 Preview** Get ready for Chapter 2. Review these skills and compare them with what you will learn in this chapter.

| What You Know | What You Will Learn |
|---|---|
| You know the order of operations.<br>1. Simplify the expressions inside grouping symbols, like parentheses.<br>2. Find the value of all powers.<br>3. Multiply and divide in order from left to right.<br>4. Add and subtract in order from left to right.<br>**Example:**<br>$7 + 16 \div 8 = 7 + 2$ Divide.<br>$= 9$ Add. | *Lesson 2-1*<br>Follow the order of operations in reverse to solve a multi-step equation.<br>**Example:**<br>$3x + 2 = 17$ Given equation.<br>$-2 \quad -2$ Subtract.<br>$3x = 15$ Simplify.<br>$\frac{3x}{3} = \frac{15}{3}$ Divide.<br>$x = 5$ Simplify. |
| You know how to solve an equation.<br>**Example:** $4 \cdot 6 = 3 \cdot x$<br>$24 = 3x$<br>$\frac{24}{3} = \frac{3x}{3}$<br>$8 = x$<br>**TRY IT!**<br>1 $15 \cdot 7 = n \cdot 3$ $n =$ ______<br>2 $p \cdot 8 = 32 \cdot 2$ $p =$ ______ | *Lesson 2-3*<br>The **cross products** of a proportion are equal.<br>**Example:** $\frac{3}{4} = \frac{6}{8}$<br>$3 \cdot 8 = 4 \cdot 6$<br>$24 = 24$<br>You can use this fact to solve for an unknown in a proportion. |
| You know how to graph an inequality.<br>**Example:** $b \leq 3$<br>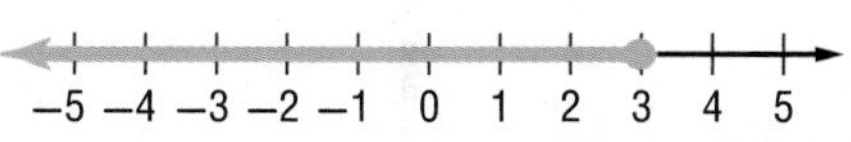 | *Lesson 2-5*<br>You can use inverse operations to solve multi-step inequalities.<br>$2x - 3 > 13$ Given equation.<br>$+3 \quad +3$ Add.<br>$2x > 16$ Simplify.<br>$\frac{2x}{2} > \frac{16}{2}$ Divide.<br>$x > 8$ Simplify. |

# Lesson 2-1 Solve Multi-Step Equations

## KEY Concept

You can model equations with algebra tiles.

$$4x + 5 = 21$$

> Use yellow tiles for positive numbers and red tiles for negative numbers.

Take away 5 tiles from each side.

Arrange the tiles into 4 equal groups.

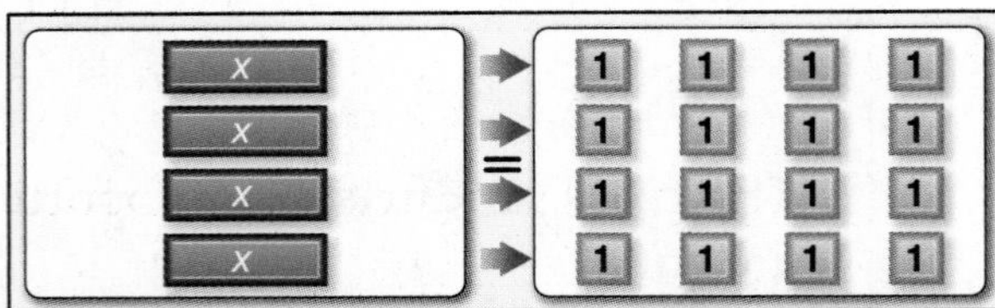

$$x = 4$$

Without a model, follow the order of operations in reverse to solve a multi-step equation. Use inverse operations to "undo" the operations in equations.

| | |
|---|---|
| $4x + 5 = 21$ | Given equation. |
| $\underline{-5 \quad -5}$ | Subtract to undo addition. |
| $4x = 16$ | Simplify. |
| $\frac{4x}{4} = \frac{16}{4}$ | Divide to undo multiplication. |
| $x = 4$ | Simplify. |

Check your solution to an equation by substituting your answer into the original equation for the variable.

### VOCABULARY

**equation**
a mathematical sentence that contains an equal sign, =, indicating that an expression on the left side of the equal sign has the same value as an expression on the right side

**inverse operations**
operations that undo each other

**order of operations**
rules that tell what order to use when evaluating expressions

## Example 1

**Solve $2x - 5 = -1$ using algebra tiles. Check the solution.**

1. Model the equation.

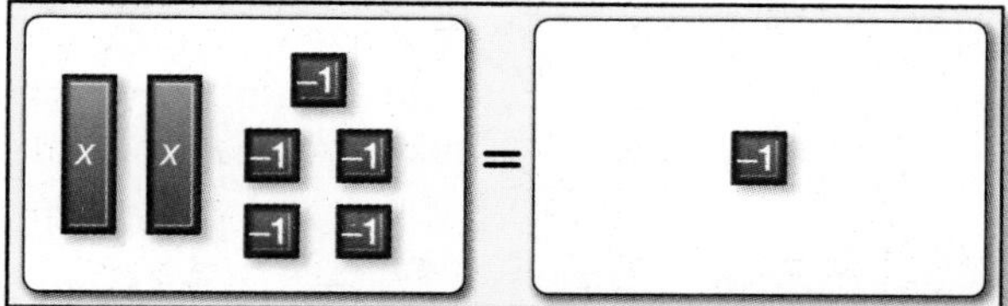

2. You need to remove 5 negative tiles from the left mat, but you do not have 5 negative tiles on the right mat to remove. To create zero pairs, add 5 positive tiles to each mat.

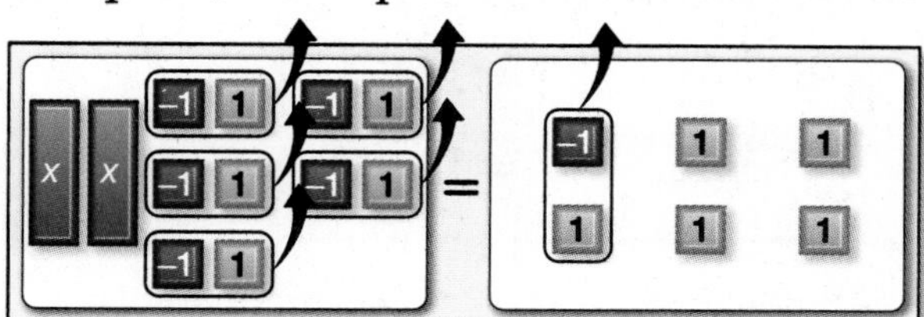

3. Remove any zero pairs. There are 2 $x$-tiles, so arrange the remaining tiles into 2 equal groups.

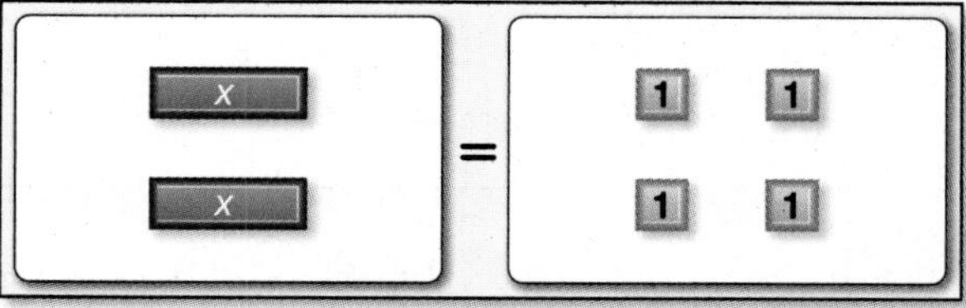

4. Write the solution.

   $x = 2$

5. Check.

   $2x - 5 = -1$

   $2(2) - 5 = -1$

   $4 - 5 = -1$ ✓

**YOUR TURN!**

**Solve $3n + 1 = 10$ using algebra tiles. Check the solution.**

1. Model the equation.

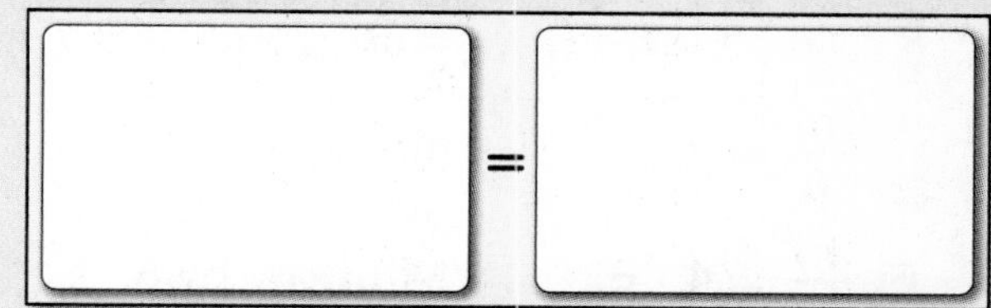

2. Remove ____ positive tile from each mat.

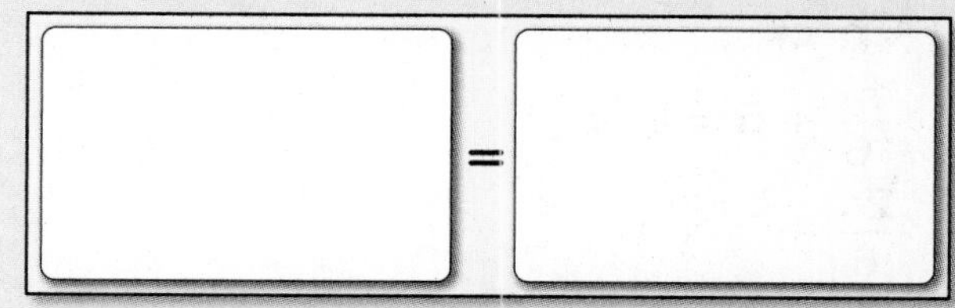

3. Remove any zero pairs. There are ____ $x$-tiles, so arrange the remaining tiles into ____ equal groups.

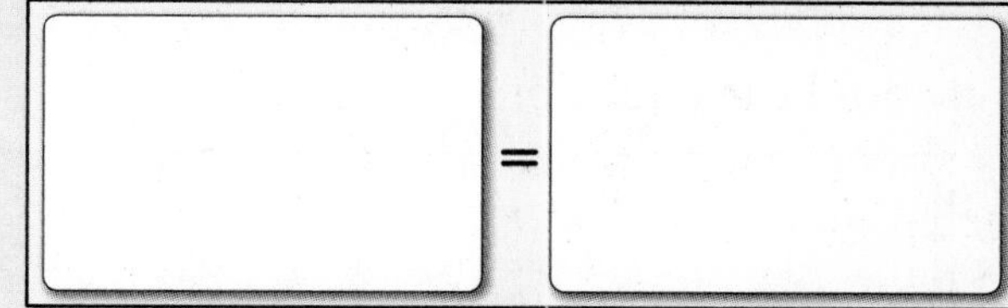

4. Write the solution.

   $n =$ ____

5. Check.

   $3n + 1 = 10$

   3____ $+ 1 = 10$

   ____ $+ 1 = 10$



GO ON

## Example 2

**Solve $\frac{m}{6} + 5 + 9$. Check the solution.**

1. Undo the operations in the equation.

| | |
|---|---|
| $\frac{m}{6} + 5 = 9$ | Given equation. |
| $-5 \quad -5$ | Subtract 5. |
| $\frac{m}{6} = 4$ | Simplify. |
| $6 \cdot \frac{m}{6} = 4 \cdot 6$ | Multiply by 6. |
| $m = 24$ | Simplify. |

2. Check.

$\frac{m}{6} + 5 = 9$

$\frac{24}{6} + 5 = 9$

$4 + 5 = 9$ ✓

**YOUR TURN!**

**Solve $-2a - 7 = 13$. Check the solution.**

1. Undo the operations in the equation.

| | |
|---|---|
| $-2a - 7 = 13$ | Given equation. |
| ________ | ________ |
| $-2a =$ ______ | Simplify. |
| $\frac{-2a}{\square} = \frac{20}{\square}$ | ________ |
| $a =$ ______ | Simplify. |

2. Check.

$-2a - 7 = 13$

$-2($______$) - 7 = 13$

______ $- 7 = 13$

## Guided Practice

**Solve each equation using algebra tiles.**

**1** $4x + 4 = 16$

Model the equation.

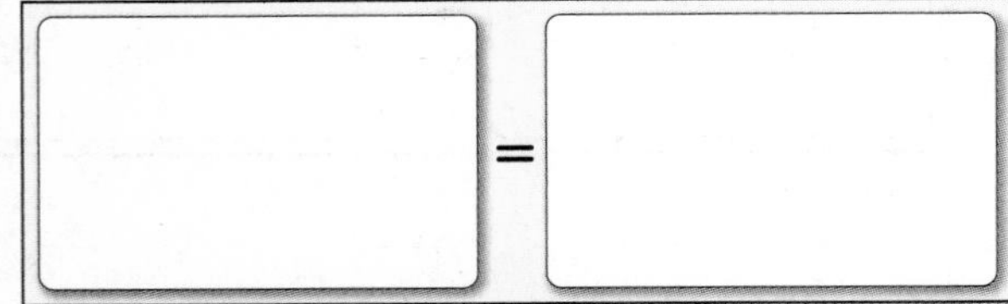

Remove 4 positive tiles from each mat.

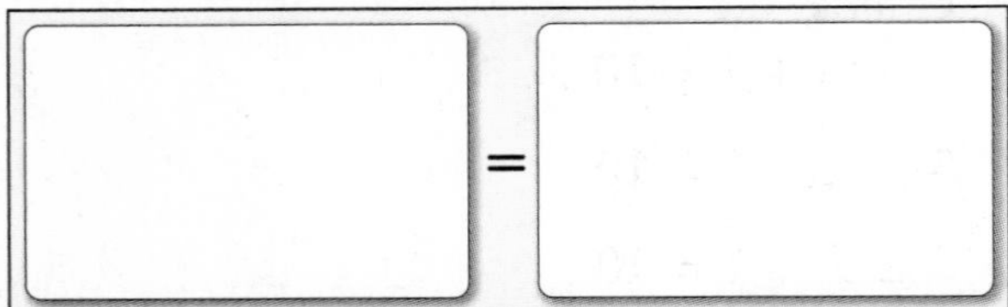

Remove any zero pairs. Arrange the remaining tiles into ____ equal groups.

Write the solution. $x =$ ____

**2** $3x - 5 = 10$

Model the equation.

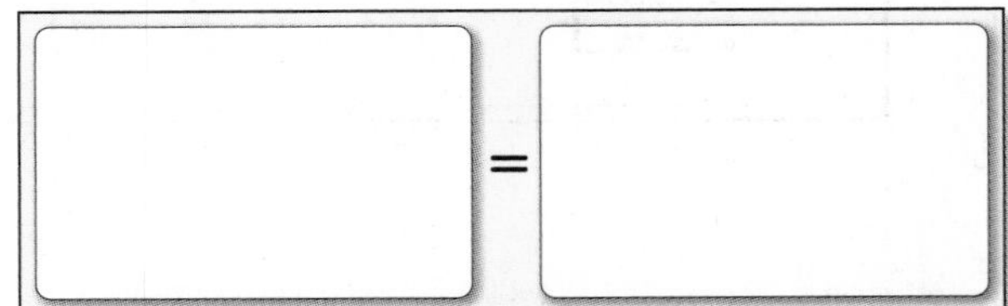

Add 5 positive tiles to each mat.

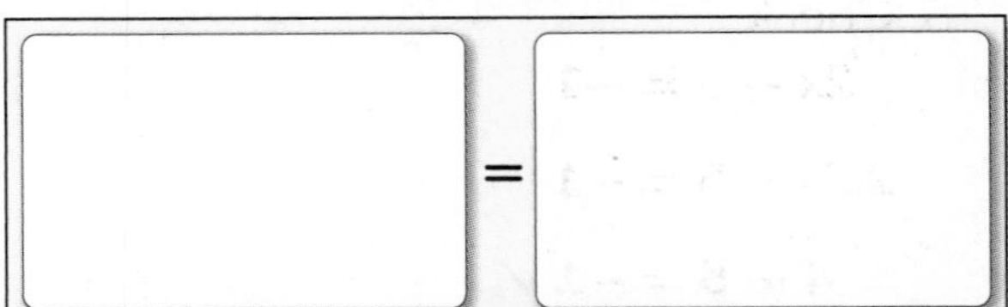

Remove any zero pairs. Arrange the remaining tiles into ____ equal groups.

Write the solution. $x =$ ____

## Step by Step Practice

**3** Solve $16 + 7z = 100$.

**Step 1** Write the given equation. $16 + 7z = 100$

**Step 2** Subtract 16 from each side of the equation. ____________

**Step 3** Simplify. ________ = ________

**Step 4** Divide each side of the equation by 7. $\frac{7z}{\square} = \frac{84}{\square}$

**Step 5** Simplify. $z =$ ________

**Solve each equation.**

**4** $8b + 4 = 60$

| | |
|---|---|
| $8b + 4 = 60$ | Given equation. |
| ____________ | ____________ |
| $8b =$ ______ | Simplify. |
| $\frac{8b}{\square} = \frac{56}{\square}$ | ____________ |
| $b =$ ______ | Simplify. |

**5** $\frac{k}{-12} + 20 = 17$

| | |
|---|---|
| $\frac{k}{-12} + 20 = 17$ | Given equation. |
| ____________ | ____________ |
| $\frac{k}{-12} =$ ______ | Simplify. |
| ______ $\frac{k}{-12} = -3$ ______ | ____________ |
| $k =$ ______ | Simplify. |

## Step by Step Problem-Solving Practice

**Solve.**

**6** **FLOWERS** Anne wants to send flowers to each of her four assistants. Anne spends \$87 on four identical arrangements, including a delivery charge of \$15. How much did each arrangement cost?

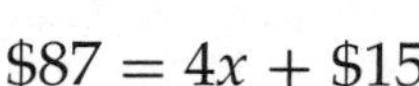

$\$87 = 4x + \$15$

Check off each step.

______ **Understand: I underlined key words.**

______ **Plan: To solve the problem, I will** ____________________.

______ **Solve: The answer is** ____________________.

______ **Check: I checked my answer by** ____________________.

GO ON

## Skills, Concepts, and Problem Solving

**Solve each equation. Check the solution.**

**7** $\frac{x}{2} - 6 = 2$

**8** $7 + 3y = 28$

**9** $\frac{b}{4} + 8 = 20$

**10** $-5 + 9g = 13$

**11** $\frac{m}{-7} - 5 = -8$

**12** $1 - \frac{k}{5} = 0$

**13** $35 = \frac{h}{9} + 23$

**14** $15 - 10y = -15$

**15** $\frac{b}{3} + 7 = 4$

**Solve.**

**16** **GROCERIES** The table at the right shows the prices of meat at a deli. Brandy purchases three pounds of turkey and one pound of roast beef. If she spends $25 on lunch meat, how much does the turkey cost per pound?

| Meat | Price per Pound |
|---|---|
| Ham | $5.50 |
| Roast Beef | $7.00 |
| Turkey | *p* |

_______________

**17** **NUMBERS** Five less than half a number is thirteen. What is the number? _______________

**Vocabulary Check** **Write the vocabulary word that completes each sentence.**

**18** Subtraction is the _______________ of addition because they undo each other.

**19** A(n) _______________ is a mathematical sentence in which the left side of the equal sign has the same value as the right side of the equal sign.

**20** **Reflect** Explain why addition and subtraction is usually the first step to solving a two-step equation.

_______________

_______________

_______________

STOP

Lesson 2-2

# Solve Equations with Variables on Each Side

## KEY Concept

To solve equations with variables on both sides, first add or subtract to get the variable terms on the same side of the equation.

| | |
|---|---|
| $-x + 3 = -2x + 2$ | Given equation. |
| $+2x \quad +2x$ | Add $2x$. |
| $x + 3 = 2$ | Simplify. |
| $-3 \quad -3$ | Subtract 3. |
| $x = -1$ | Simplify. |

Check your solution.

$$-x + 3 = -2x + 2$$
$$-(-1) + 3 = -2(-1) + 2$$
$$1 + 3 = 2 + 2$$
$$4 = 4 \checkmark$$

### VOCABULARY

**Distributive Property**
to multiply a sum by a number, you can multiply each addend by the number and add the products

**equation**
a mathematical sentence that contains an equal sign, =, indicating that an expression on the left side of the equal sign has the same value as an expression on the right side

**like terms**
terms that have the same variables to the same power

## Example 1

**Solve the equation $3s + 5 = s - 3$.**

1. Subtract $s$ from each side to keep the variable positive.

$$3s + 5 = s - 3$$
$$-s \qquad -s$$
$$2s + 5 = -3$$

2. Subtract 5 from each side.

$$2s + 5 = -3$$
$$-5 \quad -5$$
$$2s = -8$$

3. Divide by 2 on each side.

$$\frac{2s}{2} = \frac{-8}{2}$$
$$s = -4$$

## YOUR TURN!

**Solve the equation $-2 + 4d = 2d + 4$.**

1. ________________ from each side to keep the variable positive.

$$-2 + 4d = 2d + 4$$

______________

_____ + _____ = _____

2. ____________ to each side.

$$-2 + 2d = 4$$

______________

_____ = _____

3. ____________ on each side.

$$\frac{2d}{\square} = \frac{6}{\square}$$

$d =$ _____

GO ON

## Example 2

**Solve $12 - 2w = -5(w + 3)$.**

1. Use the Distributive Property on the right side of the equation. Combine like terms if possible.

$12 - 2w = -5(w + 3)$

$12 - 2w = -5w - 15$

2. Use inverse operations to solve.

$12 - 2w = -5w - 15$

$\underline{+2w \quad +2w}$ Add $2w$.

$12 = -3w - 15$ Simplify.

$\underline{+15 \qquad +15}$ Add 15.

$\frac{27}{-3} = \frac{-3w}{-3}$ Simplify. Divide by $-3$.

$-9 = w$ Simplify.

3. Check the solution.

$12 - 2w = -5(w + 3)$

$12 - 2(-9) = -5(-9 + 3)$

$12 + 18 = -5(-6)$

$30 = 30$ ✓

**YOUR TURN!**

**Solve $-7q + 17 = 5 - 6(q - 1)$.**

1. Use the Distributive Property on the right side of the equation. Combine like terms if possible.

$-7q + 17 = 5 - 6(q - 1)$

$-7q + 17 = 5$ ________

$-7q + 17 =$ _____ $- 6q$

2. Use inverse operations to solve.

$-7q + 17 = 11 - 6q$

__________ __________

$17 = 11$ _____ Simplify.

__________ __________

_____ $= q$ Simplify.

3. Check the solution.

$-7q + 17 = 5 - 6(q - 1)$

$-7($_____$) + 17 = 5 - 6($_____ $- 1)$

_____ $+ 17 = 5 - 6($_____$)$

_____ = _____

## Guided Practice

**Name the operation for each step.**

**1** $3 + 2x = 4x - 5$

$3 = 2x - 5$ Subtract $2x$.

$8 = 2x$ __________

$4 = x$ __________

**2** $3h - 5 = -2h + 10$

$5h - 5 = 10$ Add $2h$.

$5h = 15$ __________

$h = 3$ __________

**Solve each equation. Name the operation for each step.**

**3** $-8 + 7x = 3x + 16$

______ ______

$-8 + \_\_\_ = 16$ Simplify.

______ ______

$\frac{4x}{\square} = \frac{\_\_\_}{\square}$ Simplify.

______

$x = \_\_\_$ Simplify.

**4** $6m - 4 = 3m + 32$

______ ______

$\_\_\_ - 4 = 32$ Simplify.

______ ______

$\frac{3m}{\square} = \frac{\_\_\_}{\square}$ Simplify.

______

$m = \_\_\_$ Simplify.

## Step by Step Practice

**5** Solve $-\frac{1}{2}(y - 4) = -y + 5$.

**Step 1** Use the distributive property.

$$-\frac{1}{2}(y - 4) = -y + 5$$

$$\frac{-y}{\square} + \_\_\_ = -y + 5$$

Recall that $\frac{1}{2} \cdot y = \frac{1}{2} \cdot \frac{y}{1} = \frac{y}{2}$.

**Step 2** Use inverse operations to solve. Name the operation for each step.

$$\frac{-y}{\square} + \_\_\_ = -y + 5$$

______ ______

$\frac{y}{\square} + \_\_\_ = \_\_\_$ Simplify.

______ ______

$\frac{y}{\square} = \_\_\_$ Simplify.

$\_\_\_ \quad \frac{y}{\square} = \_\_\_$ ______

$y = \_\_\_$ Simplify.

**Step 3** Check your solution.

$$-\frac{1}{2}(y - 4) = -y + 5$$

$$-\frac{1}{2}(\_\_\_ - 4) = -\_\_\_ + 5$$

$$-\frac{1}{2}(\_\_\_) = \_\_\_$$

$$-1 = -1$$

GO ON

**Solve each equation.**

6 $-3(2x - 1) = 11 + 2x$

$\_\_\_\_ + \_\_\_\_ = 11 + 2x$

$+ 6x \qquad\qquad + 6x$

$3 = 11 + \_\_\_\_$

$\_\_\_\_\_\_\_\_$

$\_\_\_\_ = \_\_\_\_$

$\_\_\_\_ = \_\_\_\_$

7 $\frac{x}{15} - 10 = 2(\frac{x}{15} - 6)$

$\frac{x}{15} - 10 = \_\_\_\_ - \_\_\_\_$

$\_\_\_\_\_\_\_\_$

$\_\_\_\_ = \_\_\_\_ - 12$

$\_\_\_\_\_\_\_\_$

$\_\_\_\_ = \_\_\_\_$

$\_\_\_\_ = \_\_\_\_$

## Step by Step Problem-Solving Practice

**Solve.**

8 **EMPLOYMENT** This week, Ming and Jack worked the same number of hours and earned the same amount of money. Ming earned \$10 an hour and received a \$100 bonus. Her salary is modeled by the expression $10h + 100$, where $h$ is the number of hours she worked. Jack earned a rate of \$15 an hour, modeled by the monomial $15h$. How many hours did Ming and Jack each work this week?

Ming's earnings = Jack's earnings

________ = ________

________________

________ = ________

________ = ________

Check off each step.

____ **Understand: I underlined key words.**

____ **Plan: To solve the problem, I will** ________.

____ **Solve: The answer is** ________.

____ **Check: I checked my answer by** ________.

## Skills, Concepts, and Problem Solving

**Solve each equation.**

9 $3x + 6 = 4x - 2$

10 $36 + s = 8(2s - 3)$

11 $-2q + 42 = 4 - 6(3q - 1)$

12 $2b - 25 = -3b + 35$

13 $-3t + 8 = 2t - 22$

14 $9 - 3(2 + 2d) = 3d - 24$

15 $-\frac{1}{2}(a + 12) = 4 - a$

16 $20 - 5w = 6(w - 4)$

17 $\frac{1}{4}(z - 16) = \frac{z}{2} - 10$

**Solve.**

18 **GEOMETRY** The perimeters of the figures shown at the right are equal. What is the perimeter of each figure?

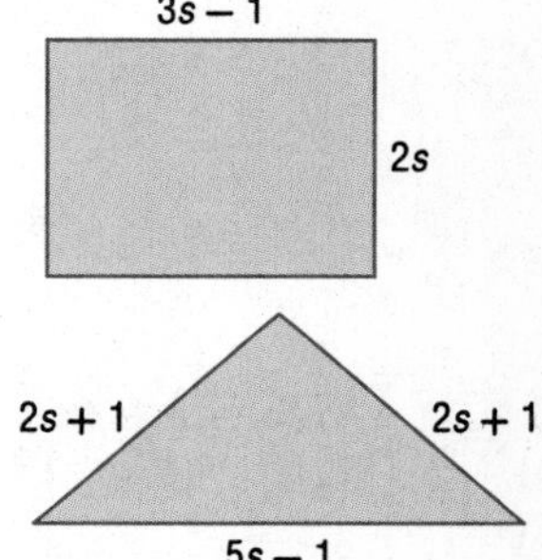


______________________________

______________________________

19 **TRUCK RENTAL** The Movers Moving Company charges $30 for the day and $0.40 per mile. This charge is modeled by the expression $30 + $0.40*m*, where *m* is the number of miles the truck is driven. The Luxury Moving Company charges $50 for the day and $0.30 per mile, modeled by the expression $50 + $0.30*m*. For what mileage will the charges of the companies be equal?

______________________________

**Vocabulary Check** **Write the vocabulary word that completes each sentence.**

20 Terms like $3b$ and $-2b$, that have the same variables to the same exponent, are called ____________.

21 In a(n) ____________ the expressions on either side of the equal sign have the same value.

22 **Reflect** Instead of using the Distributive Property as your first step in solving the equation $-\frac{1}{4}(x - 16) = x - 2$, what could you do to eliminate the fraction in the problem?

______________________________

Chapter 2

# Progress Check 1 (Lessons 2-1 and 2-2)

**Solve each equation.**

1 $50 + 2d = 100$

2 $7 - \frac{x}{8} = 5$

3 $-8 = 12 - 4n$

4 $-w + 17 = 20$

5 $24 = -6 - 5z$

6 $18 - y = -5$

7 $-3g - 12 = -6(2g + 5)$

8 $18 - 4(p - 4) = 5p - 2$

9 $8 - x = 2x - 13$

10 $7v - 6 = 3v + 6$

11 $-26 + 2y = 4y - 20$

12 $2 - 5(6 + d) = 2 + d$

**Solve.**

13 **GEOMETRY** The perimeters of the figures shown are equal. What is the perimeter of each figure?

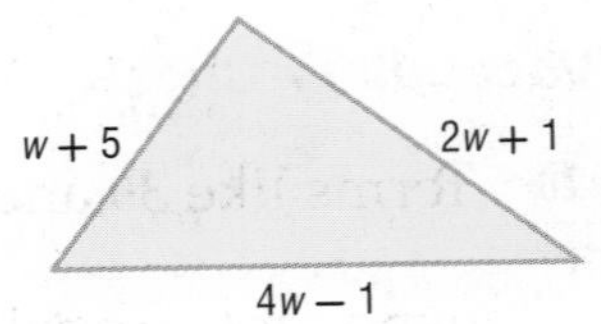


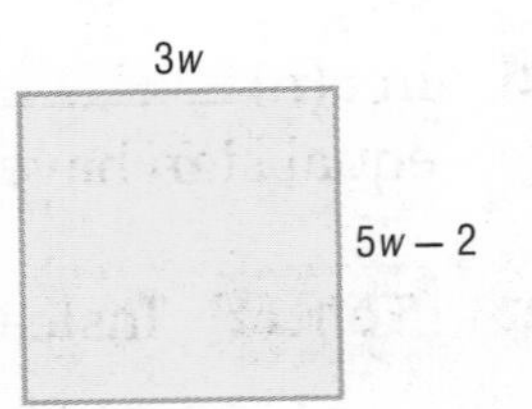


______

14 **TEMPERATURE** To convert a temperature from Celsius to Fahrenheit, use the formula $F = 1.8C + 32$, where $F$ is the temperature in Fahrenheit, and $C$ is the Celsius temperature. If the outside temperature is 87°F, what is the Celsius temperature to the nearest degree?

______

Lesson 2-3

# Proportions

## KEY Concept

Proportions are related to rates and ratios.

**Ratios** compare two different values. Ratios can be written in three forms.

4 to 5     4:5     $\frac{4}{5}$

A **rate** is a special type of ratio that compares units of measurements or amounts.

64 feet in 4 minutes     $45 for 3 shirts

A **unit rate** is a rate where one of the quantities is a single unit.

16 feet in 1 minute     $15 for 1 shirt

A **proportion** is is an equation that shows two ratios or rates are equivalent.

$$\frac{a}{b} = \frac{c}{d}$$

The **cross products** of a proportion are equal. You can use this fact to solve for an unknown in a proportion.

$\frac{a}{b} \times \frac{c}{d}$     $a \cdot d = b \cdot c$

$\frac{5}{8} \times \frac{10}{16}$     $5 \cdot 16 = 8 \cdot 10$

$80 = 80$

### VOCABULARY

**cross product**
in a proportion, the product of the numerator of one ratio and the denominator of the other ratio

**proportion**
an equation stating that two ratios or rates are equivalent

**rate**
a ratio of two measurements or amounts made with different units

**ratio**
a comparison of two quantities by division

**unit rate**
a rate that describes how many units of the first type of quantity are equal to 1 unit of the other type of quantity

## Example 1

**Determine whether the ratios $\frac{2}{3}$ and $\frac{8}{12}$ form a proportion.**

$$\frac{2}{3} \stackrel{?}{=} \frac{8}{12}$$

$$2 \cdot 12 \stackrel{?}{=} 3 \cdot 8$$

$$24 = 24 \checkmark$$

The ratios form a proportion because the cross products are equal.

### YOUR TURN!

**Determine whether the ratios $\frac{4}{7}$ and $\frac{6}{11}$ form a proportion.**

$$\frac{4}{7} \stackrel{?}{=} \frac{6}{11}$$

$$4 \cdot 11 \stackrel{?}{=} 7 \cdot 6$$

_____ ◯ _____

The ratios do not form a proportion because the cross products are not equal.

GO ON

## Example 2

**Solve $\frac{9}{5} = \frac{x}{15}$.**

Proportions are solved like equations. Isolate the variable.

$$\frac{9}{5} = \frac{x}{15}$$

1. Cross multiply. $9 \cdot 15 = 5 \cdot x$
2. Simplify. $135 = 5x$
3. Divide by 5. $\frac{135}{5} = \frac{5x}{5}$
4. Simplify. $27 = x$

**YOUR TURN!**

**Solve $\frac{12}{x} = \frac{6}{3}$.**

$$\frac{12}{x} = \frac{6}{3}$$

1. Cross multiply. $12 \cdot 3 = x \cdot 6$
2. Simplify. ______ $= 6x$
3. Divide by ______. $\frac{\square}{\square} = \frac{6x}{\square}$
4. Simplify. ______ $= x$

## Guided Practice

**Determine whether each set of ratios form a proportion. Write = or ≠ to make each a true statement.**

1 $\frac{3}{8}$ and $\frac{5}{12}$

$$\frac{3}{8} \stackrel{?}{=} \frac{5}{12}$$

______ $\cdot\ 12 \stackrel{?}{=} 8\ \cdot$ ______

The ratios ____________ form a proportion.

2 $\frac{4}{5}$ and $\frac{12}{15}$

$$\frac{4}{5} \stackrel{?}{=} \frac{12}{15}$$

______ $\cdot\ 15 \stackrel{?}{=} 5\ \cdot$ ______

The ratios ____________ form a proportion.

## Step by Step Practice

3 Solve $\frac{12}{21} = \frac{4}{c}$.

**Step 1** Cross multiply. ______ $\cdot$ ______ $=$ ______ $\cdot$ ______

______ $=$ ______

**Step 2** Divide by ______. $\frac{\square}{\square} = \frac{\square}{\square}$

**Step 3** Simplify. ______ $=$ ______

**Solve each proportion.**

4 $\frac{10}{x} = \frac{5}{4}$

_____ $\cdot$ 4 = _____ $\cdot$ 5 Cross multiply.

_____ = _____ Simplify.

$\frac{40}{\square} = \frac{5x}{\square}$ Divide by _____.

_____ $= x$ Simplify.

5 $\frac{2}{8} = \frac{x}{20}$

_____ $\cdot$ 20 = _____ $\cdot\, x$ Cross multiply.

_____ = _____ Simplify.

$\frac{40}{\square} = \frac{8x}{\square}$ Divide by _____.

_____ $= x$ Simplify.

6 $\frac{6}{x} = \frac{12}{3}$

_____ $\cdot$ 3 = _____ $\cdot$ 12

_____ = _____

$\frac{18}{\square} = \frac{12x}{\square}$

_____ $= x$

7 $\frac{x}{9} = \frac{4}{6}$

_____ $\cdot$ 6 = _____ $\cdot$ 4

_____ = _____

$\frac{6x}{\square} = \frac{36}{\square}$

$x =$ _____

8 $\frac{8}{20} = \frac{20}{x}$

9 $\frac{x}{9} = \frac{15}{27}$

## Step by Step Problem-Solving Practice

**Solve.**

10 **HEALTH** Angela's heart rate is 68 beats per minute. How many times does Angela's heart beat in 3 minutes?

(Hint: Set up fractions as $\frac{\text{beats}}{\text{minutes}}$.)

Check off each step.

_____ **Understand: I underlined key words.**

_____ **Plan: To solve the problem, I will** _____________________.

_____ **Solve: The answer is** _____________________.

_____ **Check: I checked my answer by** _____________________.

GO ON

## Skills, Concepts, and Problem Solving

**Solve each proportion.**

11 $\frac{16}{x} = \frac{4}{5}$

12 $\frac{12}{4} = \frac{3}{x}$

13 $\frac{9}{11} = \frac{x}{44}$

14 $\frac{4.5}{12} = \frac{x}{36}$

15 $\frac{6}{10} = \frac{x}{14}$

16 $\frac{6}{x} = \frac{8}{16}$

**Solve.**

17 **NATURE** In a certain rural area, the population of cicadas during the year of their emergence is 0.25 million per acre. On an 80 acre farm, what is the expected population of cicadas?

_______________

**SURVEY** A survey of 50 students from the ninth grade is a representative sample of the 800 students in the entire ninth grade class.

| Question | No | Yes |
|---|---|---|
| Do you ride the bus to school? | 18 | 32 |
| Do you have more than two siblings? | 29 | 21 |
| Do you have a pet at home? | 13 | 37 |
| Are you involved in a school-sponsored extracurricular activity? | 24 | 26 |

18 Use the table to create a proportion to predict the number of students in the entire ninth grade who are involved in a school-sponsored extracurricular activity. _______________

19 Predict the number of students who have more than two siblings. _______________

**Vocabulary Check** **Write the vocabulary word that completes each sentence.**

20 A(n) _______________ shows that two ratios or rates are equivalent.

21 A comparison of two numbers by division is a(n) _______________.

22 **Reflect** Suppose you have a list of twenty unit rates. If you write each of them as a ratio, what will each ratio have in common? Explain.

_______________

_______________

Lesson

# 2-4 Solve for a Specific Variable

## KEY Concept

When solving for a specific variable in an equation, there may be more variables than just the one for which you are solving.

Use **inverse operations** to solve an equation for a specific variable. Inverse operations help isolate the term or terms with the specific variable. The following are common inverse operations used in solving equations.

| Operation | Inverse Operation |
|---|---|
| Addition (+) | Subtraction (−) |
| Subtraction (−) | Addition (+) |
| Multiplication (×) | Division (÷) |
| Division (÷) | Multiplication (×) |

**VOCABULARY**

**equation**
a mathematical sentence that contains an equal sign, =

**inverse operations**
operations that undo each other

## Example 1

**Solve $h = 4k + 3$ for $k$.**

1. Locate the term with the variable $k$.
   $4k$
2. What do you need to do to get the $k$ term by itself on one side of the equal sign?
   Subtract 3.
   $$\begin{aligned} h &= 4k + 3 \\ -3 & \quad\ \ -3 \\ \hline h - 3 &= 4k \end{aligned}$$
3. What do you need to do to get the $k$ by itself?
   Divide by 4.
   $$\frac{h-3}{4} = \frac{4k}{4}$$
4. Simplify.
   $$\frac{h-3}{4} = k$$

**YOUR TURN!**

**Solve $c = by - d$ for $b$.**

1. Locate the term with the variable $b$.

   ______________

2. What do you need to do to get the $b$ term by itself on one side of the equal sign? ______________
   $$c = by - d$$
   ______________
   $$c \text{ \_\_\_\_\_\_ } = by$$
3. What do you need to do to get $b$ by itself?
   $$\frac{c+d}{\square} = \frac{by}{\square}$$
   ______________
4. Simplify.
   ______________ $= b$

GO ON

## Example 2

Solve $\frac{a + 5}{6} = c$ for $a$.

1. Locate the term with the variable *a*.
   a

2. Can you get the *a* term by itself on one side of the equal sign with one inverse operation? no

3. Multiply each side of the equation by 6 so that you can isolate the *a* term.

$$\frac{a + 5}{6} = c$$

$$6 \cdot \frac{a + 5}{6} = 6 \cdot c$$

$$a + 5 = 6c$$

4. Perform the inverse operations.

$$a + 5 = 6c$$
$$-5 \quad -5$$
$$a = 6c - 5$$

**YOUR TURN!**

Solve $m = \frac{n - 2}{-7p}$ for $n$.

1. Locate the term with the variable *n*.

   _____

2. Can you get the *n* term by itself on one side of the equal sign with one inverse operation? _____

3. ________ each side of the equation by ________ so that you can isolate the *n* term.

$$m = \frac{n - 2}{-7p}$$

________ $m = \frac{n - 2}{-7p}$ ________

________ = ________

4. Perform the inverse operations.

________ = ________

________________

________ $= n$

## Guided Practice

**Solve for *x*. Name the inverse operation for each step.**

1. $c = 7x + 9$

   $c - 9 = 7x$ Subtract 9.

   $\frac{c - 9}{7} = x$ ____________

2. $y = 10x + z$

   $y - z = 10x$ Subtract z.

   $\frac{y - z}{10} = x$ ____________

3. $\frac{x + 12}{5} = p$

   $x + 12 = 5p$ ____________

   $x = 5p - 12$ ____________

4. $w = \frac{x - 9}{11y}$

   $11yw = x - 9$ ____________

   $11yw + 9 = x$ ____________

## Step by Step Practice

**5** Solve $ab = -3d + 2c$ for $d$.

**Step 1** Locate the term with the variable $d$.

______________

**Step 2** What is the first inverse operation to perform? $ab = -3d + 2c$

______________ ______________

______________ = ______________

**Step 3** What is the next inverse operation to perform? ______________ = ______________

______________ ______________ = ______________

**Step 4** Simplify. ______________ = ______________

**Identify the two inverse operations needed to solve for the given variable. Solve each equation for the given variable.**

**6** $x = -8z + 3y$ for $z$

first inverse operation: ______________

next inverse operation: ______________

$z =$ ______________

**7** $-a = 3b - c$ for $b$

first inverse operation: ______________

next inverse operation: ______________

$b =$ ______________

**8** $3r = -2s - 5q$ for $q$

first inverse operation: ______________

next inverse operation: ______________

$q =$ ______________

**9** $\frac{k - 7}{10} = j$ for $k$

first inverse operation: ______________

next inverse operation: ______________

$k =$ ______________

GO ON

## Step by Step Problem-Solving Practice

**Solve.**

**10** **BIKING** Kendrick is racing his younger brother Jordan on their new bikes. The distance $d$ (in yards) is modeled by the equation $d = vt$, where $v$ is the velocity (in yards per second), and $t$ is the time (in seconds). If Kendrick gives Jordan a 20-yard head start, the distance equation becomes $d = vt + 20$. Solve the new distance equation for $v$.

$$d = vt + 20$$

______________

__________ = __________

__________ = __________

Check off each step.

_____ **Understand: I underlined key words.**

_____ **Plan: To solve the problem, I will** ____________________.

_____ **Solve: The answer is** ____________________.

_____ **Check: I checked my answer by** ____________________.

## Skills, Concepts, and Problem Solving

**Solve each equation for *x*.**

**11** $m = -3x - 19$

**12** $-5r = 6x + 9q$

**13** $w = \dfrac{x + 20}{-10y}$

**14** $-35e = \dfrac{13 + x}{f}$

**15** $18w = 5x + 11q$

**16** $2j = \dfrac{-18 + x}{2k}$

**Solve each equation for $x$.**

17 $12k = 7j - x$

18 $u = \frac{x - 1}{-17v}$

19 $9x - 23b = -14d$

**Solve. Write the answer in simplest form.**

20 **CIRCLES** The diameter of a circle pictured at the right is twice the radius. If 14 inches is added to the diameter of a circle, then the formula for the diameter is given by the equation $d = 2r + 14$. Solve the new equation for the radius $r$.

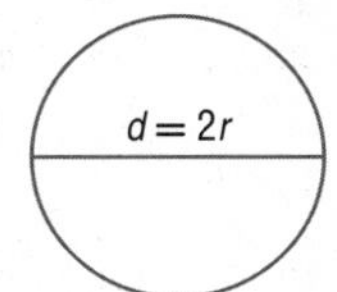


______________________________

21 **DISTANCE** Grace is controlling a remote control car. When the car is 15 feet from Grace, it begins to travel at a constant velocity $v$ (ft/sec). At time $t$ (sec), the distance $d$ (ft) that the car is from Grace is given by the formula $d = vt + 15$. Solve this equation for $t$. Using the new equation, at what time will the car be 100 feet from Grace if its velocity is 8 ft/sec? Round your answer to the nearest tenth of a second.

______________________________

**Vocabulary Check** **Write the vocabulary word that completes each sentence.**

22 Operations that undo each other, such as multiplication and division, are called ______________________.

23 A(n) ______________________ is a mathematical sentence that contains an equal sign.

24 **Reflect** Exercise 20 used a formula from geometry that contains multiple variables. What is another formula for a geometric measurement that contains multiple variables? Solve this formula for a different variable.

______________________________

______________________________

STOP

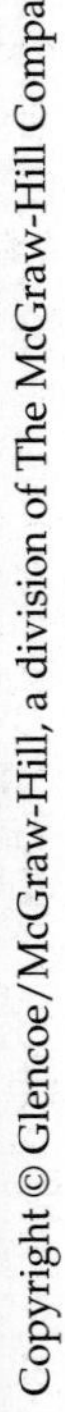

# Chapter 2 Progress Check 2 (Lessons 2-3 and 2-4)

**Solve each proportion.**

1 $\frac{8}{x} = \frac{12}{30}$

2 $\frac{b}{9} = \frac{15}{27}$

3 $\frac{7}{6} = \frac{14}{y}$

4 $\frac{11}{10} = \frac{z}{5}$

5 $\frac{w}{30} = \frac{3}{8}$

6 $\frac{10}{x} = \frac{4}{7}$

**Solve each equation for *x*.**

7 $-3z = \frac{x - 2}{9}$

8 $-16m - x = 6h$

9 $8g = -4h - 3x$

10 $m = -3x - 19$

11 $-5r = 6x + 9q$

12 $w = \frac{6 + x}{10 - y}$

**Solve.**

13 **GEOMETRY** The formula for the perimeter of a rectangle is $P = 2\ell + 2w$. Suppose you know the perimeter and the width, but need to find the length. Solve this formula for $\ell$ so that you have a formula for finding the length when the perimeter and width are given.

____________________________

14 **GROCERIES** Selma spent \$7.08 on 12 pounds of bananas for the cross country team to eat after a race. Use a proportion to determine the price of bananas per pound.

____________________________

Lesson 2-5

# Solve Multi-Step Inequalities

## KEY Concept

Solving inequalities differs from solving equations.

- An **inequality** symbol is used in place of an equal sign.
- You change the direction of the inequality symbol when you multiply or divide by a negative number.
- The solution is a range of numbers.

**Inequality Symbols**

| | | | |
|---|---|---|---|
| $<$ | is less than | $\leq$ | is less than or equal to |
| $>$ | is greater than | $\geq$ | is greater than or equal to |
| $\neq$ | is not equal to | | |

**Graphing Inequalities**

The number line below shows $-1 < x$ or $x \geq 2$.

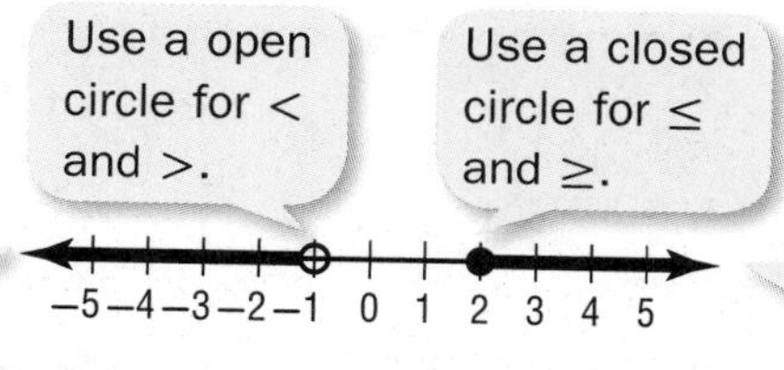


### VOCABULARY

**inequality**
a number sentence that compares two unequal expressions and uses $<$, $>$, $\leq$, $\geq$, or $\neq$

**inverse operations**
operations that undo each other

## Example 1

**Solve $\frac{x}{2} - 3 \leq 3$. Graph the solution.**

1. Use inverse operations to solve.

| | |
|---|---|
| $\frac{x}{2} - 3 \leq 3$ | Given inequality. |
| $+3 + 3$ | Add 3. |
| $\frac{x}{2} \leq 6$ | Simplify. |
| $2 \cdot \frac{x}{2} \leq 2 \cdot 6$ | Multiply by 2. |
| $x \leq 12$ | Simplify. |

2. Graph the solution. Draw a closed circle at 12 and shade to the left.

5 6 7 8 9 10 11 12 13 14 15

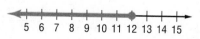

## YOUR TURN!

**Solve $6y + 4 > -20$. Graph the solution.**

1. Use inverse operations to solve.

| | |
|---|---|
| $6y + 4 > -20$ | Given inequality. |
| ________ | ________ |
| $\frac{6y}{\square} > \frac{\_\_\_\_}{\square}$ | Simplify. |
| | ________ |
| $y >$ ____ | Simplify. |

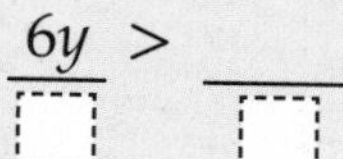

2. Graph the solution. Draw ____________ at ________ and shade to the ________.

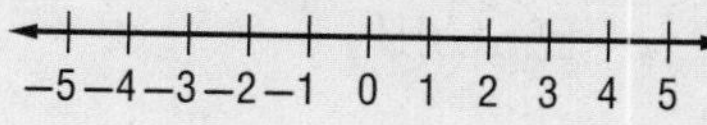


GO ON

## Example 2

**Solve $\frac{d+7}{-2} < -4$. Graph the solution.**

1. Use inverse operations to solve.

| | |
|---|---|
| $\frac{d+7}{-2} < -4$ | Given. |
| $-2 \cdot \frac{d+7}{-2} > -4 \cdot -2$ | Multiply by −2. |
| $d + 7 > 8$ | Simplify. |
| $\underline{-7 \quad -7}$ | Subtract 7. |
| $d > 1$ | Simplify. |

Multiplying or dividing by a negative number flips the inequality symbol.

2. Graph the solution. Draw an open circle at 1 and shade to the right.

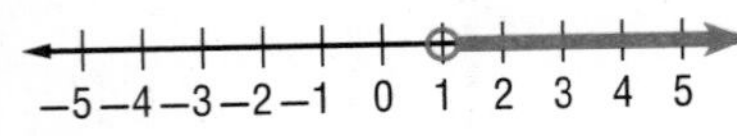


**YOUR TURN!**

**Solve $-6a - 3 \leq 15$. Graph the solution.**

1. Use inverse operations to solve.

| | |
|---|---|
| $-6a - 3 \leq 15$ | Given. |
| ______ | ______ |
| $\frac{-6y}{\square} > \frac{\_\_\_\_}{\square}$ | Simplify. |
| | ______ |
| $a \geq$ ______ | Simplify. |

2. Graph the solution. Draw ______ at ______ and shade to the ______.

## Guided Practice

**Solve each inequality. Graph the solution.**

1. $7y + 9 < -12$

______ ______

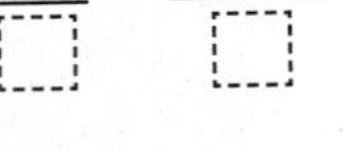

$\frac{7y}{\square} < \frac{\_\_\_\_}{\square}$ Simplify.

______

$y <$ ______ Simplify.

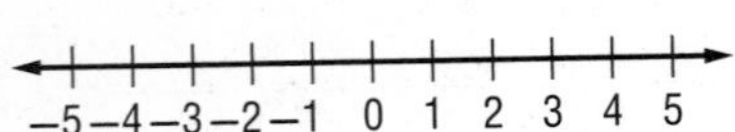


2. $\frac{x}{4} - 9 \geq -6$

______ ______

$\frac{x}{4} \geq$ ______ Simplify.

______ $\frac{x}{4} \geq$ ______ 3 ______

$x \geq$ ______ Simplify.

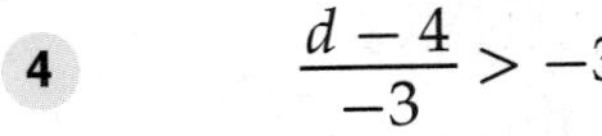


3. $3y - 12 \geq 6$

______ ______

$\frac{3y}{\square} \geq \frac{\_\_\_\_}{\square}$ Simplify.

______

$y \geq$ ______ Simplify.

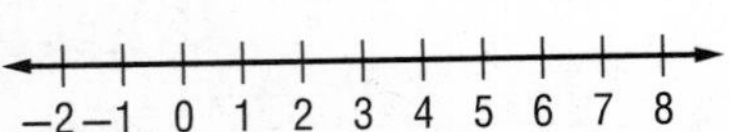


4. $\frac{d-4}{-3} > -3$

______ $\frac{d-4}{-3} <$ ______ −3 ______

$d - 4 <$ ______ Simplify.

______ ______

$d <$ ______ Simplify.

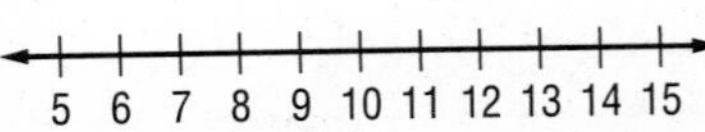

## Step by Step Practice

**5** Solve $-5a + 7 \leq 17$. Graph the solution.

**Step 1** Use inverse operations to solve. $-5a + 7 \leq 17$ Given.

______ ______

**Step 2** Change the direction of the ______ because you divided by a ______.

$\frac{\_\_\_}{[\ ]} \geq \frac{\_\_\_}{[\ ]}$ Simplify.

______

$a \geq$ ______ Simplify.

**Step 3** Graph the solution. Draw ______ at ______ and shade to the ______.

–5 –4 –3 –2 –1 0 1 2 3 4 5

**6** $3b - 6 \geq -3b$ $b \geq$ ______

–5 –4 –3 –2 –1 0 1 2 3 4 5

**7** $\frac{g}{8} + 2 \leq 9$ $g \leq$ ______

48 49 50 51 52 53 54 55 56 57 58

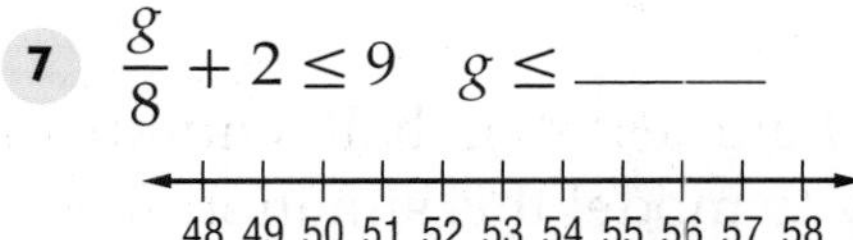

## Step by Step Problem-Solving Practice

**8** **NUMBERS** Five more than twice a number is less than thirteen. Write an inequality to model the situation. Solve and graph the inequality.

| twice a number | plus | five | less than | thirteen |
|---|---|---|---|---|
| ______ | ______ | ______ | ◯ | ______ |

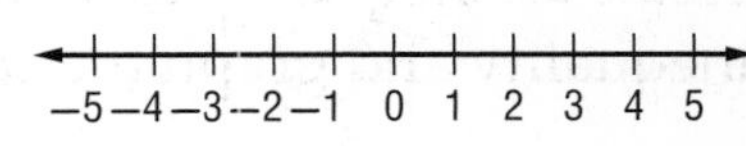

–5 –4 –3 –2 –1 0 1 2 3 4 5

Check off each step.

______ **Understand: I underlined key words.**

______ **Plan: To solve the problem, I will** ______.

______ **Solve: The answer is** ______.

______ **Check: I checked my answer by** ______.

GO ON

## Skills, Concepts, and Problem Solving

**Solve each inequality. Graph the solution.**

**9** $6a + 2 < 14$

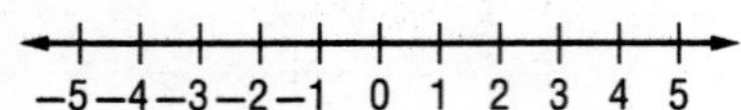


**10** $-15 - 3x > 6$

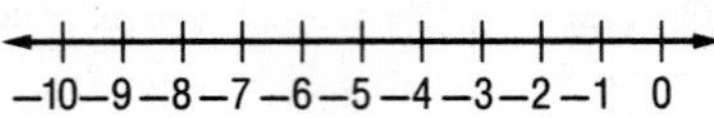


**11** $-1 + \frac{y}{10} > 0$

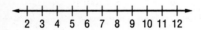

2 3 4 5 6 7 8 9 10 11 12

**12** $\frac{s}{6} + 9 \leq 8$

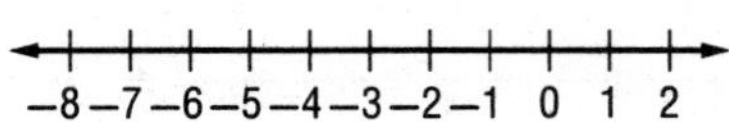


**13** **NUMBERS** Four less than half a number is greater than six. Write an inequality to model this situation. Then solve it and graph the solution.

________________________

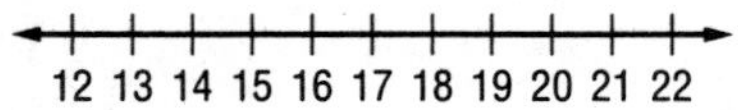


**14** **MONEY** The table shows the prices for items being sold at the sidewalk sale for the Big Book Store. John has only $20 to spend. John buys 1 DVD and some paperback books. If $n$ represents the number of paperback books John buys, create an inequality to model this situation. Solve the inequality and graph the solution.

| Item | Price |
|---|---|
| Paperback Books | $3 |
| Hardcover Books | $5 |
| CDs | $6 |
| DVDs | $8 |

________________________

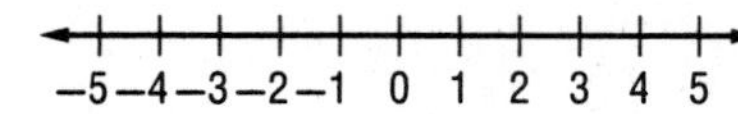


**Vocabulary Check** **Write the vocabulary word that completes each sentence.**

**15** A number sentence that compares two unequal expressions is a(n) ____________________.

**16** **Reflect** Many word problems use the phrase "at most." What inequality symbol represents this phrase? What inequality symbol represents the phrase "at least?"

________________________________________________

________________________________________________

STOP

Lesson 2-6

# Solve Inequalities with Variables on Each Side

## KEY Concept

To solve inequalities with variables on both sides, use the same process you use to solve equations, with one exception:

Change the direction of the inequality symbol when you multiply or divide by a negative number.

$-y < 2 \rightarrow y > -2$

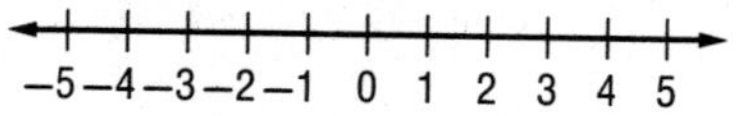


$-y > 2 \rightarrow y < -2$

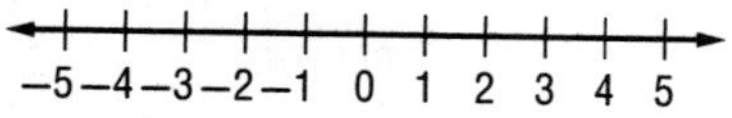


$-y \leq -2 \rightarrow y \geq 2$

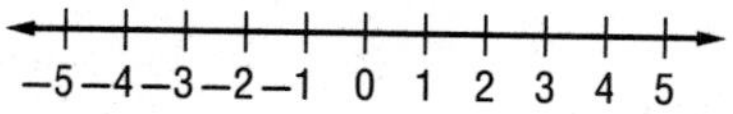


$-y \geq -2 \rightarrow y \leq 2$

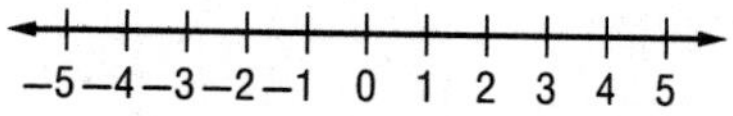


Recall that $-y$ is the same as $-1 \cdot y$ and to isolate the variable, divide by $-1$.

### VOCABULARY

**Distributive Property**
to multiply a sum by a number, you can multiply each addend by the number and add the products

**inequality**
a number sentence that compares two unequal expressions and uses <, >, ≤, ≥, or ≠

**inverse operations**
operations that undo each other

## Example 1

**Solve $-7x - 4 > 12 + x$.**
**Graph the solution.**

1. Use inverse operations to solve.

| | |
|---|---|
| $-7x - 4 > 12 + x$ | Given. |
| $-x \qquad\qquad -x$ | Subtract $x$. |
| $-8x - 4 > 12$ | Simplify. |
| $+4 \quad +4$ | Add 4. |
| $\frac{-8x}{-8} < \frac{16}{-8}$ | Simplify. Divide by −8. |
| $x < -2$ | Simplify. |

Remember, multiplying or dividing by a negative (−) number flips the inequality symbol.

2. Graph the solution. Draw an open circle at −2 and shade to the left.

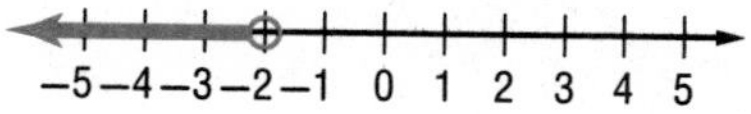


### YOUR TURN!

**Solve $m + 10 \geq -6m + 31$.**
**Graph the solution.**

1. Use inverse operations to solve.

| | |
|---|---|
| $m + 10 \geq -6m + 31$ | Given. |
| ________ | ________ |
| ____ $\geq$ ____ | Simplify. |
| ________ | ________ |
| $\frac{____}{\square} \geq \frac{____}{\square}$ | Simplify. |
| | ________ |
| $m \geq$ ____ | Simplify. |

2. Graph the solution. Draw ________ at ________ and shade to the ________.

GO ON

## Example 2

**Solve $-3(t - 8) > 14 - t$.**
**Graph the solution.**

1. Distribute and combine like terms.

$-3(t - 8) > 14 - t$

$-3t + 24 > 14 - t$

2. Use inverse operations to solve.

| | |
|---|---|
| $-3t + 24 > 14 - t$ | |
| $+t \qquad\qquad +t$ | Add $t$. |
| $-2t + 24 > 14$ | Simplify. |
| $-24 \; -24$ | Subtract 24. |
| $\frac{-2t}{-2} < \frac{-10}{-2}$ | Simplify. Divide by $-2$. |
| $t < 5$ | Simplify. |

3. Graph the solution. Draw an open circle at 5 and shade to the left.

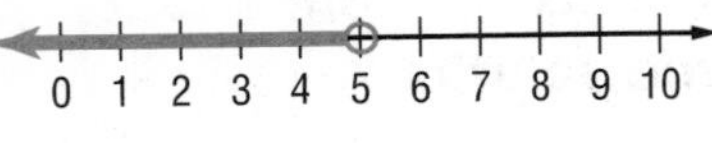


### YOUR TURN!

**Solve $-(9 - v) \leq -5(v + 9) + 2v$.**
**Graph the solution.**

1. Distribute and combine like terms.

$-(9 - v) \leq -5(v + 9) + 2v$

________ $\leq$ ________ $+ 2v$

$-9 + v \leq$ ________ $- 45$

2. Use inverse operations to solve.

| | |
|---|---|
| $-9 + v \leq -3v - 45$ | |
| ____________ | ________ |
| ______ $\leq$ ______ | Simplify. |
| ____________ | ________ |
| ____ $\leq$ ____ (over ☐ ☐) | Simplify. |
| | ________ |
| $v \leq$ ________ | Simplify. |

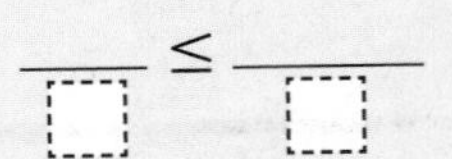

3. Graph the solution. Draw ________ at ________ and shade to the ________.

−15 −14 −13 −12 −11 −10 −9 −8 −7 −6 −5

## Guided Practice

**Solve each inequality. Graph the solution.**

**1** $3y - 5 \leq 7 + 2y$

| | |
|---|---|
| ________ | ________ |
| ______ $- 5 \leq 7$ | Simplify. |
| ________ | ________ |
| $y \leq$ ______ | Simplify. |

5 6 7 8 9 10 11 12 13 14 15

**2** $7b + 12 > -15 - 2b$

| | |
|---|---|
| ____________ | ________ |
| ______ $+ 12 > -15$ | Simplify. |
| ________ | ________ |
| $\frac{9b}{☐} >$ ____ (over ☐) | Simplify. |
| | ________ |
| $b >$ ______ | Simplify. |

−5 −4 −3 −2 −1 0 1 2 3 4 5

## Step by Step Practice

**3** Solve $5w + 4(w - 2) \leq 4 + 3(2 + w)$. Graph the solution.

$5w + 4(w - 2) \leq 4 + 3(2 + w)$

**Step 1** Use the ________________ to eliminate parentheses. Then combine like terms.

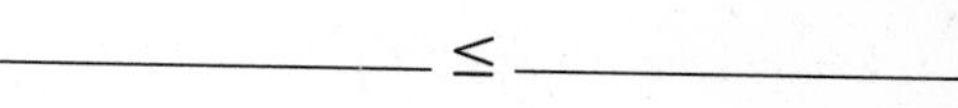

**Step 2** Use inverse operations to solve.

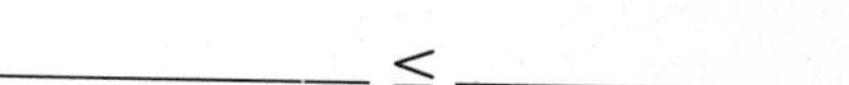

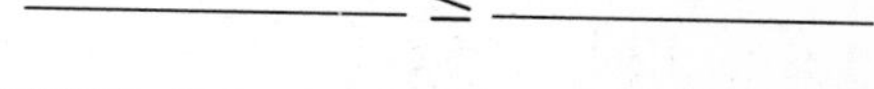

**Step 3** Graph the solution. Draw ________________ at __________ and shade to the __________.

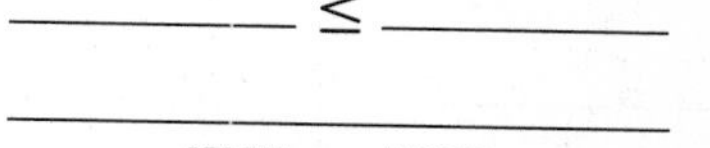

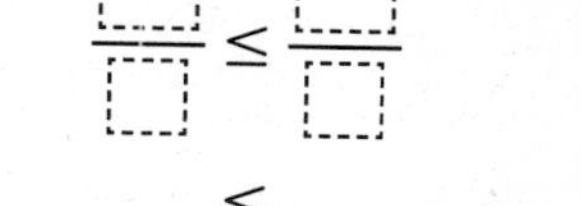

________ $\leq$ ________

−5 −4 −3 −2 −1 0 1 2 3 4 5

**Solve each inequality. Graph the solution.**

**4** $-(3p + 6) < 4(p - 5)$

−5 −4 −3 −2 −1 0 1 2 3 4 5

**5** $v + 9 \geq 18 - 2(v - 3)$

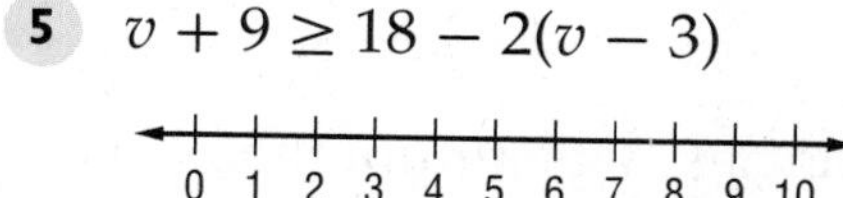


## Step by Step Problem-Solving Practice

**6** **SHOPPING** Pete is at a sale where shirts are \$10 and pants are \$15. Pete buys six items and spends more money on shirts than pants. Pete's purchase can be modeled by $10s \geq 15(6 - s)$, where $s$ is the number of shirts Pete buys. Solve the inequality to find the least number of shirts Pete bought.

$10s \geq 15(6 - s)$

________ $\geq$ ________

________ $\geq$ ________

________ $\geq$ ________

Check off each step.

______ **Understand: I underlined key words.**

______ **Plan: To solve the problem, I will** ________________________________.

______ **Solve: The answer is** ________________________________.

______ **Check: I checked my answer by** ________________________________.

GO ON

## Skills, Concepts, and Problem Solving

**Solve each inequality. Graph the solution.**

7 $4 - 2s > 6 - s$ ________

−5 −4 −3 −2 −1 0 1 2 3 4 5

8 $-3(2p - 4) < 5p - 10$ ________

−5 −4 −3 −2 −1 0 1 2 3 4 5

9 $\frac{10k + 10}{-3} < 3(-2k - 2)$ ________

−5 −4 −3 −2 −1 0 1 2 3 4 5

10 $5(4y + 6) \geq 4(2y + 3) - 18$ ________

−5 −4 −3 −2 −1 0 1 2 3 4 5

**Solve.**

11 **GEOMETRY** The perimeter of the isosceles triangle shown below is less than 45 cm. Write an inequality. Solve and graph the solution.

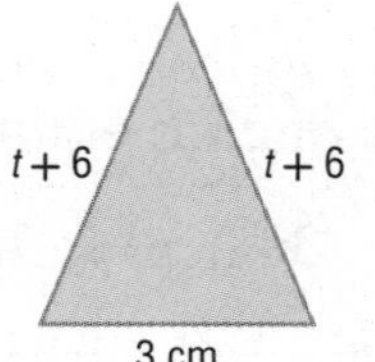


________________

7 8 9 10 11 12 13 14 15 16 17

12 **JOBS** Travis has two summer jobs. He earns \$7 per hour babysitting and \$9 per hour doing dishes at a local bakery. One week, Travis works 20 hours. If he earns more money babysitting that week than doing dishes, the situation can be modeled by the inequality $7b > 9(20 - b)$, where $b$ is the number of hours Travis worked babysitting that week. Solve and graph the solution.

________________

7 8 9 10 11 12 13 14 15 16 17

**Vocabulary Check** **Write the vocabulary word that completes each sentence.**

13 When you use the ________________ to multiply a sum by a number, you can multiply each addend by the number and add the products.

14 **Reflect** Explain how you can check your solution to an inequality.

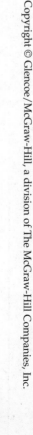

________________________________

________________________________

________________________________

STOP

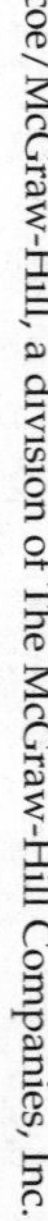

# Chapter 2 Progress Check 3 (Lessons 2-5 and 2-6)

**Solve each inequality. Graph the solution.**

1 $-9z + 20 < -16$

−5 −4 −3 −2 −1 0 1 2 3 4 5

2 $15 - x > 19$

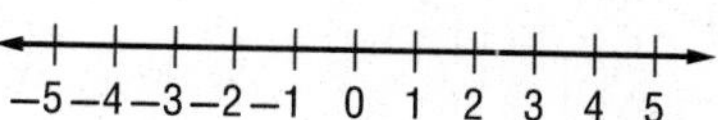

3 $\frac{k + 10}{-4} > 2$

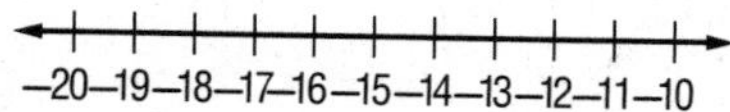

4 $\frac{p - 12}{-5} \leq 4$

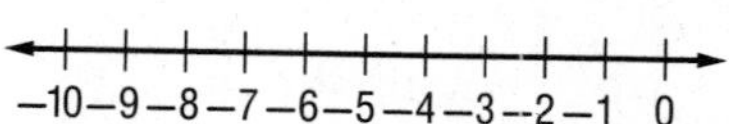

5 $\frac{t - 3}{-2} < 2(2t + 1)$

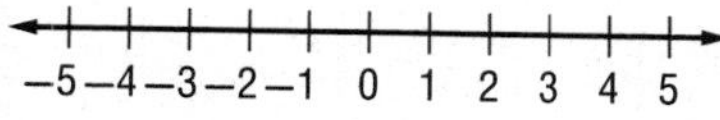

6 $-5(k + 2) < 3(4 - k) + 8$

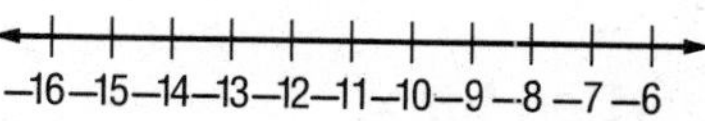

7 **GRADES** Alicia will need to score at least 80 points on the final Algebra test of the year to be exempt from final exams. She will receive 2 points for each question answered correctly and will get 20 points for a bonus she has already completed. Alicia's needed score is modeled by $2q + 20 \geq 80$, where $q$ is the number of questions Alicia answers correctly on the test.

Solve and graph the inequality to determine how many questions Alicia needs to answer correctly so that she can be exempt from the exam.

22 23 24 25 26 27 28 29 30 31 32

Chapter 2

# Chapter Test

**Solve each equation.**

1 $-c - 25 = 18$

2 $18 = -4 + 2r$

3 $-113 - p = -8$

4 $18 - y = y + 2$

5 $-3a - 7 = 5a + 1$

6 $15(b + 1) = 3b - 45$

7 $-2q + 42 = 4 - 6(3q - 1)$

8 $\frac{1}{4}(z - 16) = \frac{z}{2} - 10$

**Solve each proportion for *x*.**

9 $\frac{x}{36} = \frac{3.6}{12}$

10 $\frac{10}{3} = \frac{32}{x}$

11 $\frac{1.8}{2} = \frac{x}{10}$

**Solve each equation for *x*.**

12 $-35e = \frac{13 + x}{f}$

13 $18w = 5x + 11q$

14 $w = \frac{x + 13}{-4y}$

**Solve each inequality. Graph the solution.**

**15** $\frac{h-13}{-8} < 3$

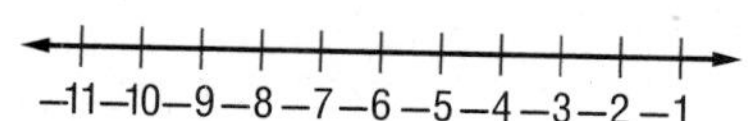


**16** $20 - 4t \leq -12$

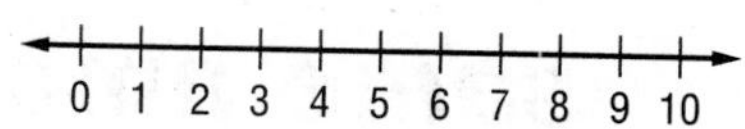


**17** $-2v - 6 < -4 - 3(3v - 4)$

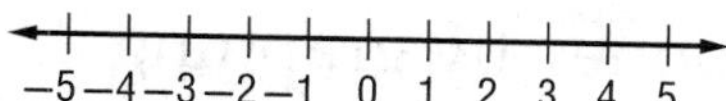


**18** $2 + 2(4x - 1) \geq \frac{-6x - 5}{-2}$

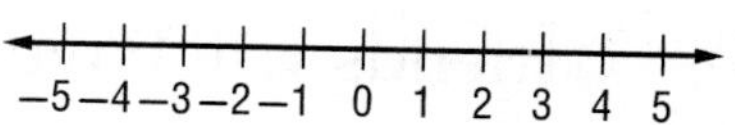


**Solve.**

**19** **GROCERIES** Sari spent $12.56 on 8 pairs of socks. Use a proportion to determine the price 20 pairs of socks.

______________________________

**20** **GEOMETRY** The perimeter of the smaller rectangle is one-half the perimeter of the larger rectangle. What is the perimeter of each rectangle?

______________________________

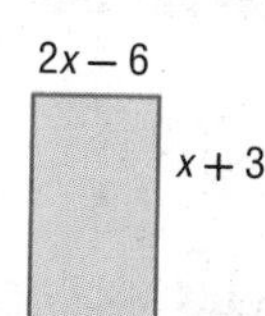


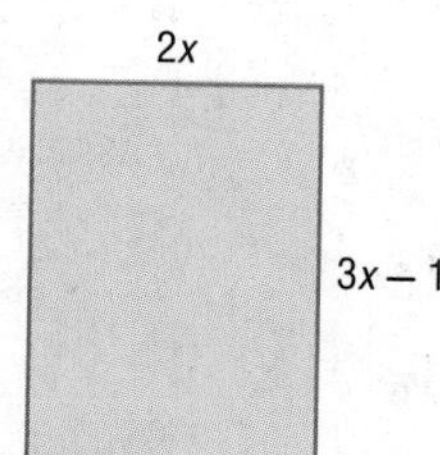


**Correct the mistake.**

**21** Raphael solved the equation $\frac{r-1}{15s} = 15t$ for $r$. In his first step, he subtracted 1 and multiplied by $15s$. Olivia told him that he should multiply first and then add. Who is correct and what is the equation solved for $r$?

______________________________

STOP

# Graphs

## *When will they have the same amount?*

Two friends open savings accounts. The first friend starts with no initial deposit and deposits $10 each week. The second friend starts with a deposit of $20 and deposits $5 each week. You can figure out in which week the two friends will have the same amount of money by graphing the system of equations.

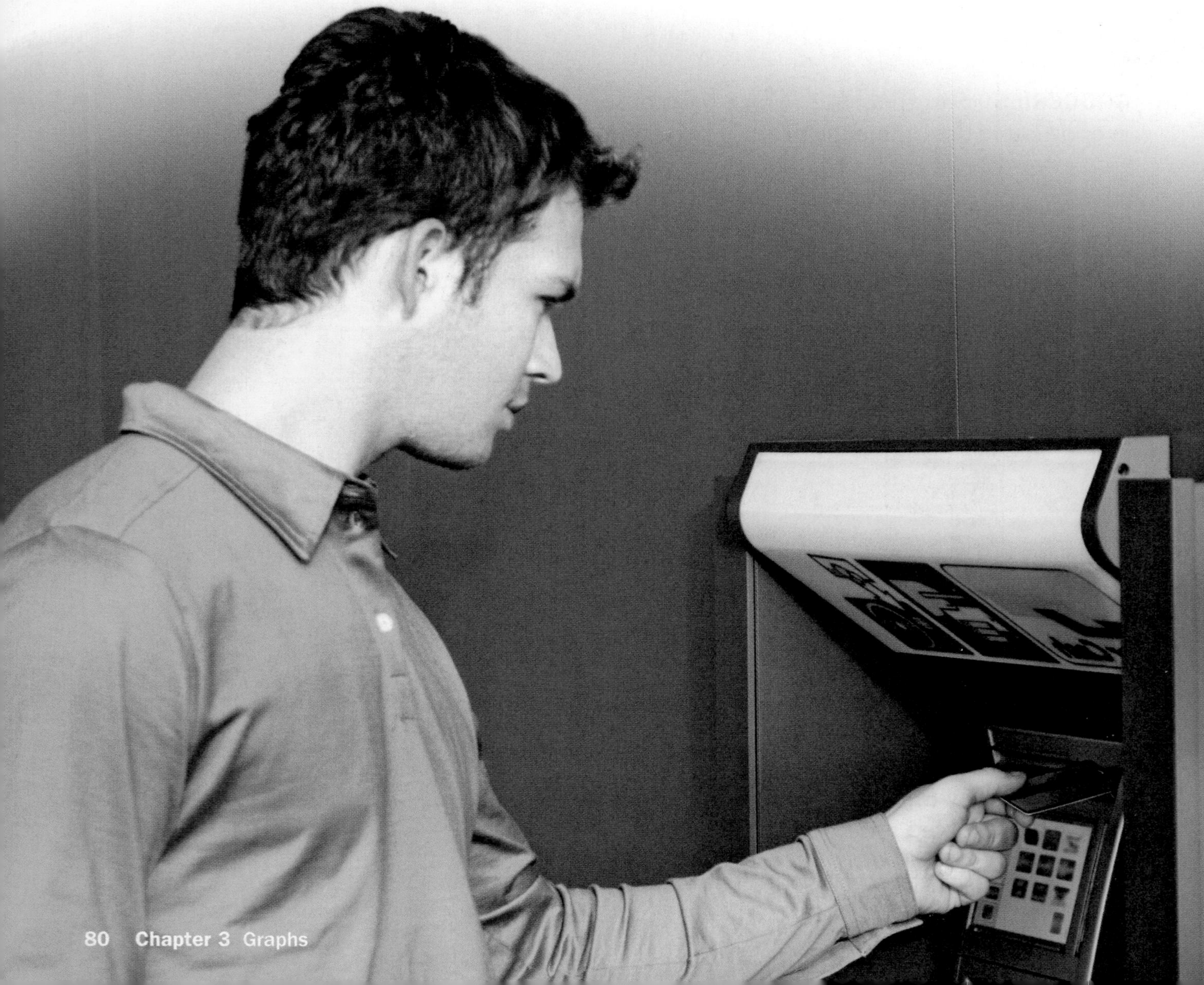

**STEP 1 Chapter Pretest** Are you ready for Chapter 3? Take the Chapter 3 Pretest to find out.

**STEP 2 Preview** Get ready for Chapter 3. Review these skills and compare them with what you will learn in this chapter.

| What You Know | What You Will Learn |
|---|---|
| You know how to graph points. **Example:** Graph and label point $S(4, 3)$. 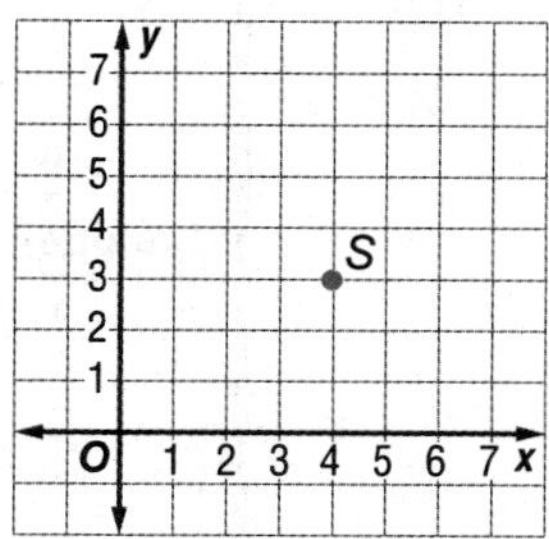 **TRY IT!** Graph and label point $T(2, 5)$. 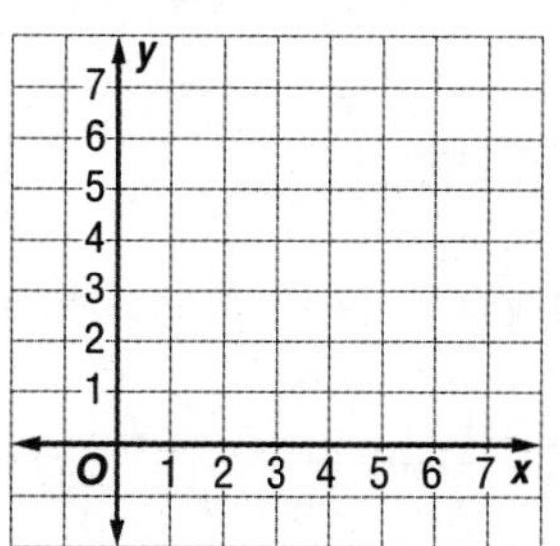 | *Lesson 3-1* You can graph linear equations using a table. **Example:** Graph $y = x + 2$ using a table. <table><tr><th>x</th><th>y = x + 2</th><th>y</th></tr><tr><td>0</td><td>0 + 2</td><td>2</td></tr><tr><td>1</td><td>1 + 2</td><td>3</td></tr><tr><td>2</td><td>2 + 2</td><td>4</td></tr></table> 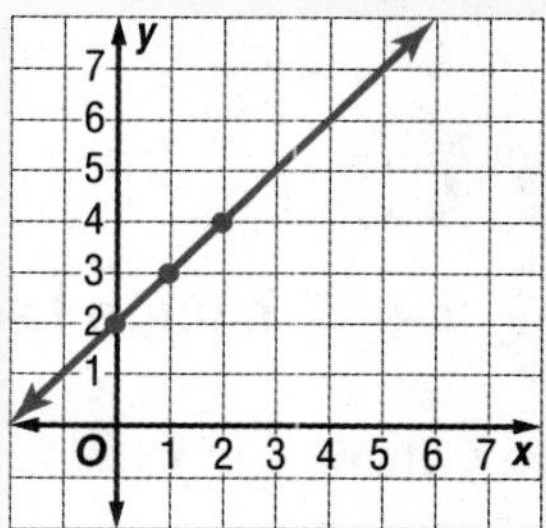 |
| You know that intersecting lines are lines that meet or cross at a point. 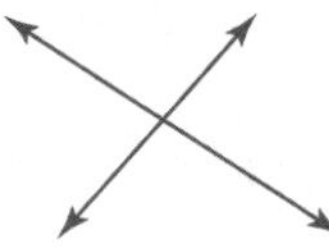 | *Lesson 3-2* You can graph a system of linear equations by graphing both equations on the same graph. When the graph of a system of equations shows intersecting lines, the system has one solution. 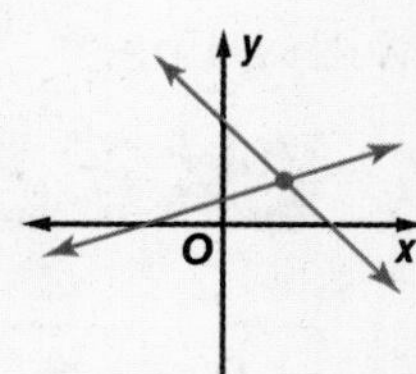 |

Lesson 3-1

# Graph Linear Equations

## KEY Concept

A **linear equation** has a graph that is a straight line. Every non-vertical line has a unique **slope** and passes through the *y*-axis at exactly one point. The slope and that point of intersection can be used to graph the line.

### Slope-Intercept Form

Linear equations can be placed in the form $y = mx + b$. This helps show the slope and the *y*-intercept.

$$y = \frac{2}{3}x - 1$$

slope (points to $\frac{2}{3}$) *y*-intercept (points to $-1$)

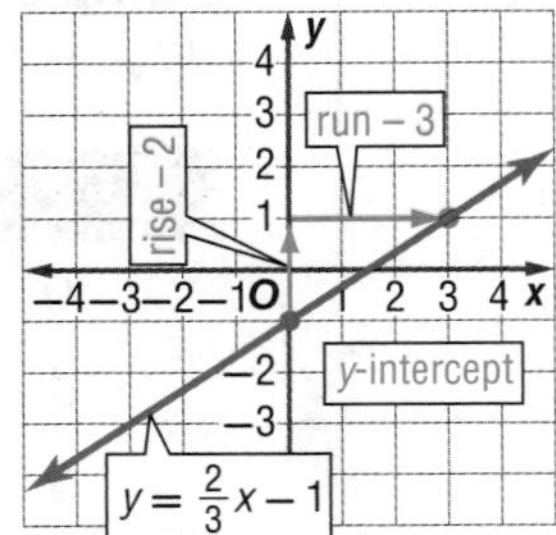


### VOCABULARY

**linear equation**
an equation for which the graph is a straight line

**slope**
the rate of change between any two points on a line; the ratio of vertical change to horizontal change (rise over run)

**slope-intercept form**
a linear equation in the form $y = mx + b$, where *m* is the slope and *b* is the *y*-intercept

***y*-intercept**
the point where a line crosses the *y*-axis

## Example 1

**Graph $y = \frac{1}{2}x + 3$ using a table.**

1. Substitute values for *x*. Solve for *y*.

$x = -2 \quad y = \frac{1}{2}(-2) + 3 = 2$

$x = 0 \quad y = \frac{1}{2}(0) + 3 = 3$

$x = 2 \quad y = \frac{1}{2}(2) + 3 = 4$

2. Plot the points. Then connect them with a line.

| x | y |
|---|---|
| −2 | 2 |
| 0 | 3 |
| 2 | 4 |

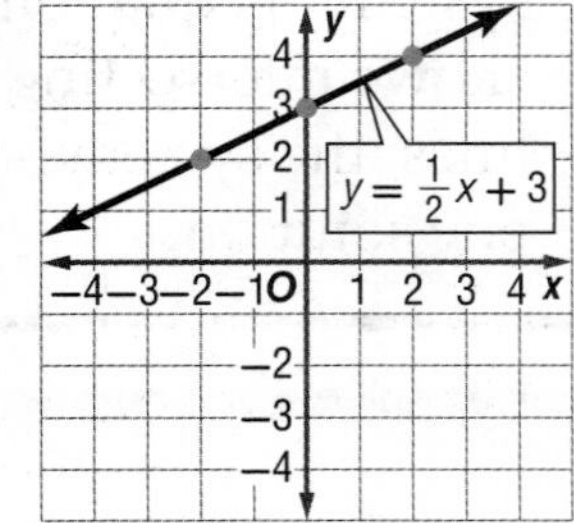


## YOUR TURN!

**Graph $y = -x - 2$ using a table.**

1. Substitute values for *x*. Solve for *y*.

$x = -1 \quad y = -(\_\_\_\_) - 2 = \_\_\_\_$

$x = 0 \quad y = -(\_\_\_\_) - 2 = \_\_\_\_$

$x = 1 \quad y = -(\_\_\_\_) - 2 = \_\_\_\_$

2. Plot the points. Then connect them with a line.

| x | y |
|---|---|
| −2 | |
| 0 | |
| 2 | |

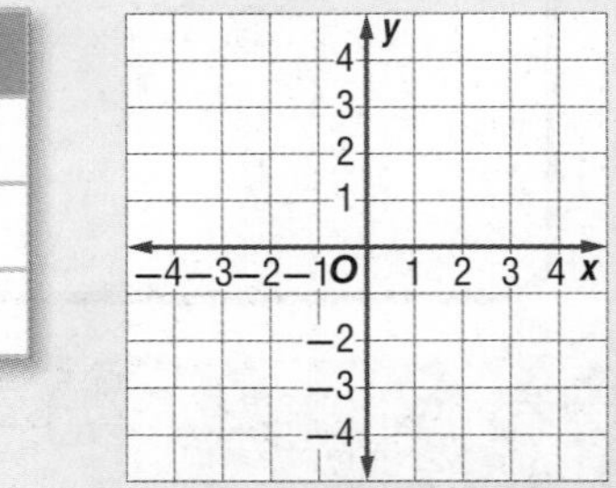

## Example 2

**Graph $y = \frac{3}{4}x - 2$ using the slope and the y-intercept.**

1 Graph the $y$-intercept.

**−2**

2. Graph another point using the slope $\frac{3}{4}$.

**rise = 3, run = 4**

3. Connect the points.

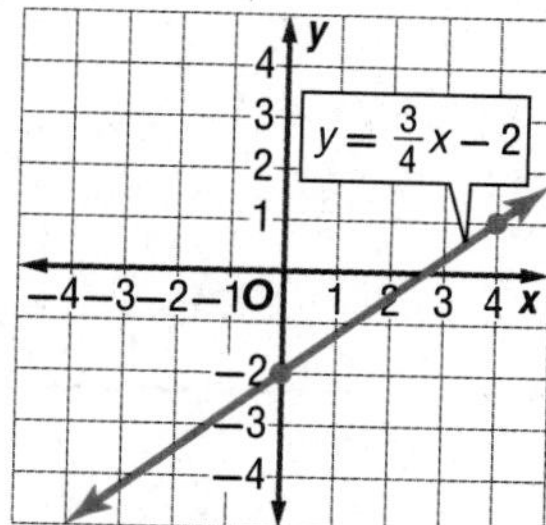


**YOUR TURN!**

**Graph $y = -\frac{5}{2}x + 1$ using the slope and the y-intercept.**

1. Graph the $y$-intercept.

_____

2. Graph another point using the slope _____.

rise = _____, run = _____

3. Connect the points.

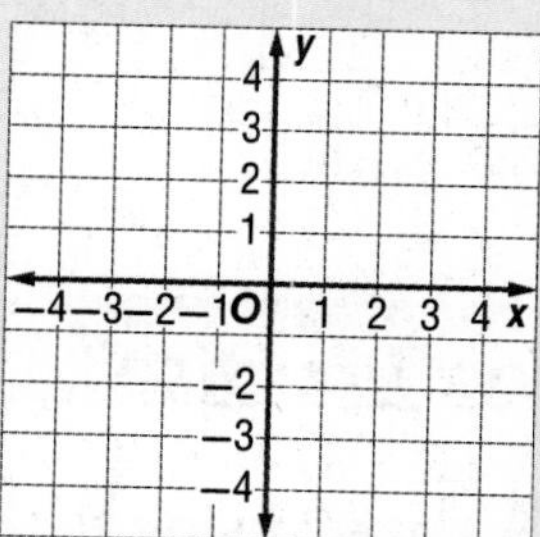


## Guided Practice

**Graph each equation using a table.**

1 $y = 5x + 5$

$x = -2$ $y = 5(\_\_\_) + 5 = \_\_\_$

$x = 0$ $y = 5(\_\_\_) + 5 = \_\_\_$

$x = 2$ $y = 5(\_\_\_) + 5 = \_\_\_$

| x | y |
|---|---|
| −2 | |
| 0 | |
| 2 | |

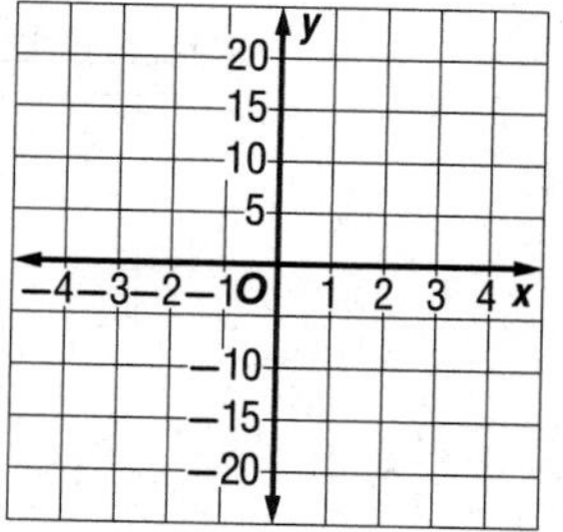


2 $y = -\frac{1}{2}x - 3$

$x = -2$ $y = -\frac{1}{2}(\_\_\_) - 3 = \_\_\_$

$x = 0$ $y = -\frac{1}{2}(\_\_\_) - 3 = \_\_\_$

$x = 2$ $y = -\frac{1}{2}(\_\_\_) - 3 = \_\_\_$

| x | y |
|---|---|
| −2 | |
| 0 | |
| 2 | |

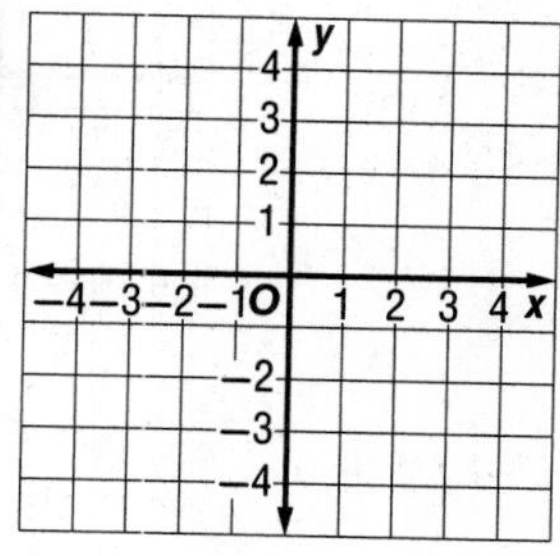


GO ON

**Graph each equation using the slope and the y-intercept.**

3 $y = -\frac{5}{6}x + 4$

y-intercept = ______ slope = ______

rise = ______, run = ______

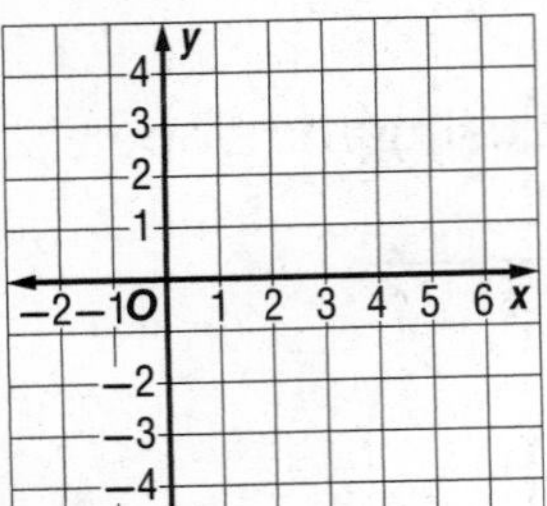

4 $y = \frac{1}{3}x$

y-intercept = ______ slope = ______

rise = ______, run = ______

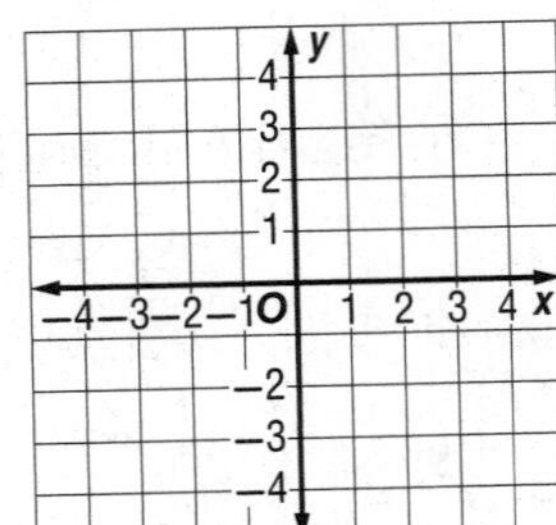

## Step by Step Practice

5 Graph $y = \frac{5}{3}x - 2$ using the slope and the y-intercept.

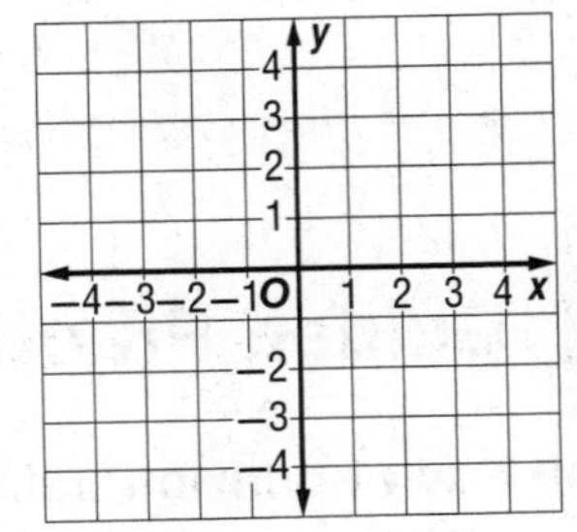

**Step 1** Graph the y-intercept. ______

**Step 2** Graph another point using the slope ______.

The rise is ______ and the run is ______.

**Step 3** Connect the points.

**Graph each equation using the slope and the y-intercept.**

6 $y = \frac{1}{4}x - 2$

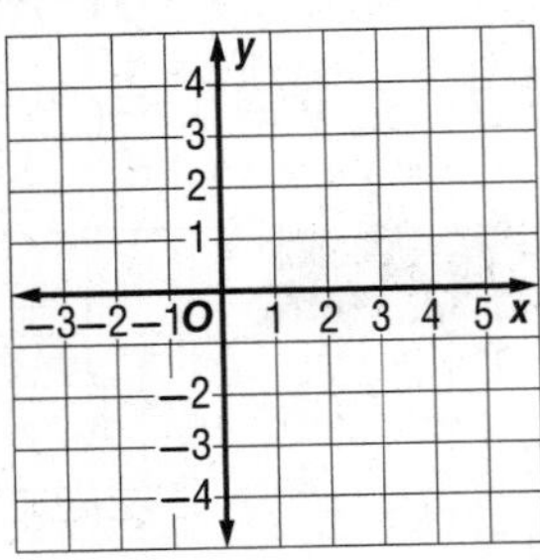

7 $y = 3x$

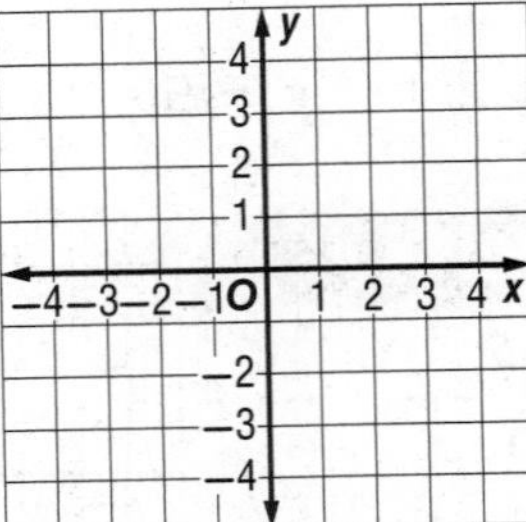

8 $y = -2x + 1$

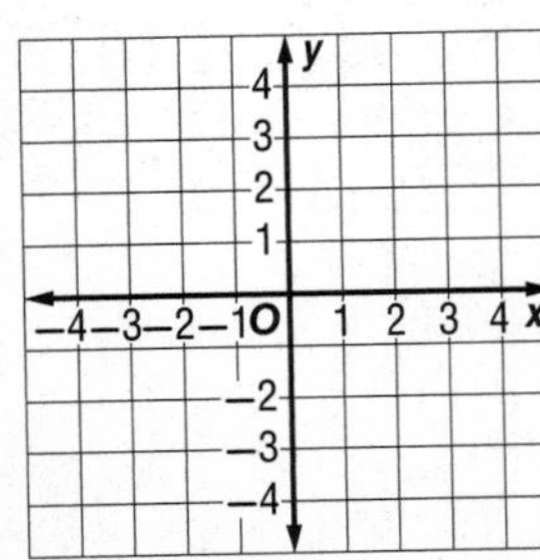

9 $y = -\frac{6}{5}x + 7$

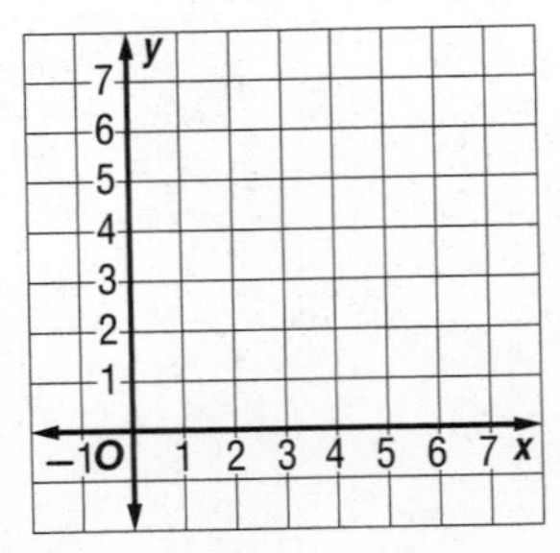

## Step by Step Problem-Solving Practice

**Solve.**

**10** **GRADES** The equation Mrs. Simpson uses to calculate each student's grade is $y = 3x + 5$, where $y$ is the total grade and $x$ is the number of questions answered correctly. Graph this equation.

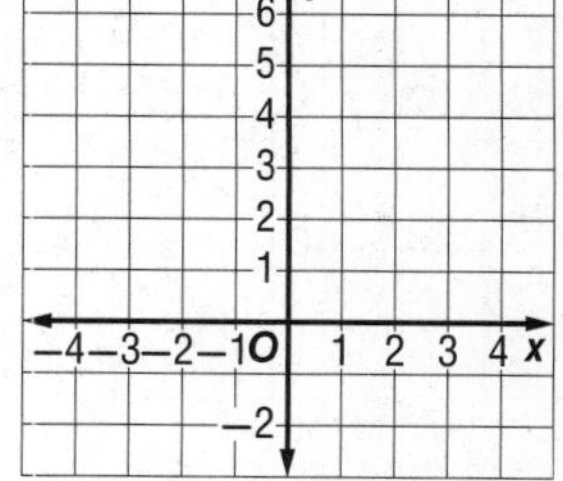

Check off each step.

_______ Understand: I underlined key words.

_______ Plan: To solve the problem, I will _______________________________.

_______ Solve: The answer is shown in the graph.

_______ Check: I checked my answer by _______________________________

_______________________________________________.

## Skills, Concepts, and Problem Solving

**Graph each equation using the slope and the $y$-intercept.**

**11** $y = -\frac{4}{3}x + 3$

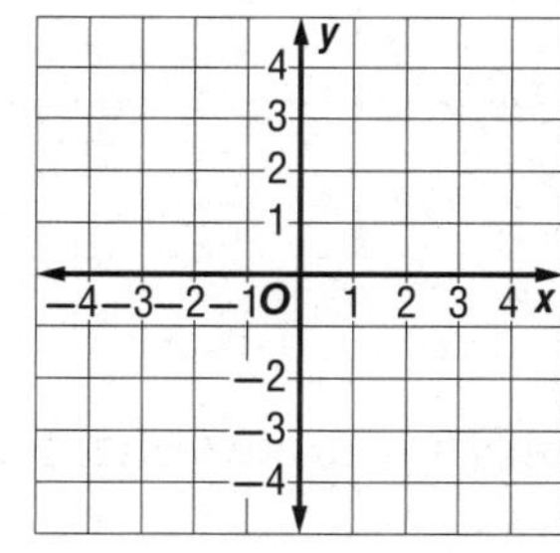

**12** $y = 4x - 5$

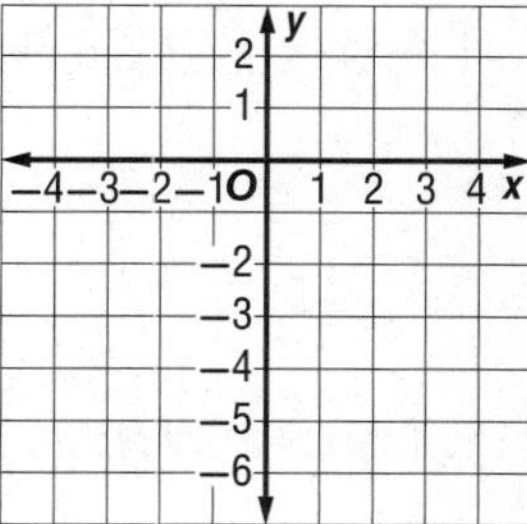

**13** $y = -2x + 1$

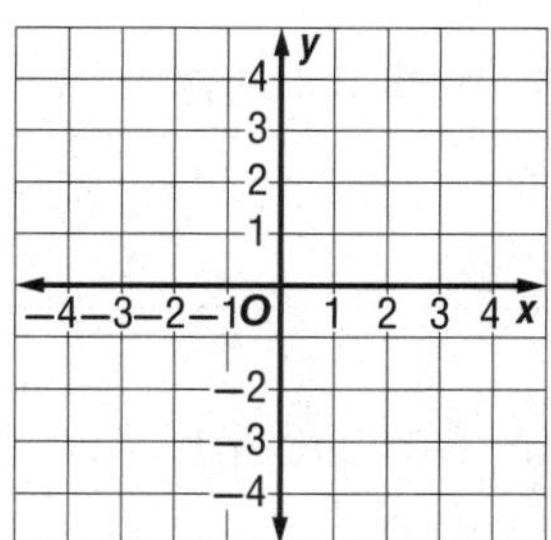

**14** $y = \frac{1}{2}x - 2$

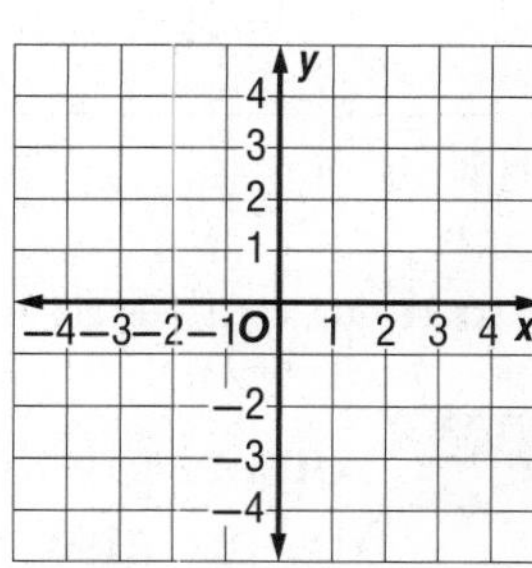



GO ON

**Graph each equation using the slope and the $y$-intercept.**

15 $y = -6x + 4$

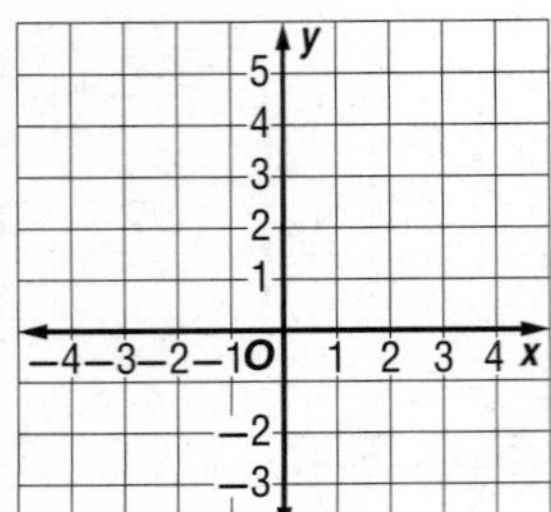

16 $y = -x$

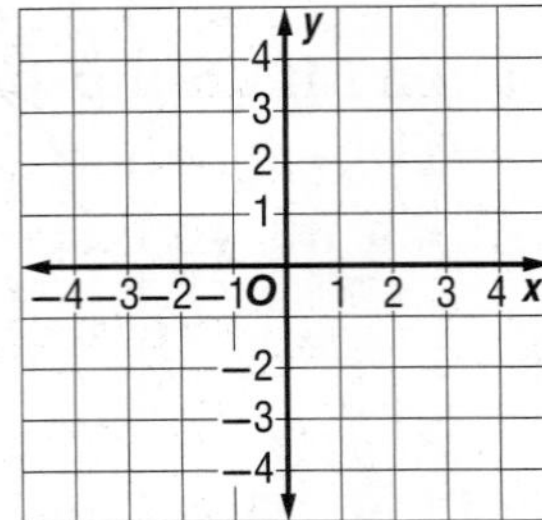

**Solve.**

17 **FOOD** The base charge for a single dip ice cream cone is \$3.00. Each additional topping costs \$0.25 each. The total cost $y$ of a cone with $x$ number of toppings is modeled by the equation $y = \frac{1}{4}x + 3$. Graph this equation.

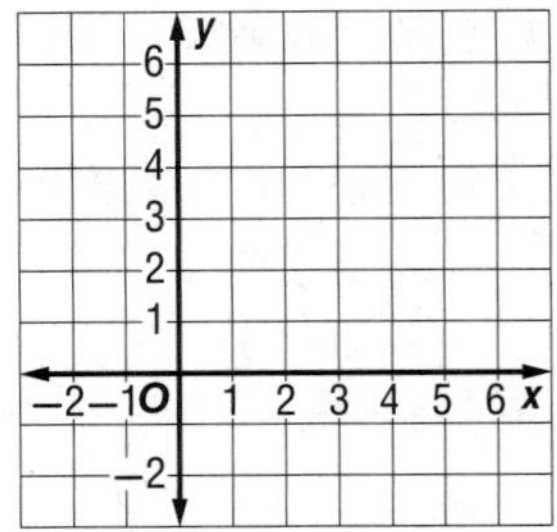

18 **ANIMALS** A dog kennel buys 400 pounds of dog food and goes through 18 pounds of food per day. The amount of food $y$ that remains after $x$ amount of days is modeled by the equation $y = -18x + 400$. Graph this equation.

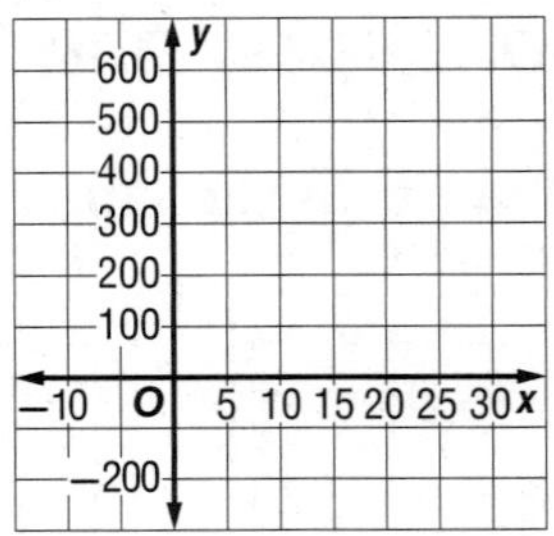

**Vocabulary Check** **Write the vocabulary word that completes each sentence.**

19 A line crosses the $y$-axis at the ____________________.

20 The ____________________ of a line is the rise of a line over the run of the line.

21 An equation that is in the form $y = mx + b$ is in ____________________.

22 **Reflect** You are given a point on a line other than the $y$-intercept. You are also given the slope of the line. Are you able to graph the line without knowing the $y$-intercept? Explain.

________________________________________

________________________________________

STOP

Lesson 3-2

# Graph Systems of Linear Equations

A **system of linear equations** is a set of two or more linear equations with the same variables. The graph of a system of equations may have intersecting lines, parallel lines, or may be the same lines.

When the graph of a system of linear equations is intersecting lines, the system has one solution.

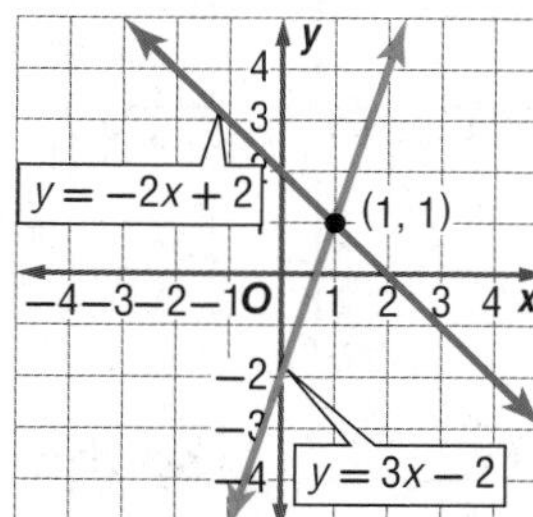


$y = 3x - 2$
$y = -2x + 2$

Notice the lines have different slopes. The lines intersect at (1, 1), so the solution is (1, 1).

When the graph of a system of linear equations is parallel lines, the system has no solution.

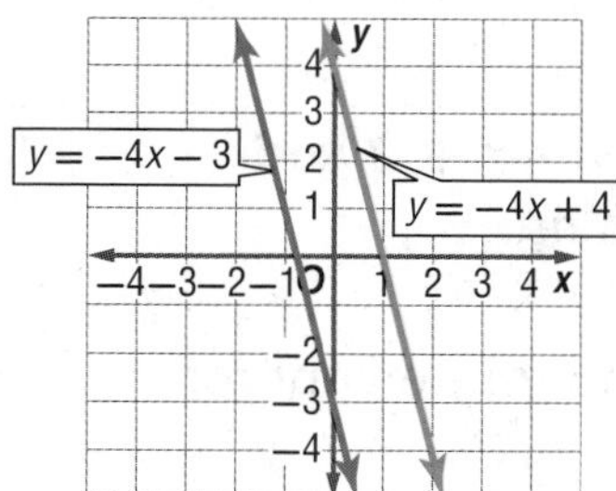


$y = -4x + 4$
$y = -4x - 3$

Notice the lines are parallel and the slopes are the same. So there is no solution.

When the graph of a system of linear equations is one line, the system has infinitely many solutions.

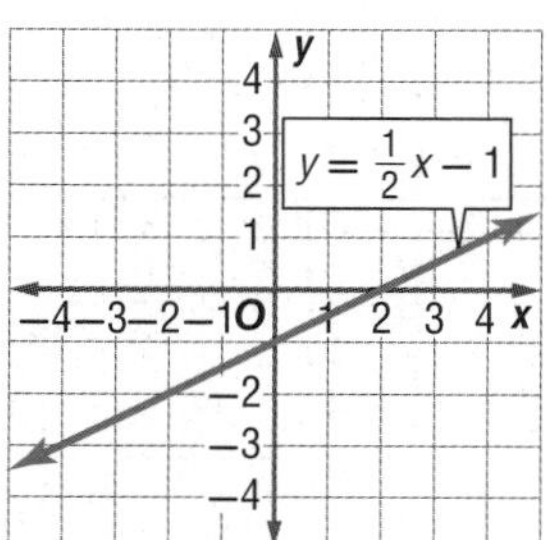


Notice the lines are the same. So there are infinitely many solutions.

## VOCABULARY

**linear equation**
an equation for which the graph is a straight line

**slope**
the rate of change between any two points on a line; the ratio of vertical change to horizontal change (rise over run)

**slope-intercept form**
a linear equation in the form $y = mx + b$, where $m$ is the slope and $b$ is the $y$-intercept

**system of linear equations**
a set of two or more linear equations with the same variables

***y*-intercept**
the point where a line crosses the $y$-axis



GO ON

## Example 1

**Solve the system $y = x - 2$ and $y = -1$ by graphing.**

1. Write the slope and $y$-intercept for each equation.

   $y = x - 2$ slope = 1, $y$-intercept = −2

   $y = -1$ slope = 0, $y$-intercept = −1

2. Graph each equation.

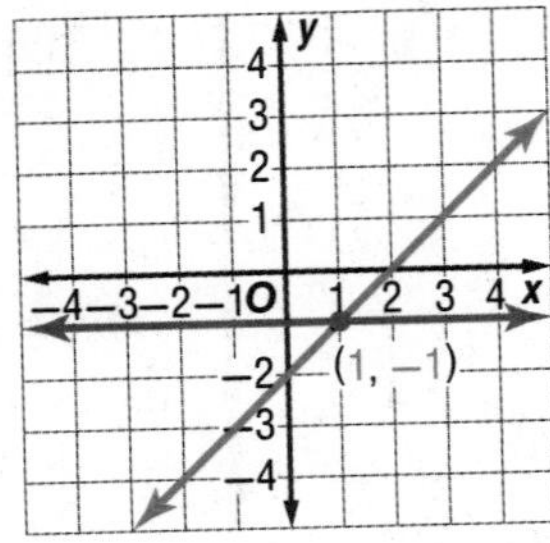


3. The lines intersect at (1, −1), so the solution is (1, −1).

**YOUR TURN!**

**Solve the system $y = -x + 3$ and $y = 2x - 3$ by graphing.**

1. Write the slope and $y$-intercept for each equation.

   $y = -x + 3$ slope = ____, $y$-intercept = ____

   $y = 2x - 3$ slope = ____, $y$-intercept = ____

2. Graph each equation.

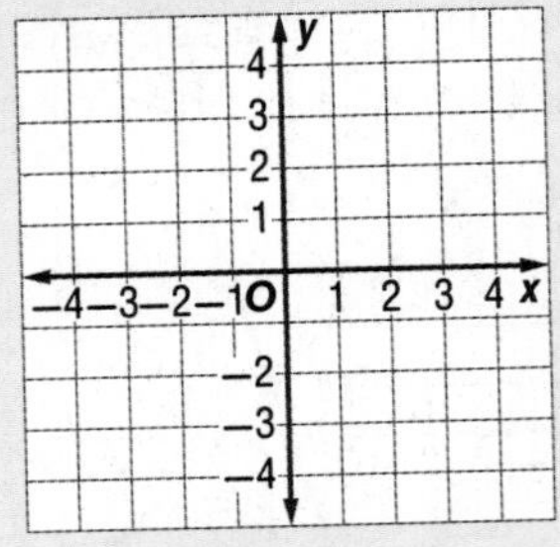


3. The lines intersect at ________, so the solution is ________.

## Example 2

**Solve the system $y = \frac{1}{4}x + 1$ and $y = \frac{1}{4}x - 2$ by graphing.**

1. Write the slope and $y$-intercept for each equation.

   $y = \frac{1}{4}x + 1$ slope = $\frac{1}{4}$, $y$-intercept = 1

   $y = \frac{1}{4}x - 2$ slope = $\frac{1}{4}$, $y$-intercept = −2

2. Graph each equation.

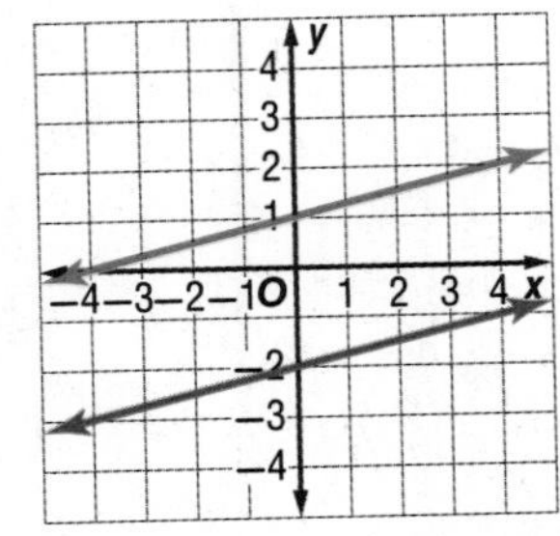


3. The lines are parallel so the slopes are the same. There is no solution.

**YOUR TURN!**

**Solve the system $y = 4x + 2$ and $y = 4x - 5$ by graphing.**

1. Write the slope and $y$-intercept for each equation.

   $y = 4x + 2$ slope = ____, $y$-intercept = ____

   $y = 4x - 5$ slope = ____, $y$-intercept = ____

2. Graph each equation.

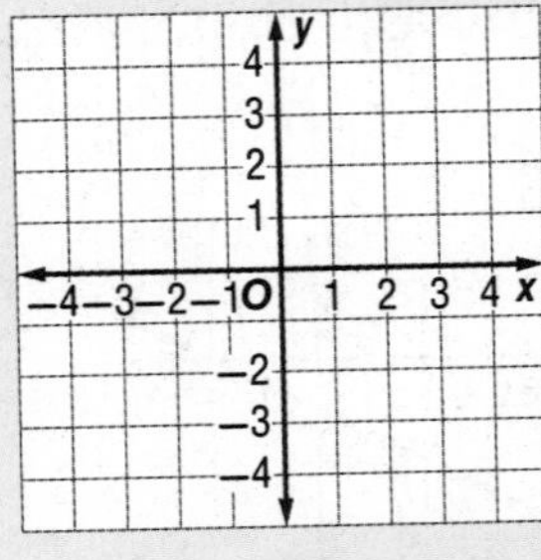


3. The lines are ________ so the slopes are the ________. There is ________.

## Guided Practice

**State the slope and *y*-intercept. Then graph each equation.**

1 $y = -\frac{1}{6}x + 3$

slope = ______ $y$-intercept = ______

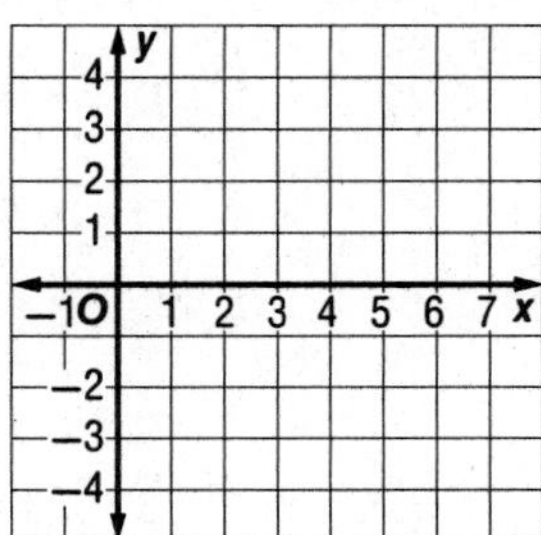

2 $y = 3x$

slope = ______ $y$-intercept = ______

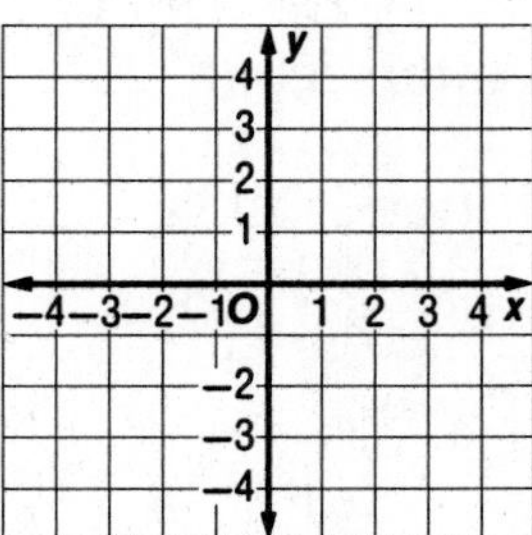

**Use the graph of system of linear equations to name the solution.**

3

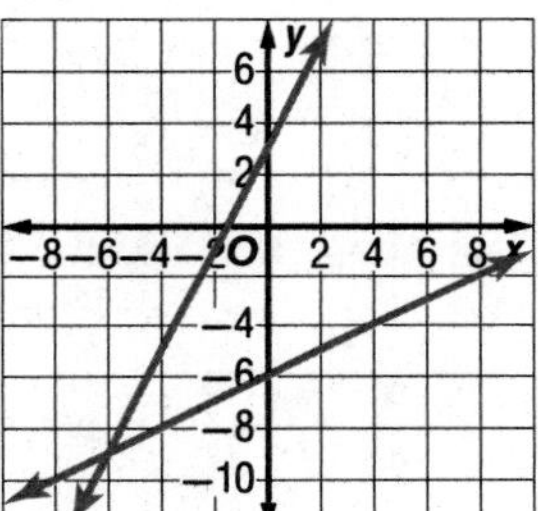

The lines intersect at the point ____________.

The solution is ____________.

4

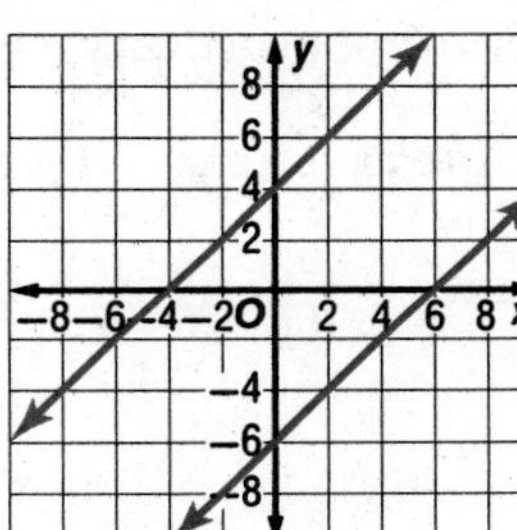

The lines are ____________.

There are ______________________.

5

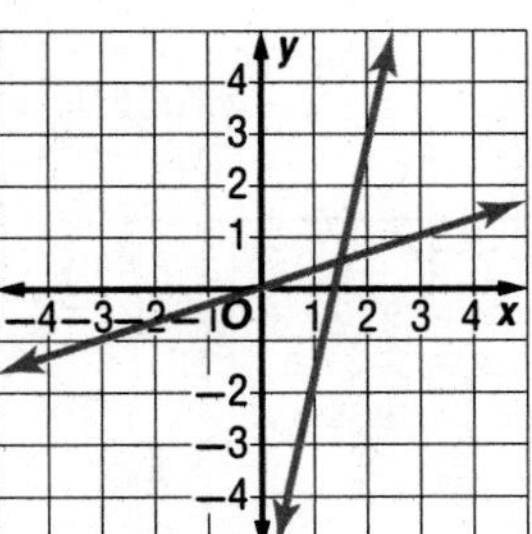

The lines intersect at the point ____________.

The solution is ____________.

6

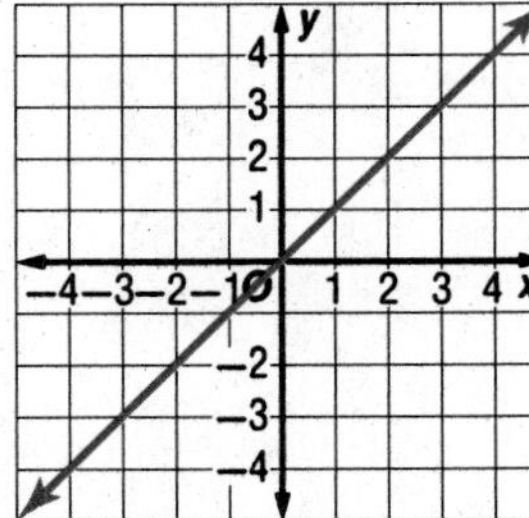

The lines are ____________.

There are ______________________.

GO ON

## Step by Step Practice

7 Solve the system $y = x + 3$ and $y = 2x + 4$ by graphing.

**Step 1** Graph $y = x + 3$. slope = ________ $y$-intercept = ________

**Step 2** Graph $y = 2x + 4$. slope = ________ $y$-intercept = ________

**Step 3** The lines intersect at ________,

so the solution is ________.

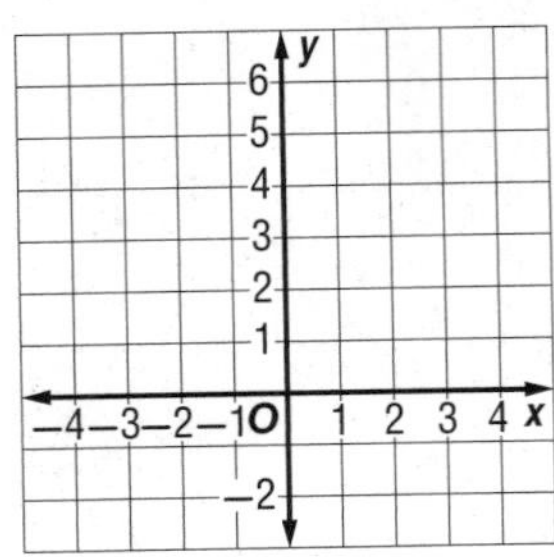

**Solve each system of linear equations by graphing.**

8 $y = -\frac{1}{3}x + 1$ slope = ________ $y$-intercept = ________

$y = x + 3$ slope = ________ $y$-intercept = ________

________

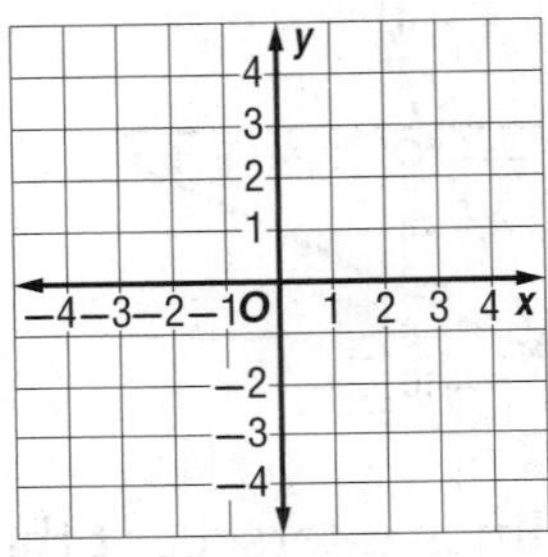

9 $y = 5x + 2$ slope = ________ $y$-intercept = ________

$y = 5x - 2$ slope = ________ $y$-intercept = ________

________

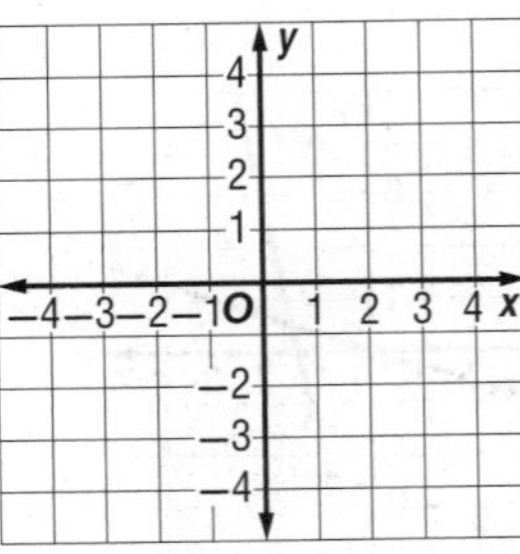

10 $y = \frac{1}{4}x$ slope = ________ $y$-intercept = ________

$y = 4x$ slope = ________ $y$-intercept = ________

________

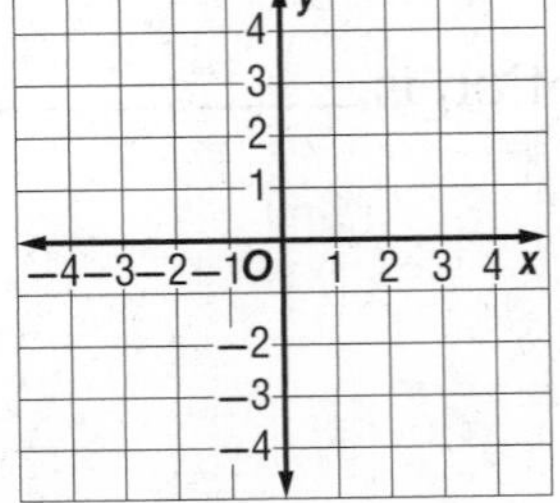

## Step by Step Problem-Solving Practice

**Solve.**

**11** **MUSIC** An online music store charges $20 to join and $1 for each song that is downloaded. The total cost $y$ for $x$ number of downloaded songs is modeled by the equation $y = x + 20$.

Another online music store charges $11 to join and $2 for each downloaded song. The total cost $y$ for $x$ songs is modeled by the equation $y = 2x + 11$. Find the number of songs that would cost the same at both sites.

$y = x + 20$

slope = ________

$y$-intercept = ________

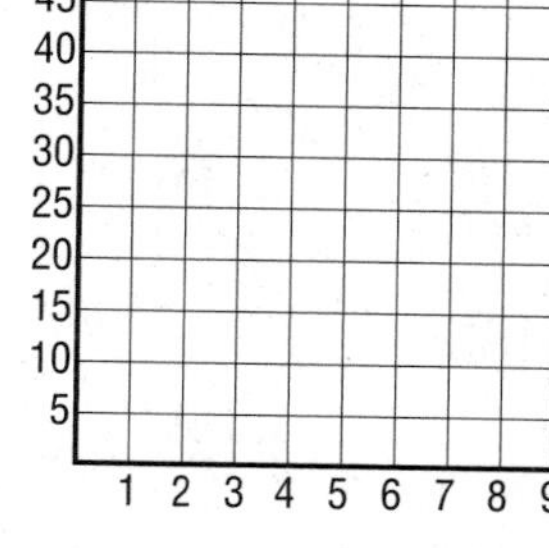


$y = 2x + 11$

slope = ________

$y$-intercept = ________

Check off each step.

______ **Understand: I underlined key words.**

______ **Plan: To solve the problem, I will** ________________________________________________
________________________________________________.

______ **Solve: The answer is** ________________________________________________.

______ **Check: I checked my answer by** ________________________________________________
________________________________________________.



GO ON

## Skills, Concepts, and Problem Solving

**Solve each system of linear equations by graphing.**

**12** $y = -x + 4$

$y = \frac{1}{2}x + 10$

________________

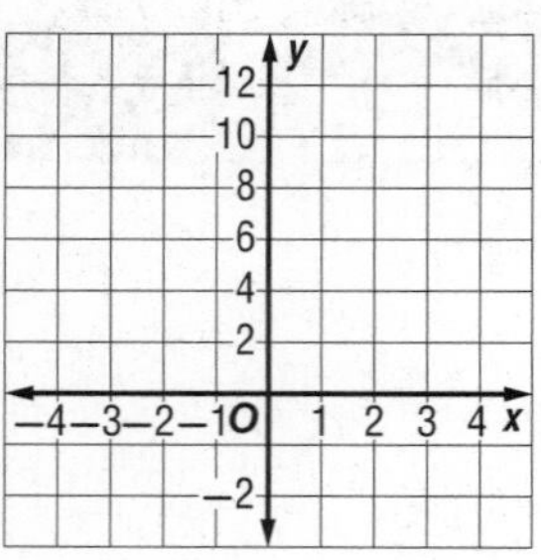


**13** $y = 3x - 5$

$y = 3x + 6$

________________

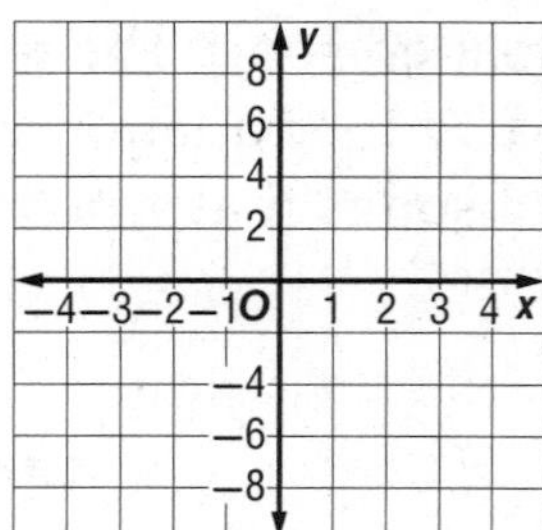


**14** $y = 2x + 7$

$y = 5x + 4$

________________

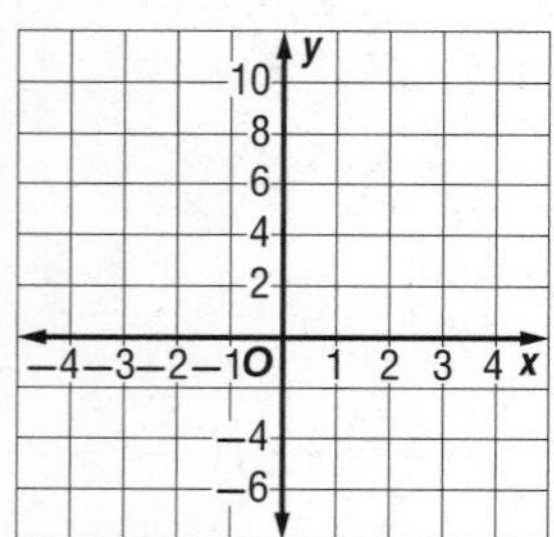


**15** $y = \frac{1}{3}x - 3$

$y = -4$

________________

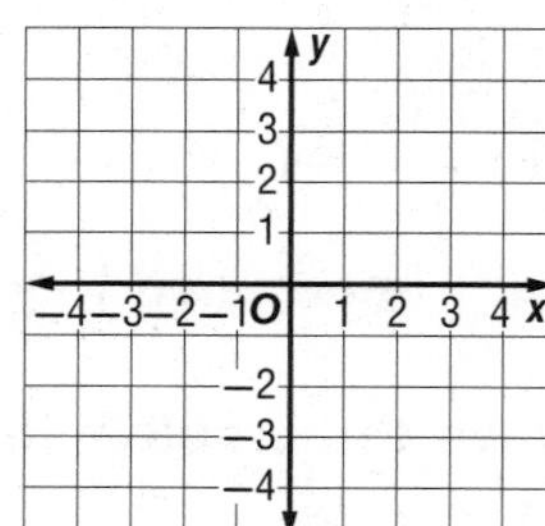


**16** $y = \frac{5}{2}x - 3$

$y = \frac{5}{2}x + 1$

________________

y
4
3
2
1
−4 −3 −2 −1 O 1 2 3 4 x
−2
−3
−4

**17** $y = 2$

$y = x$

________________

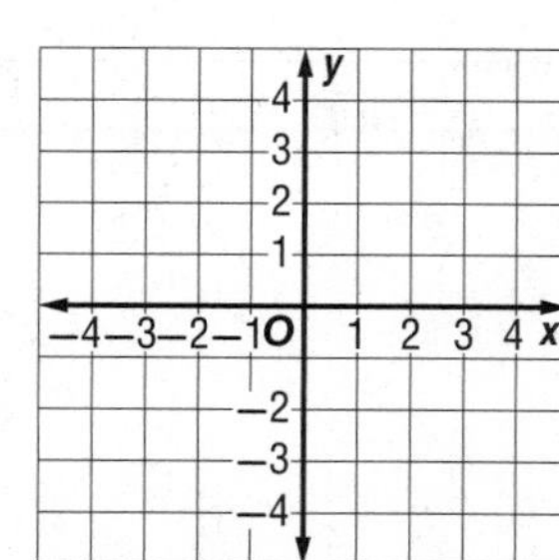

**Solve.**

**18** **SALES** Tory and Joseph both have sales positions at their jobs. Tory earns $200 per week and 15% commission on her sales. Joseph earns $150 per week and 20% commission on his sales. The graph at the right models the amount $y$ each person earns in a week when his or her sales are $x$ amount. For what amount of sales do Tory and Joseph earn the same amount of money in a week?

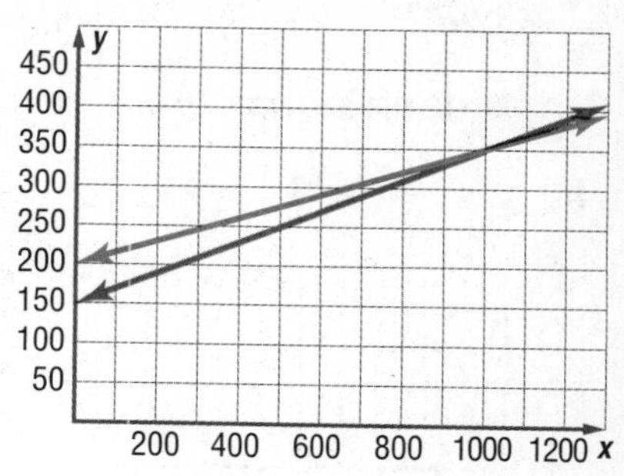


_______________

**19** **MAPS** If Ben and Mai's neighborhood were mapped on a coordinate plane, each could follow a direct line from each of their houses to the library. Ben would follow the line $y = 2x - 8$, and Mai would follow the line $y = -3x + 12$. At what point is the library located?

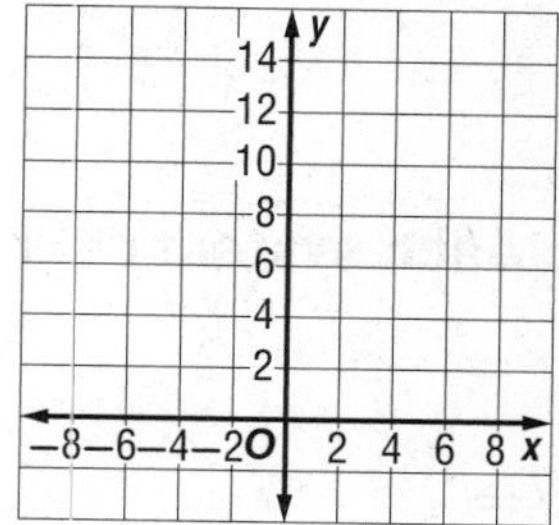


_______________

**20** **FOOD** Yummy Pizzeria charges $10 for a large cheese pizza and $0.50 for each additional topping. Paisano Pizzeria charges $7 for a large cheese pizza and $1 for each additional topping. For what number of toppings is the cost of a large pizza the same?

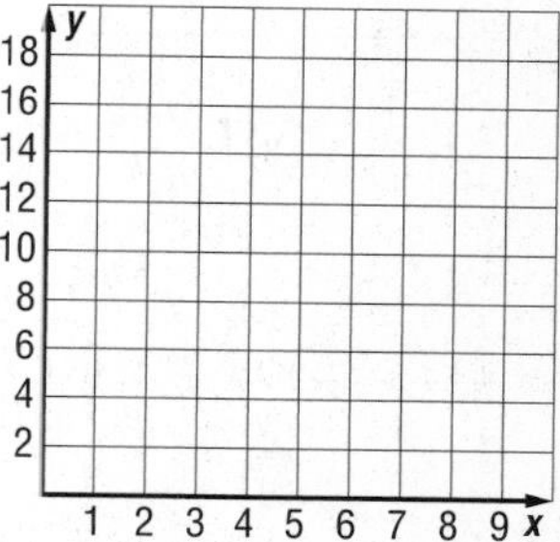


_______________

**Vocabulary Check** **Write the vocabulary word that completes each sentence.**

**21** A set of two or more linear equations with the same variables is a(n) _______________.

**22** The _______________ of a line is the rise over the run.

**23** A(n) _______________ is an equation for which the graph is a straight line.

**24** **Reflect** Given the equation $y = 12x - 9$, write a second equation so that the system has no solution. Explain why there is no solution.

_______________

_______________

_______________

STOP

# Progress Check 1 (Lessons 3-1 and 3-2)

**Graph each equation.**

1 $y = \frac{1}{2}x + 1$

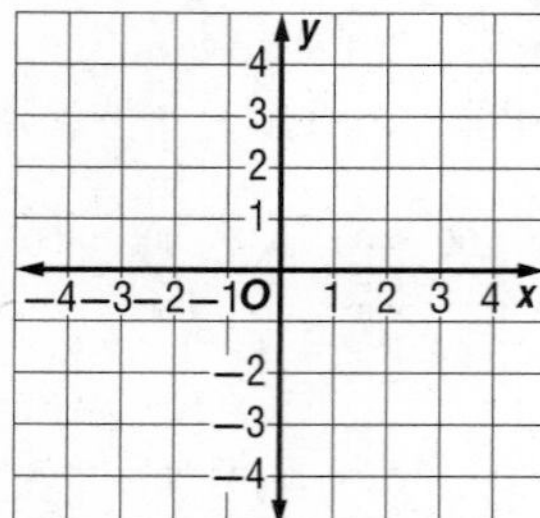

2 $y = 2x$

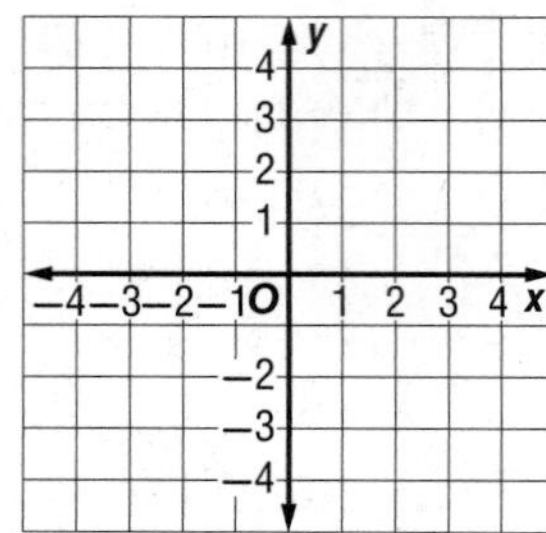

**Solve each system of linear equations by graphing.**

3 $y = -2x + 12$

$y = \frac{1}{3}x + 5$

______________

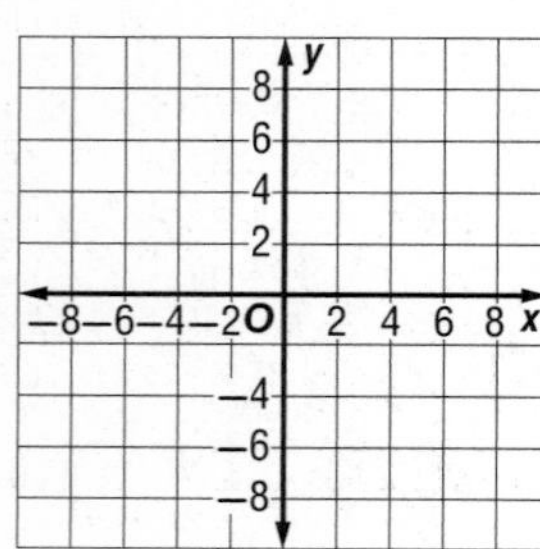

4 $y = 2x - 3$

$y = 2x + 4$

______________

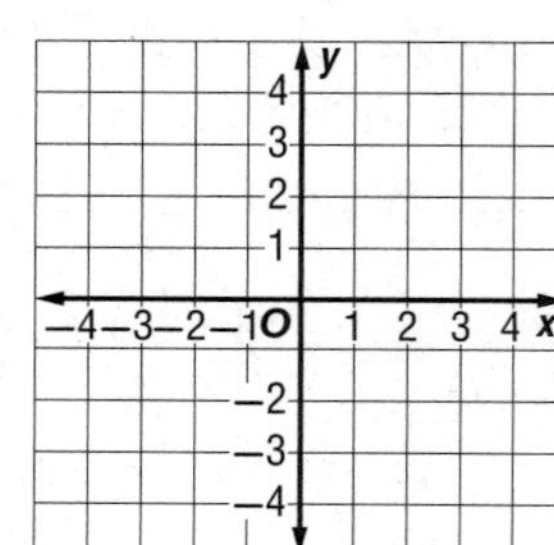

**Solve.**

5 **GARDENING** Andrea and Tim are each renting a rototiller to prepare their gardens for planting. Andrea's rototiller costs $50 plus $10 per hour of use. Tim's rototiller is a flat fee of $90 per day. At how many hours of work will Andrea and Tim pay the same amount for their equipment?

______________________________

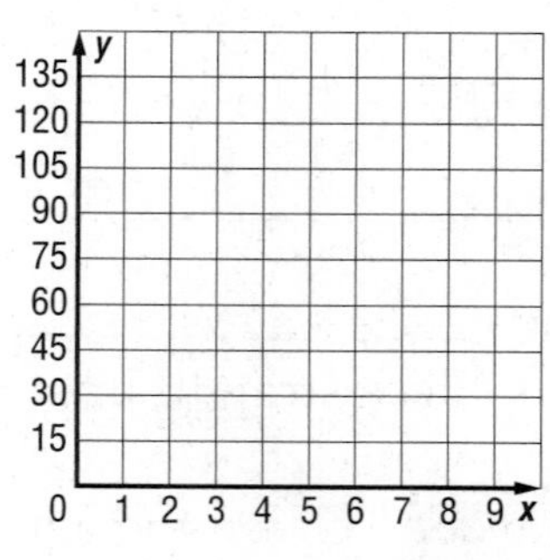

6 **BASKETBALL** Brad started practicing free throws for the basketball season. The number of free throws made, $y$, out of the number of days, $x$, he practiced is modeled by $y = x + 10$. Graph this equation.

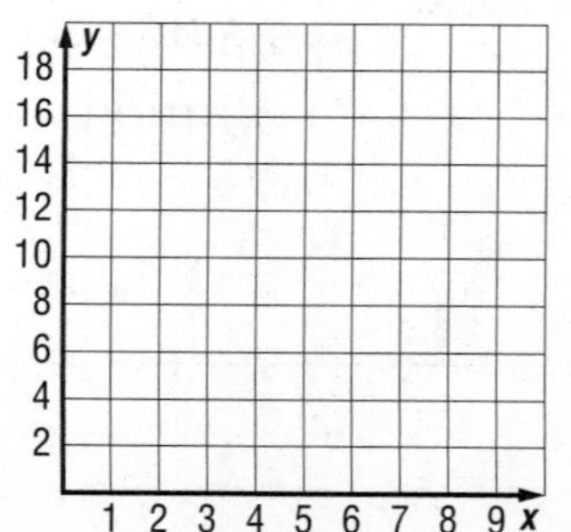

Lesson 3-3

# Graph Linear Inequalities

## KEY Concept

The graph of a **linear inequality** includes a **boundary** (either a solid or dashed line) and all ordered pairs in a region either above or below that boundary.

| Inequality Symbols | |
|---|---|
| > greater than | dashed line |
| < less than | |
| ≥ greater than or equal to | solid line |
| ≤ less than or equal to | |

The graph below shows the inequality $y > 2x - 2$.

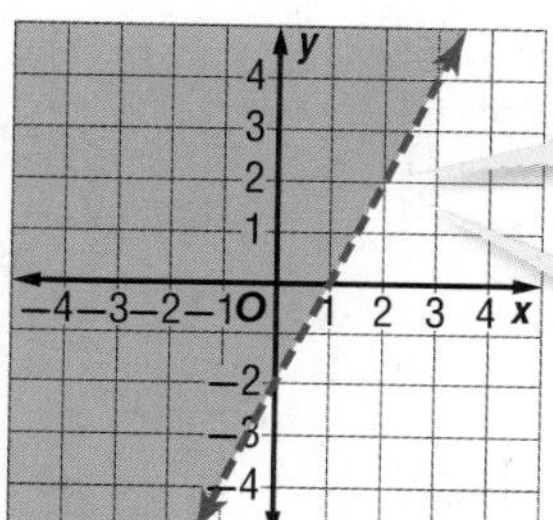


The equation of the boundary is $y = 2x - 2$.

Use a dashed line because of the inequality symbol >.

All the points in this region are solutions to the inequality. The point (0, 0) is an easy point to use to check.

**VOCABULARY**

**boundary**
the related linear equation of an inequality

**linear inequality**
a relation whose boundary is a straight line

## Example 1

**Graph $y \geq -\frac{1}{2}x - 4$.**

1. Graph the boundary. Use a solid line because the inequality symbol is ≥.
2. Substitute a point to check if it is included in the solution. Use (0, 0).

$y \overset{?}{\geq} -\frac{1}{2}x - 4$ Given.

$0 \overset{?}{\geq} -\frac{1}{2}(0) - 4$ Substitute.

$0 \geq -4$ ✓ True.

3. Shade the region that contains the point (0, 0).

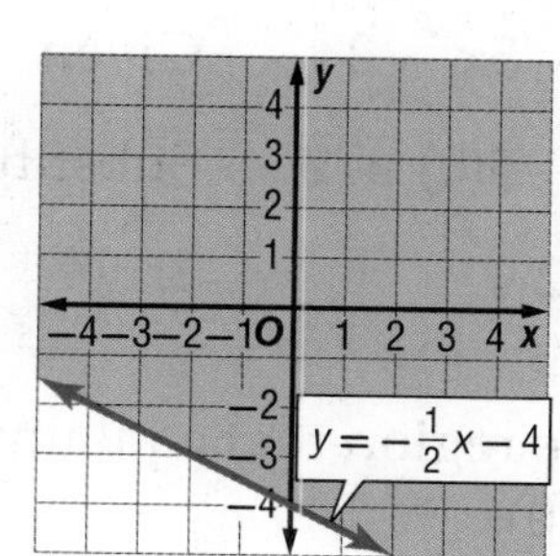


GO ON

### YOUR TURN!

**Graph $y \leq 3x - 1$.**

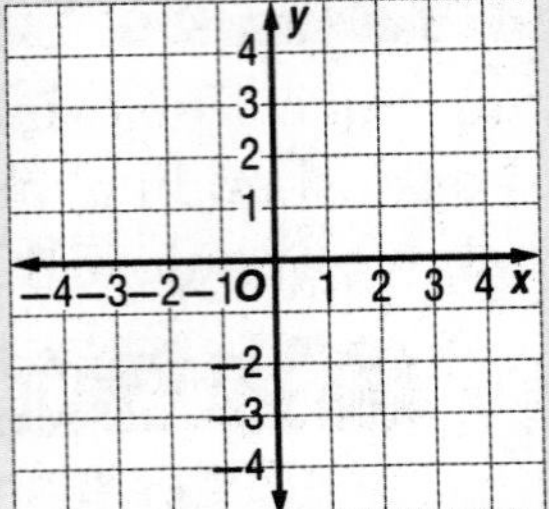


1. Graph the boundary. Use a __________ line because the inequality symbol is __________.
2. Substitute a point to check if it is included in the solution. Use __________.

| | |
|---|---|
| $y \overset{?}{\leq} 3x - 1$ | Given. |
| $\_\_\_\_ \overset{?}{\leq} 3(\_\_\_\_) - 1$ | Substitute. |
| $0 \leq \_\_\_\_$ | __________. |

3. Shade the region that __________ the point (0, 0).

## Example 2

**Graph $y < -5x + 2$.**

1. Graph the boundary. Use a dashed line because the inequality symbol is <.

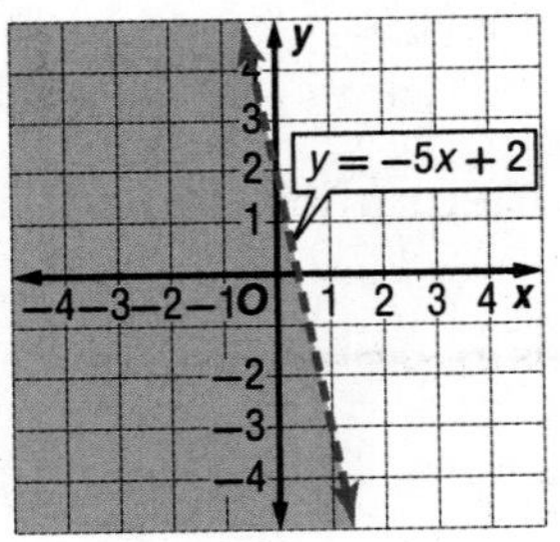


2. Substitute a point to check it it is included in the solution. Use (0, 0).

| | |
|---|---|
| $y \overset{?}{<} -5x + 2$ | Given. |
| $0 \overset{?}{<} -5(0) + 2$ | Substitute. |
| $0 < 2$ ✓ | True. |

3. Shade the region that contains the point (0, 0).

### YOUR TURN!

**Graph $y > \frac{1}{5}x - 3$.**

1. Graph the boundary.

Use a __________ line because the inequality symbol is __________.

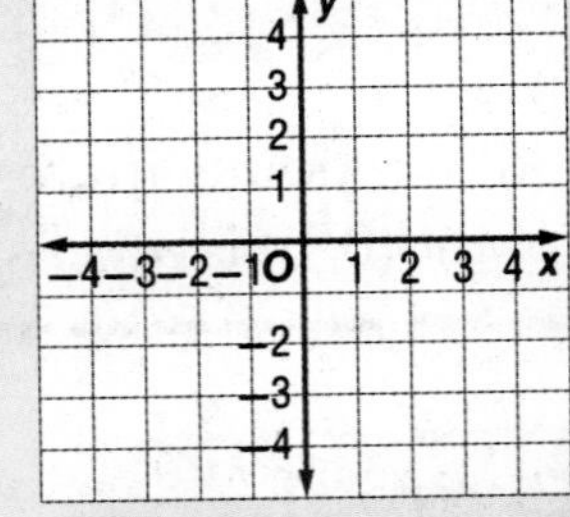


2. Substitute a point to check if it is included in the solution. Use __________.

| | |
|---|---|
| $y \overset{?}{>} \frac{1}{5}x - 3$ | Given. |
| $\_\_\_\_ \overset{?}{>} \frac{1}{5}(\_\_\_\_) - 3$ | Substitute. |
| $0 > \_\_\_\_$ | __________. |

3. Shade the region that __________ the point (0, 0).

## Guided Practice

**Graph each inequality.**

**1** $y < -3x + 5$

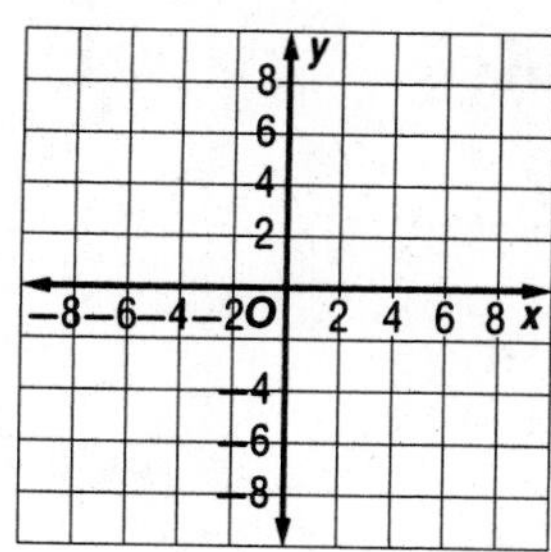

Use a ________ line because the symbol is ______.

Substitute (0, 0).

$\_\_\_\_ \overset{?}{<} -3(\_\_\_\_) + 5$

(0, 0) makes the inequality ________.

Shade the region that ________________ the point (0, 0).

**2** $y \geq \frac{1}{3}x + 7$

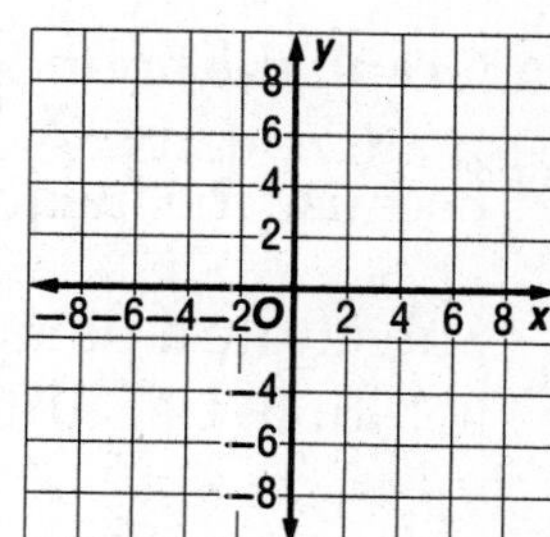

Use a ________ line because the symbol is ______.

Substitute (0, 0).

$\_\_\_\_ \overset{?}{\geq} \frac{1}{3}(\_\_\_\_) + 7$

(0, 0) makes the inequality ________.

Shade the region that ________________ the point (0, 0).

**3** $y \leq -\frac{1}{2}x - 1$

Use a ________ line because the inequality symbol is ________.

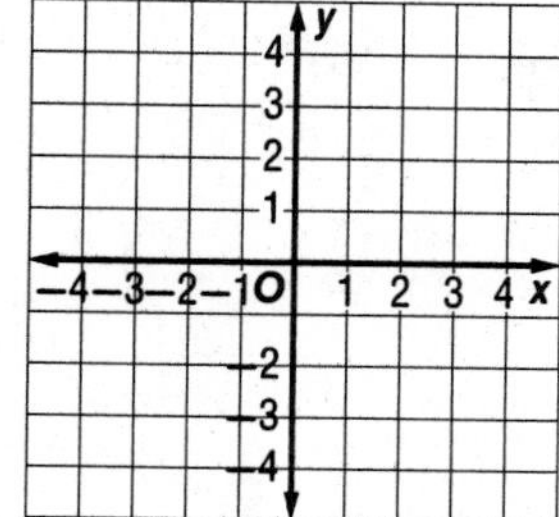

**4** $y > x + 4$

Use a ________ line because the inequality symbol is ________.

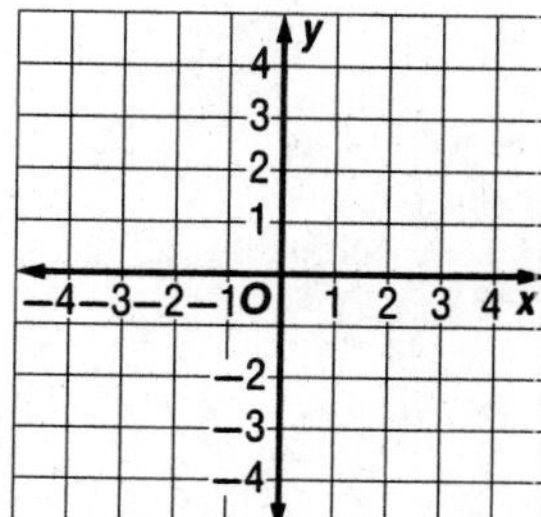



GO ON

## Step by Step Practice

**5** Graph $y < -2x + 6$.

**Step 1** Graph the boundary. Use a ______________ line because the inequality symbol is ______________.

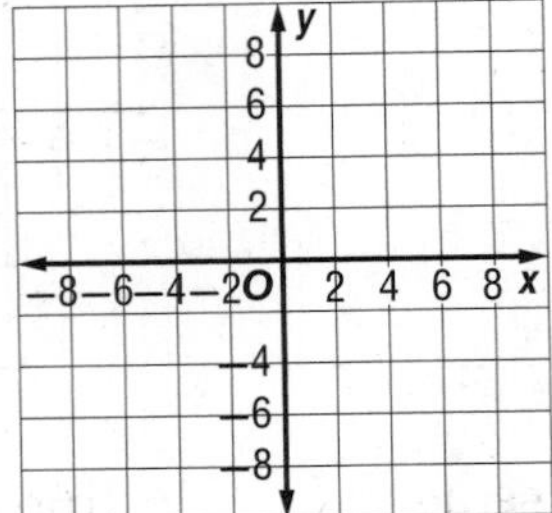


**Step 2** Substitute a point to check if it is included in the solution. Use (0, 0).

| | |
|---|---|
| $y \overset{?}{<} -2x + 6$ | Given. |
| ______ $\overset{?}{<} -2($______$) + 6$ | Substitute. |
| ______ $<$ ______ | ______. |

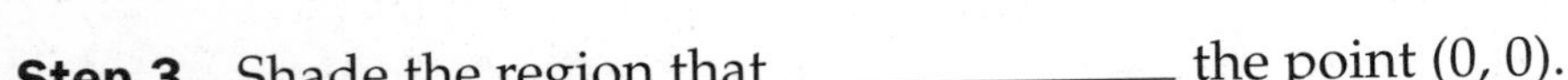

**Step 3** Shade the region that ______________ the point (0, 0).

**Graph each inequality.**

**6** $y \leq \frac{4}{5}x - 3$

Graph the boundary. Use a __________ line.

Substitute (0, 0). ______ $\leq$ ______ $- 3$

Shade the region that ______________ the point (0, 0.)

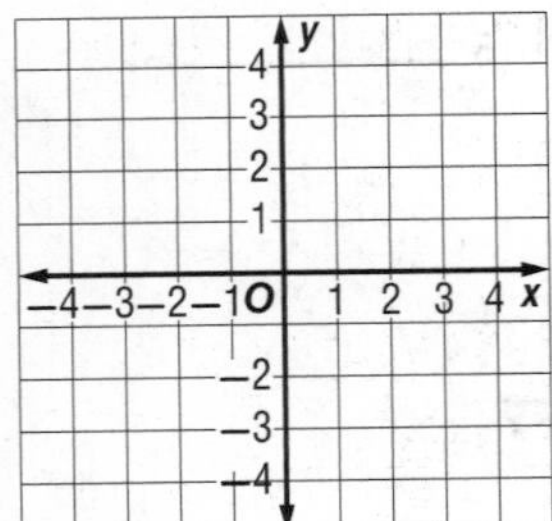


**7** $y > 6x - 9$

Graph the boundary. Use a __________ line.

Substitute (0, 0). ______ $>$ ______ $- 9$

Shade the region that ______________ the point (0, 0).

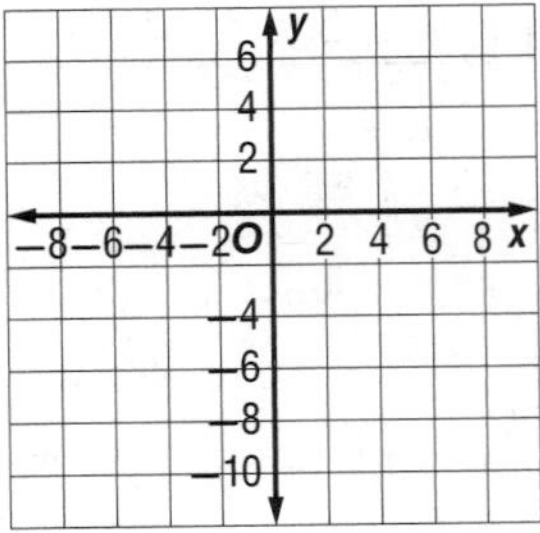


**8** $y > -x + 5$

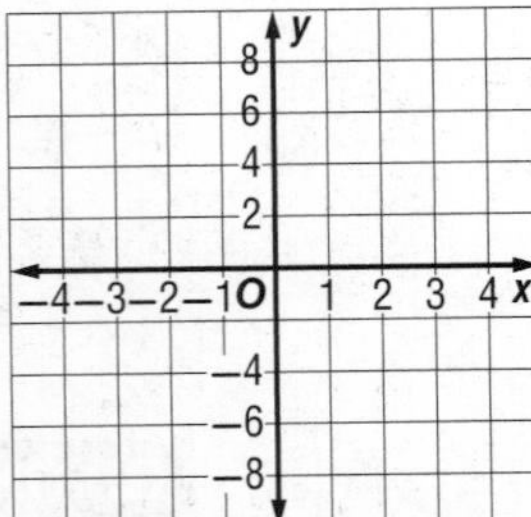


**9** $y \leq \frac{3}{8}x + 2$

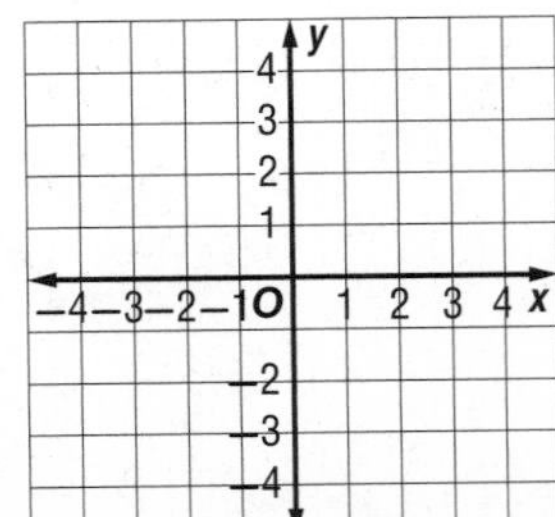

## Step by Step Problem-Solving Practice

**Solve.**

**10** BIRTHDAY PARTY Adina will purchase both paper party favors for \$3 each and plastic party favors for \$4 each. She wants to spend less than \$100. The budget for Adina's birthday party is modeled by $y < -\frac{3}{4}x + 25$, where $x$ represents the number of paper party favors, and $y$ represents the number of plastic party favors. Display the graph of Adina's budget.

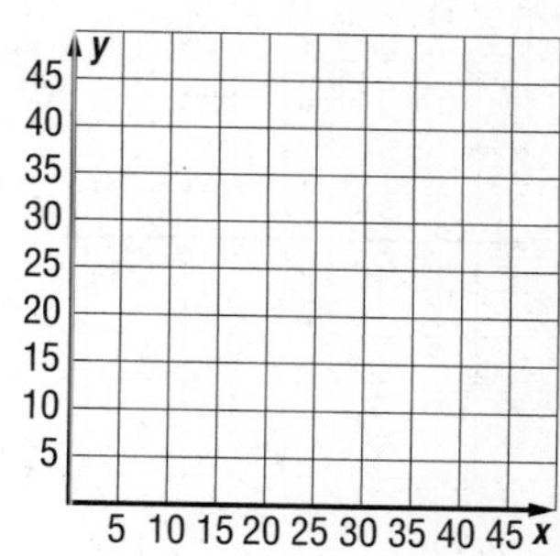

Check off each step.

_______ Understand: I underlined key words.

_______ Plan: To solve the problem, I will _______________________________.

_______ Solve: The answer is shown in the graph.

_______ Check: I checked my answer by _______________________________

_______________________________________________________________.

## Skills, Concepts, and Problem Solving

**Graph each inequality.**

**11** $y \leq -4x + 5$

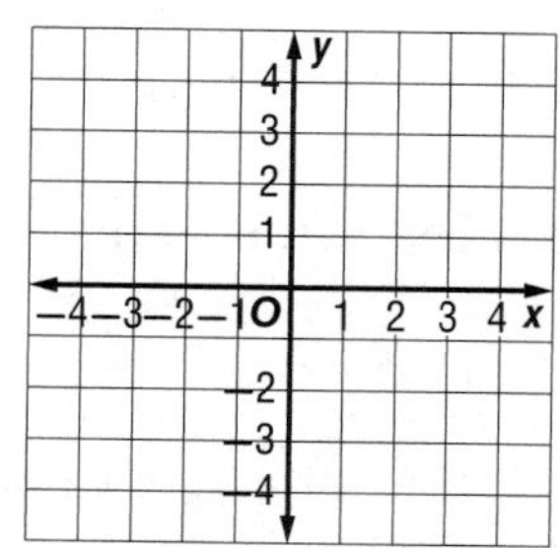

**12** $y > 2x + 3$

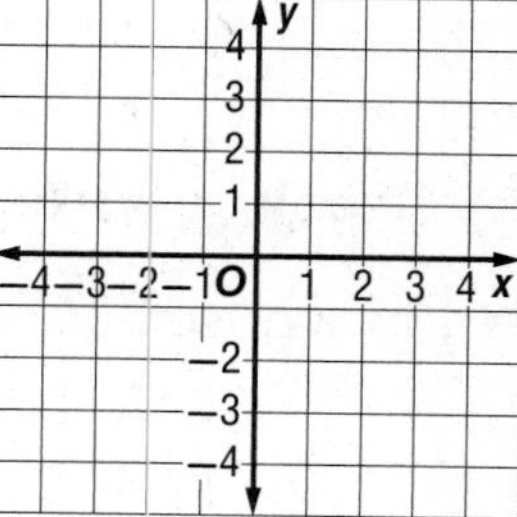

**13** $y < \frac{3}{4}x - 5$

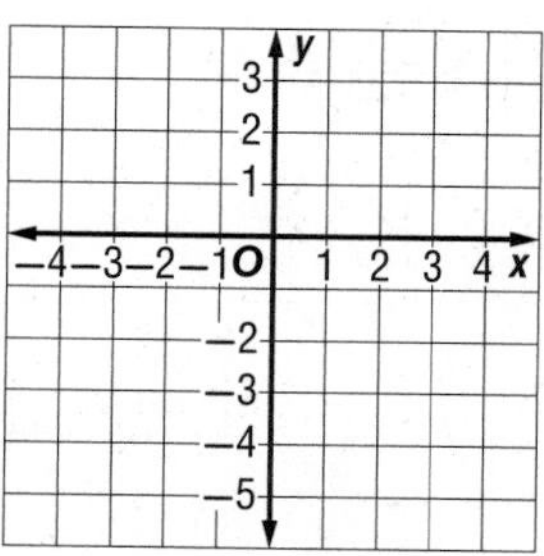

**14** $y \leq -x - 3$

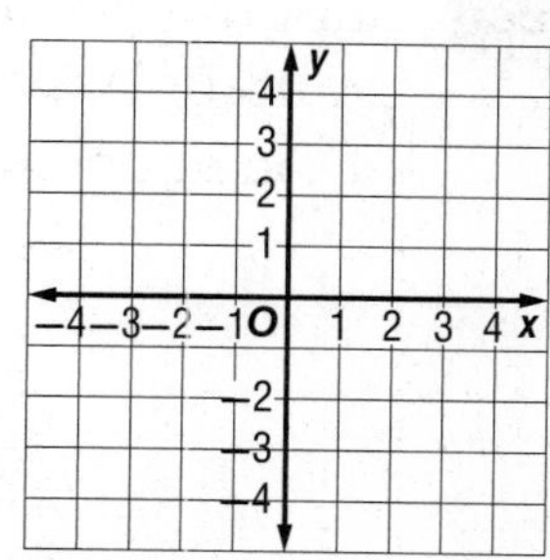

**Graph each inequality.**

15 $y \geq \frac{5}{3}x - 2$

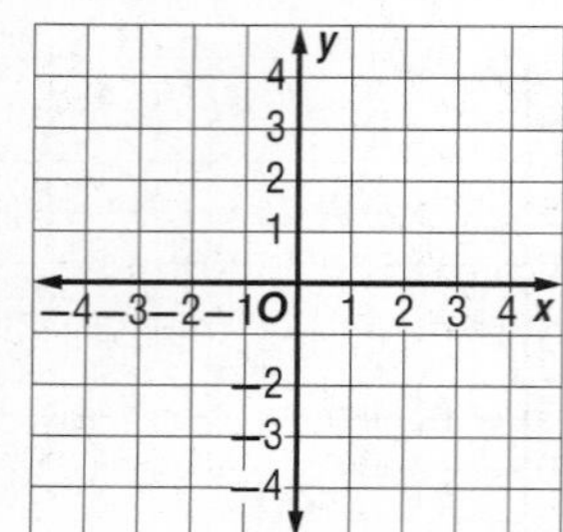

16 $y > -\frac{1}{2}x + 6$

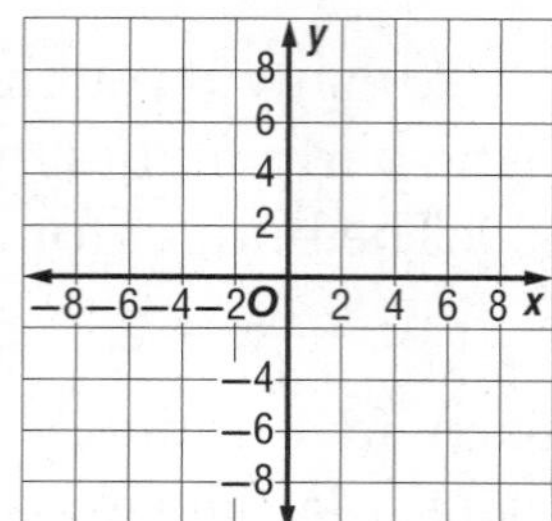

**Solve.**

17 **SALES** Paul is selling items from his closet and wants to raise at least $30. He is charging $2 for each pair of shoes and $3 for each clothing item. The amount he wants to raise from his sale is modeled by the linear inequality $y \geq -\frac{2}{3}x + 10$, where $x$ represents the number of pairs of shoes Paul sells and $y$ represents the number of clothing items that he sells. Graph this linear inequality.

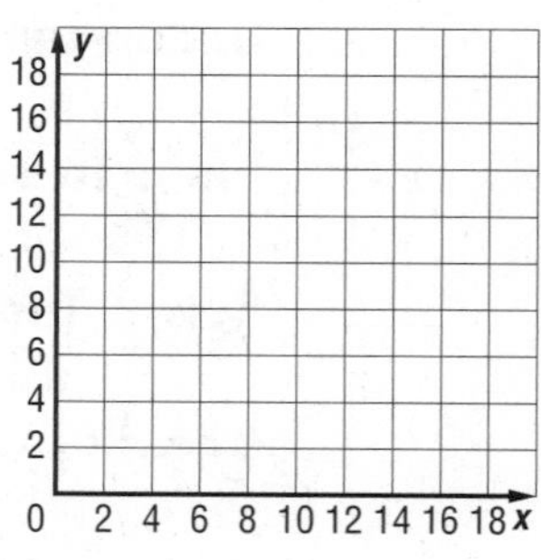

18 **BUDGETING** Ali has $8 to spend on lunch. His lunch budget is modeled by the linear inequality $y \leq 8$. Graph this linear inequality.

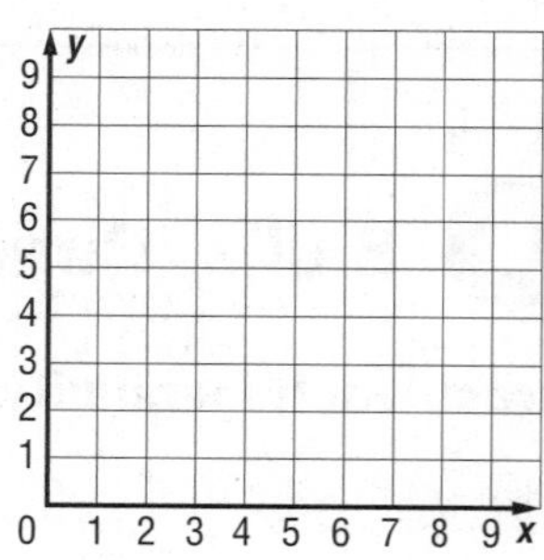

**Vocabulary Check** **Write the vocabulary word that completes each sentence.**

19 A relation whose boundary is a straight line is a(n) ______________________.

20 The ______________________ is the related linear equation of an inequality.

21 **Reflect** Suppose the boundary of a linear inequality contains the point (0, 0). What could you do to determine which side of the boundary should be shaded?

______________________________________________________________

______________________________________________________________

STOP

Lesson 3-4

# Graph Systems of Linear Inequalities

## KEY Concept

The solution to a **system of linear inequalities** is the region where the two graphs overlap.

The graph shows the solutions to the system of inequalities below.

$y > -x + 1$

$y \leq 2x - 3$

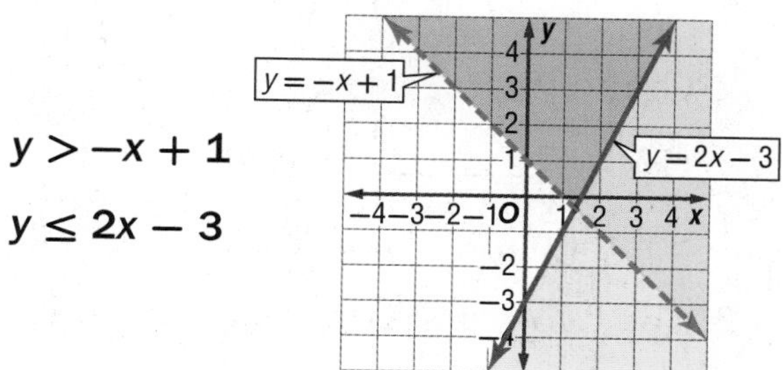


The solution of the system is the purple region.

**VOCABULARY**

**linear inequality**
a relation whose boundary is a straight line

**system of linear inequalities**
a set of two or more linear inequalities with the same variables

## Example 1

**Solve the system $y < \frac{2}{5}x + 2$ and $y > -x - 4$ by graphing.**

1. Graph $y < \frac{2}{5}x + 2$.
   Use a dashed line.
   Solution contains (0, 0).

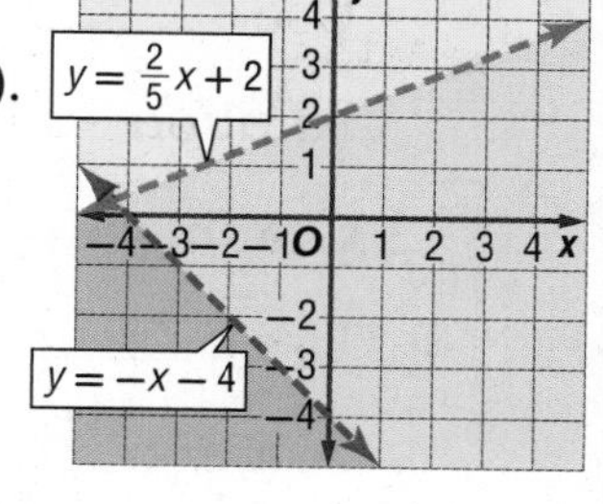


2. Graph $y > -x - 4$.
   Use a dashed line.
   Solution contains (0, 0).

3. The solution is the region where the two graphs overlap.

## YOUR TURN!

**Solve the system $y \geq x + 1$ and $y \geq -x + 2$ by graphing.**

1. Graph $y \geq x + 1$
   Use a ________ line.
   Solution ________
   ________ (0, 0).

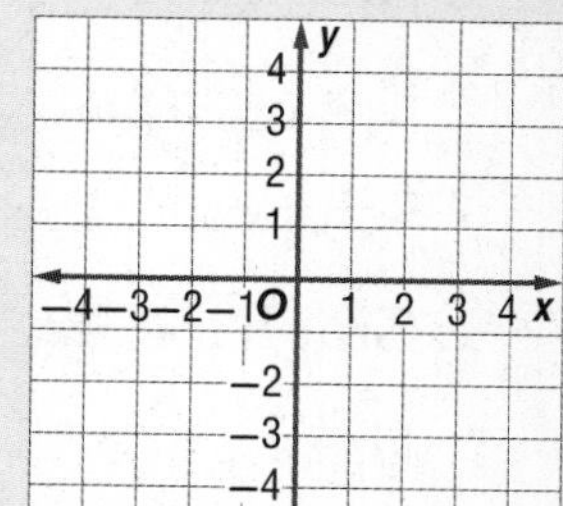


2. Graph $y \geq -x + 2$
   Use a ________ line.
   Solution ________ (0, 0).

3. The solution is the region where the two graphs ________.



GO ON

## Example 2

**Solve the system $y \geq x + 3$ and $y \geq x - 2$ by graphing.**

1. Graph $y \geq x + 3$.
   Use a solid line.
   Solution does not contain (0, 0).

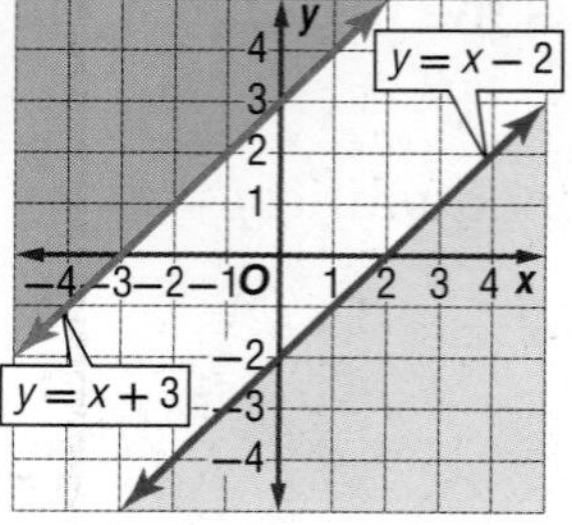


2. Graph $y \geq x - 2$.
   Use a solid line.
   Solution does not contain (0, 0).

3. The two graphs do not overlap. There is no solution.

**YOUR TURN!**

**Solve the system $y > -\frac{2}{3}x + 4$ and $y \leq -\frac{2}{3}x - 2$ by graphing.**

1. Graph $y > -\frac{2}{3}x + 4$.
   Use a ________ line.
   Solution ________ ________ (0, 0).

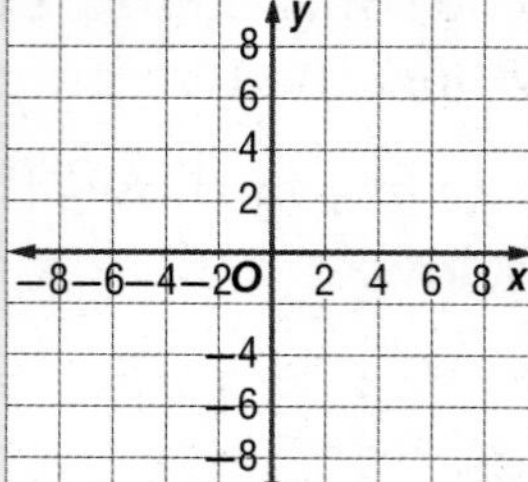

2. Graph $y \leq -\frac{2}{3}x - 2$
   Use a ________ line.
   Solution ________ (0, 0).

3. The two graphs do not overlap. There is ________.

## Guided Practice

**Solve each system of inequalities by graphing.**

**1** $y < -\frac{1}{3}x + 2$
$y \geq 2x - 5$

1. Graph $y < -\frac{1}{3}x + 2$.
   Use a ________ line.
   Solution ________ (0, 0).
2. Graph $y \geq 2x - 5$.
   Use a ________ line.
   Solution ________ (0, 0).
3. The solution is the region where the two graphs ________.

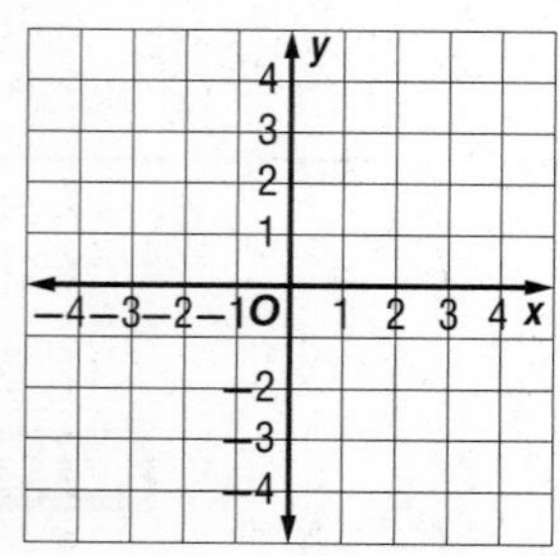

**2** $y \leq -x + 7$
$y > -3$

1. Graph $y \leq -x + 7$.
   Use a ________ line.
   Solution ________ (0, 0).
2. Graph $y > -3$.
   Use a ________ line.
   Solution ________ (0, 0).
3. The solution is the region where the two graphs ________.

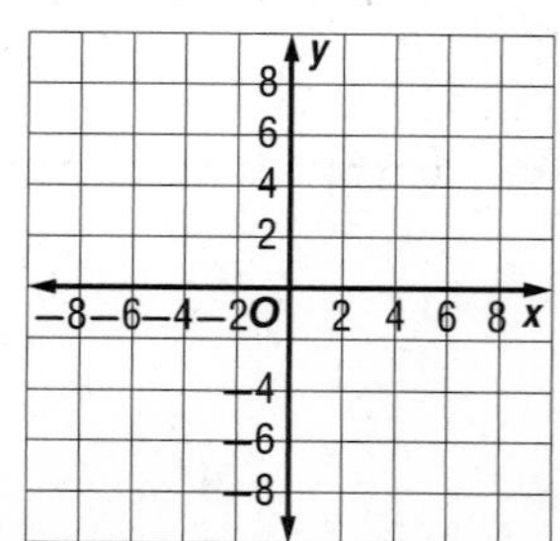

## Step by Step Practice

**3** Solve the system $y > -x + 8$ and $y \geq 4x - 3$ by graphing.

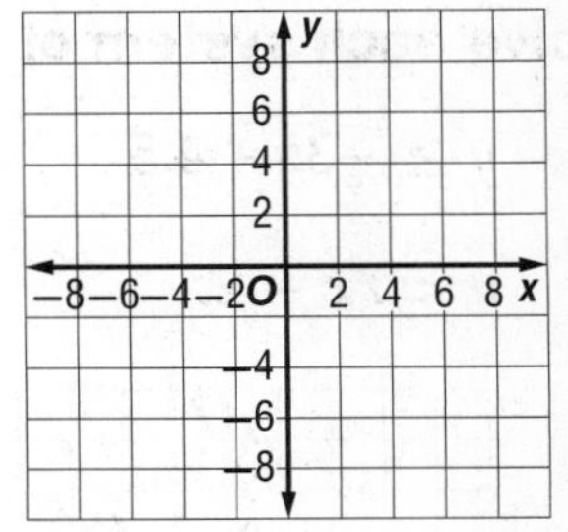

**Step 1** Graph $y > -x + 8$.

_______ line; solution _______ (0, 0)

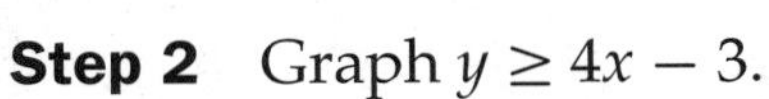

**Step 2** Graph $y \geq 4x - 3$.

_______ line; solution _______ (0, 0)

**Step 3** The solution to the system is the area where the two graphs _______.

**Solve each system of linear inequalities by graphing.**

**4** $y < 8$ _______ line; (0, 0)? _______

$y \geq -\frac{1}{5}x - 4$ _______ line; (0, 0)? _______

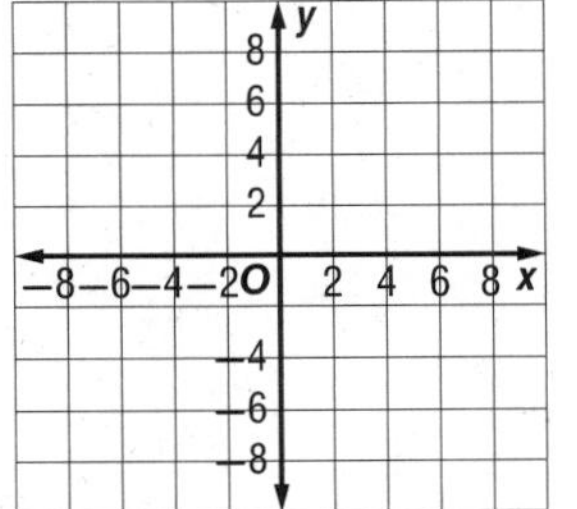

**5** $y \leq \frac{5}{3}x - 6$ _______ line; (0, 0)? _______

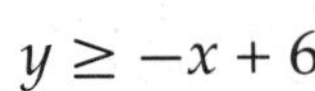

$y \geq -x + 6$ _______ line; (0, 0)? _______

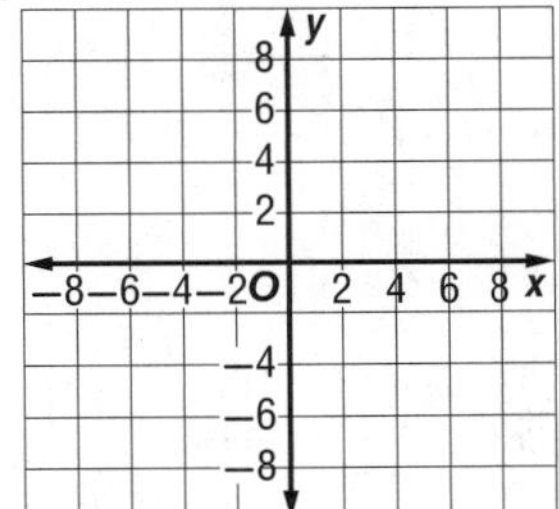

## Step by Step Problem-Solving Practice

**Solve.**

**6** Carlos is purchasing lunchmeat for a work luncheon. The budget can be modeled by the inequality $y \leq -\frac{2}{3}x + 8.35$. The total amount of meat purchased can be modeled by $y \geq -x + 10$. Graph this system to find the amount of meat Carlos can purchase.

Check off each step.

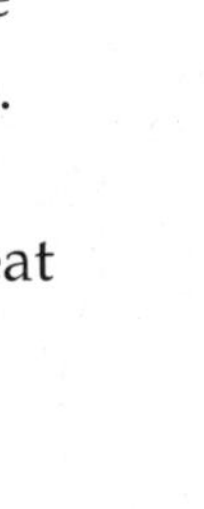

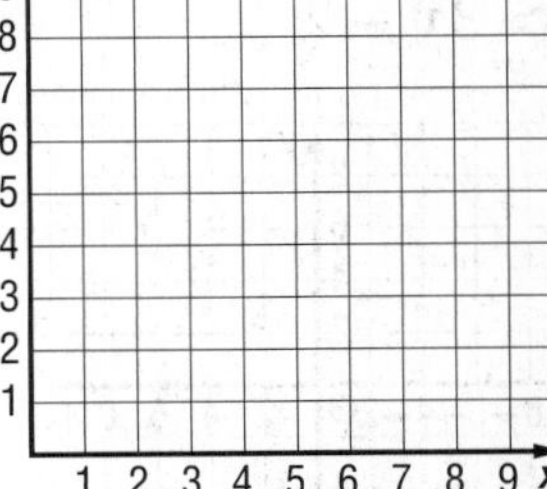

_______ **Understand: I underlined key words.**

_______ **Plan: To solve the problem, I will** _______.

_______ **Solve: The answer is shown in the graph.**

_______ **Check: I checked my answer by** _______

_______.

## Skills, Concepts, and Problem Solving

**Solve each system of linear inequalities by graphing.**

**7** $y \geq -3x + 4.5$

$y < 7x - 2$

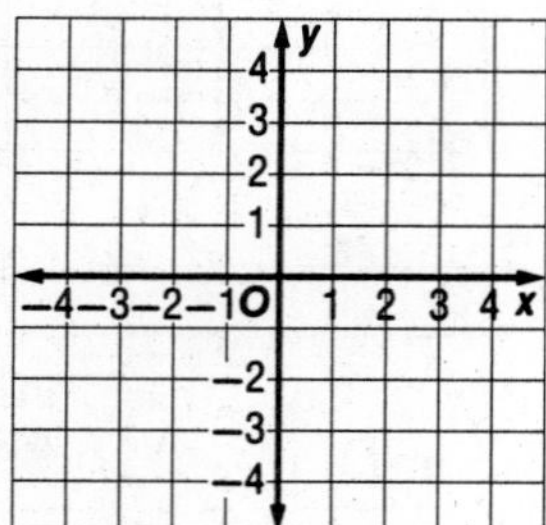

**8** $y < x + 1$

$y < 8$

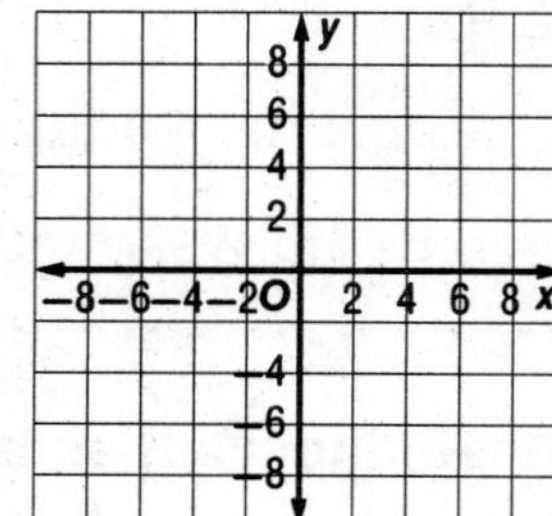

**9** $y \leq \frac{4}{5}x - 2$

$y \leq \frac{5}{4}x - 2$

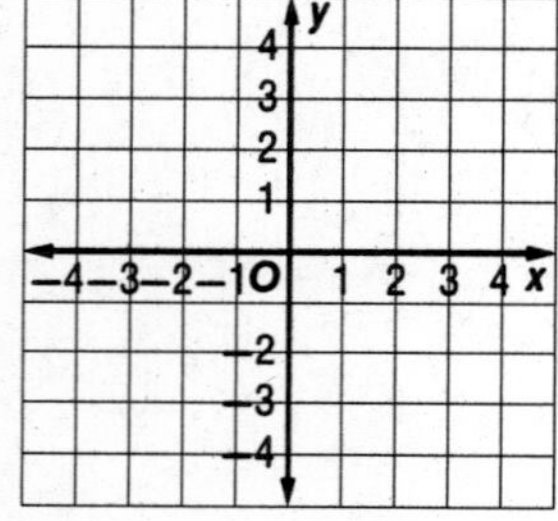

**10** $y \geq -8x + 12$

$y \leq x - 5$

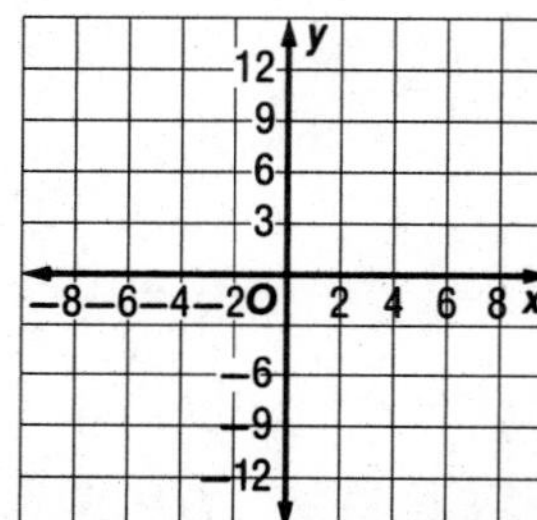

**11** $y \leq -\frac{1}{2}x + 3$

$y > 3x - 2$

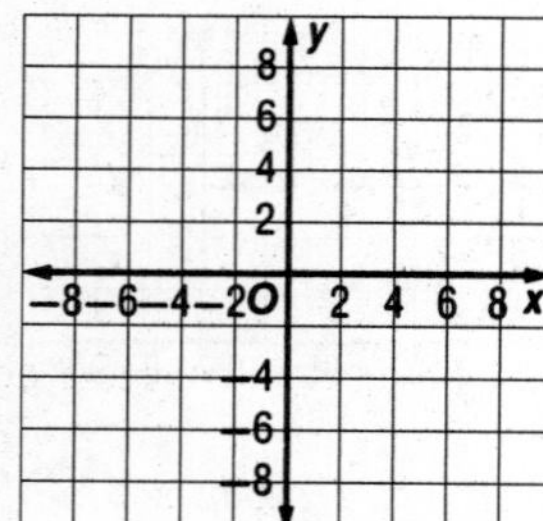

**12** $y > 5x + 4$

$y > x + 4$

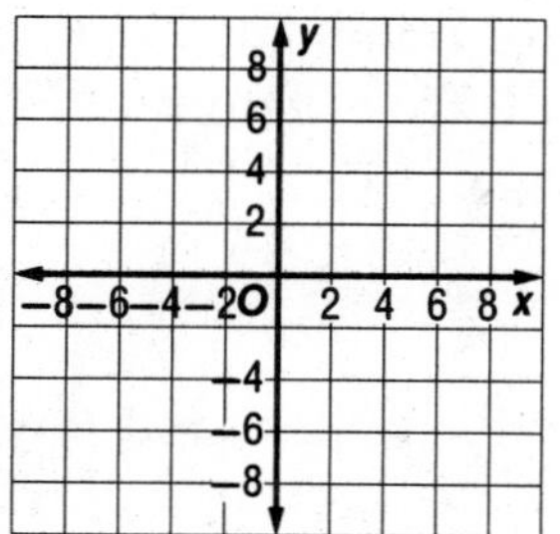

**Solve.**

13 **GEOMETRY** A rectangle can be created using the system of four linear inequalities below. Graph the system of inequalities.

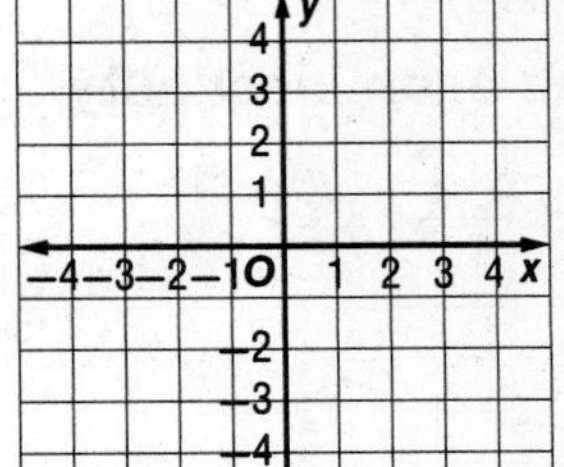

$y \geq -x - 4$

$y \leq -x + 4$

$y \leq x + 4$

$y \geq x - 4$

14 **SHOPPING** You have $40 to spend on pants and shirts at a store that is selling clearance items. You want to buy at least two pairs of pants. Let $x$ represent the number of shirts you buy, and let $y$ represent the number of pants you buy.
The inequality $y \leq -\frac{4}{5}x + 4$ models your budget of $40.
The inequality $y \geq 2$ models the fact that you will purchase at least two pairs of pants. Graph this system of inequalities.

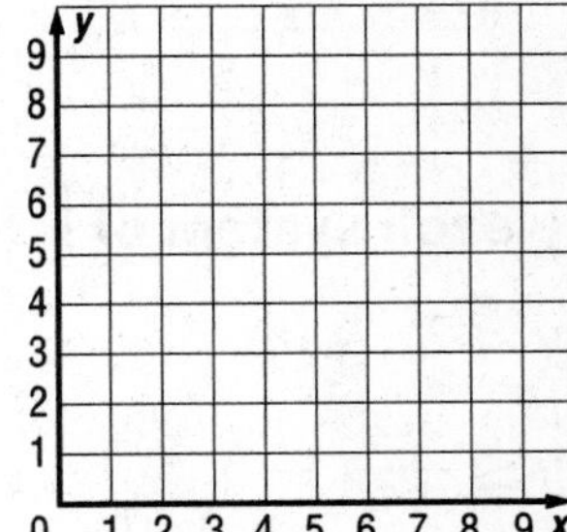

**Vocabulary Check** **Write the vocabulary word that completes each sentence.**

15 A set of two or more linear inequalities with the same variables is a(n) ________________.

16 A(n) ________________ is a relation whose boundary is a straight line.

17 **Reflect** Could there ever be a system of linear inequalities in which there is no solution, meaning the graphs would never overlap? If so, give an example?

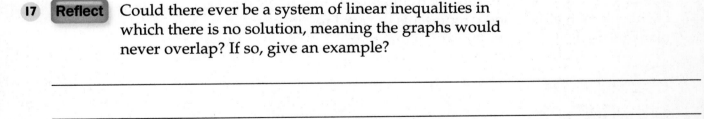

______________________________

______________________________

______________________________



STOP

# Chapter 3 Progress Check 2 (Lessons 3-3 and 3-4)

**Graph each inequality.**

1 $y \leq \frac{2}{3}x - 1$

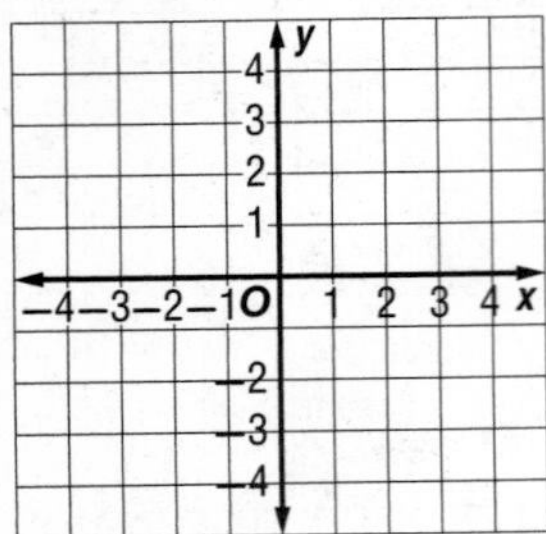

2 $y > 3x + 4$

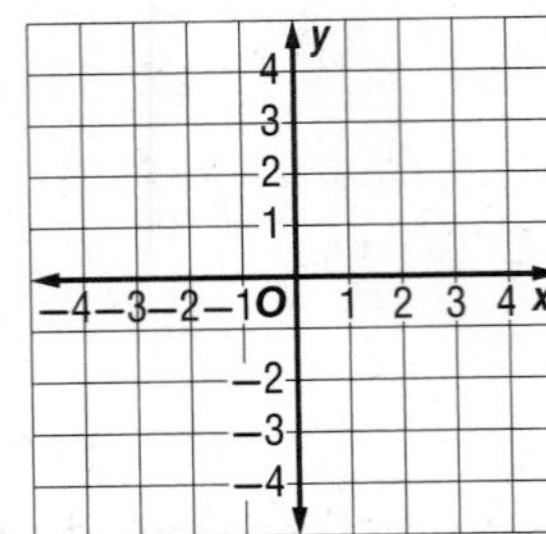

**Solve each system of inequalities by graphing.**

3 $y < 2$

$y \leq -\frac{3}{5}x - 3$

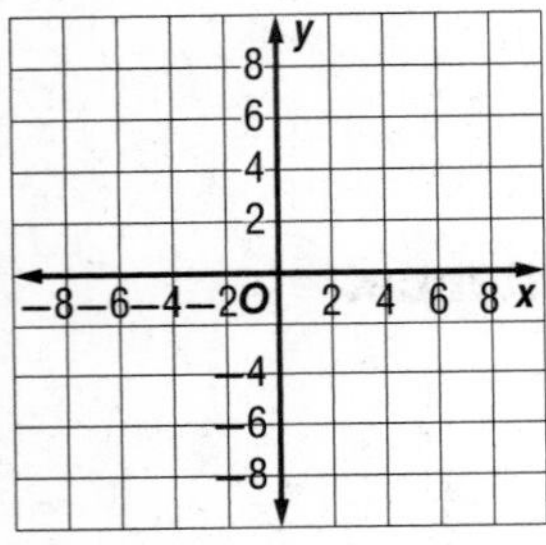

4 $y \leq \frac{3}{2}x - 4$

$y \geq -x + 2$

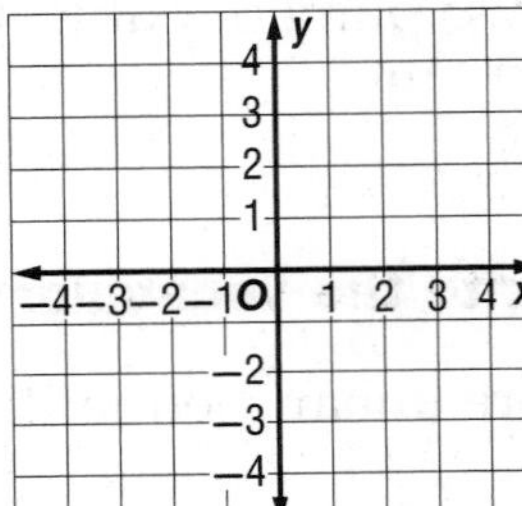

**Solve.**

5 **BLANKETS** Connie has been making blankets for many years. The number of blankets Connie has made can be modeled by $y \geq 2x + 1$, where $x$ represents the number of years Connie has been making blankets and $y$ represents the total number of blankets made. Graph this linear inequality.

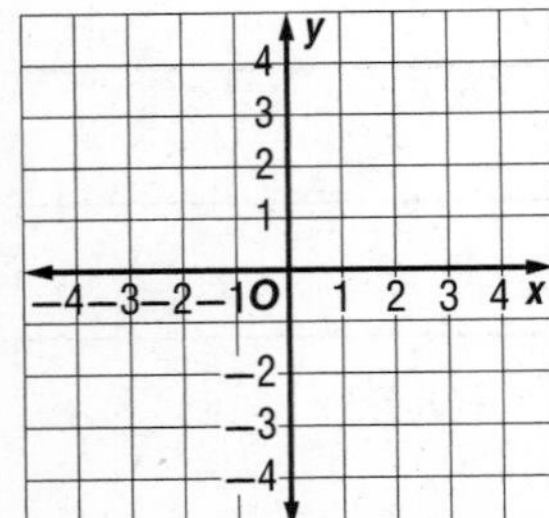

Lesson 3-5

# Graph Quadratic Functions

## KEY Concept

The graph of a **quadratic function** is a **parabola**. A parabola is a U-shaped graph that has either a **maximum** (highest point) or **minimum** (lowest point). The places where a parabola crosses the *x*-axis are the zeros, or solutions, of its related quadratic function.

### One Solution

The graph shows the quadratic function $y = x^2$. There is one solution.

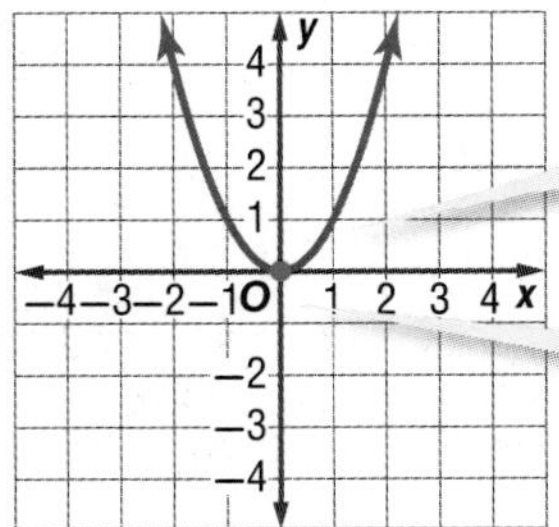

There is 1 solution, or zero, $x = 0$.

The minimum value of 0 occurs at $x = 0$.

### Two Solutions

The graph shows the quadratic function $y = x^2 - 16$. There are two solutions.

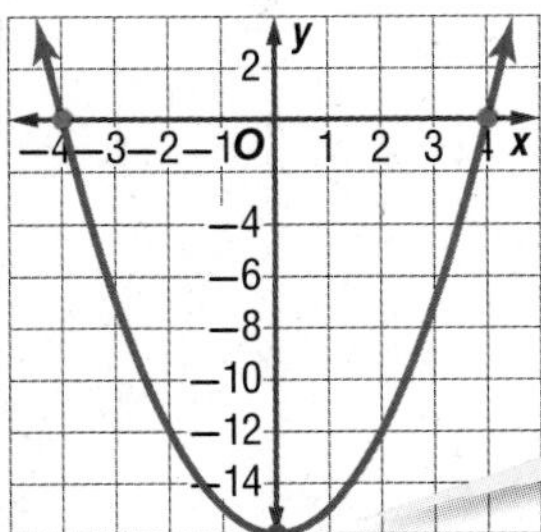

There are 2 solutions, or zeros, $x = -4$ and $x = 4$.

The minimum value of $-16$ occurs at $x = 0$.

### No Solutions

The graph shows the quadratic function $y = -x^2 - 1$. There are no solutions.

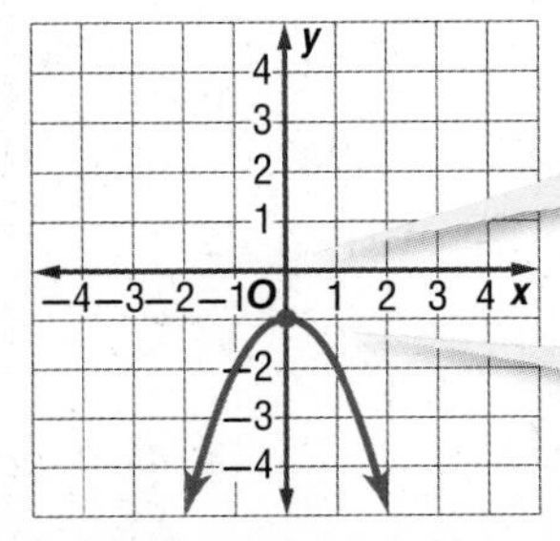

There are 0 solutions, or zeros.

The maximum value of $-1$ occurs at $x = 0$.

## VOCABULARY

**maximum**
the highest point on a parabola

**minimum**
the lowest point on a parabola

**parabola**
the U-shaped graph of a quadratic function that can open either up or down

**quadratic function**
an equation that can be written in the form $ax^2 + bx + c = 0$, where $a \neq 0$

**zeros of a function**
the solutions of a quadratic function

GO ON

## Example 1

**Use a table to graph the quadratic function $y = x^2 - 4$.**

1. Make a table.

| $x$ | −3 | −2 | 0 | 2 | 3 |
|---|---|---|---|---|---|
| $f(x)$ | 5 | 0 | −4 | 0 | 5 |

2. Plot the points and graph the parabola.

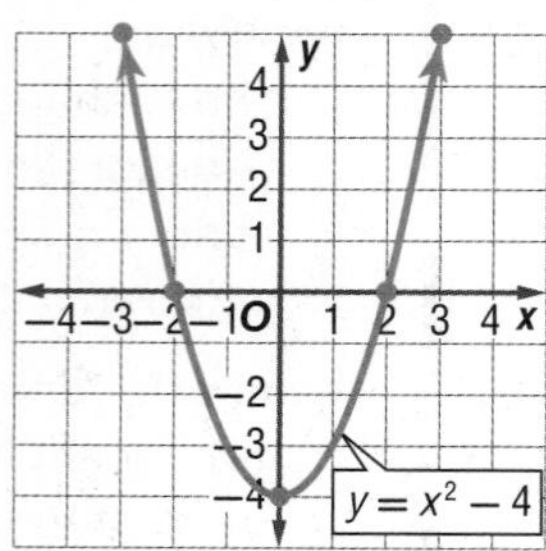


3. There are 2 solutions, or zeros.

$x = -2$ and $x = 2$

### YOUR TURN!

**Use a table to graph the quadratic function $y = 2x^2$.**

1. Make a table.

| $x$ | −2 | −1 | 0 | 1 | 2 |
|---|---|---|---|---|---|
| $f(x)$ | | | | | |

2. Plot the points and graph the parabola.

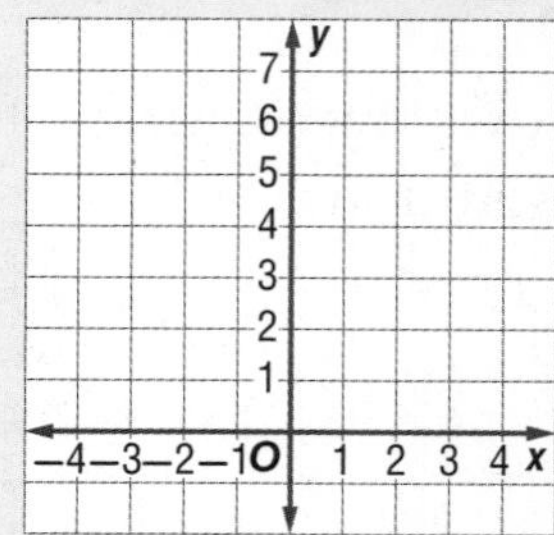

3. There is _____ solution, or zero.

$x =$ _____

## Example 2

**Identify any zeros, minimums, or maximums in the graph.**

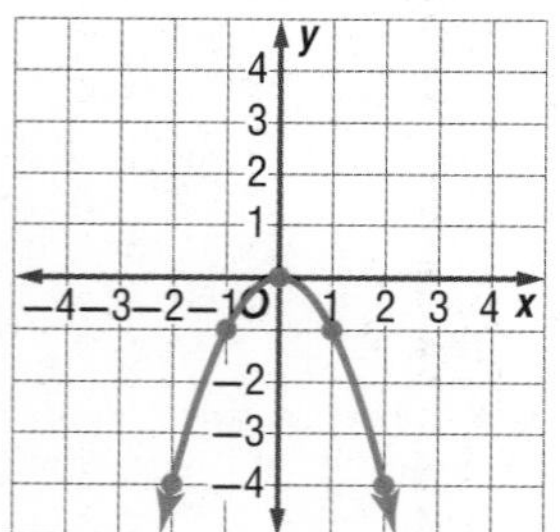

1. Identify any zeros.

The graph crosses the $x$-axis at $x = 0$. There is one zero.

2. Identify the maximum or minimum.

The graph opens downward. The maximum value of 0 occurs at $x = 0$.

### YOUR TURN!

**Identify any zeros, minimums, or maximums in each graph.**

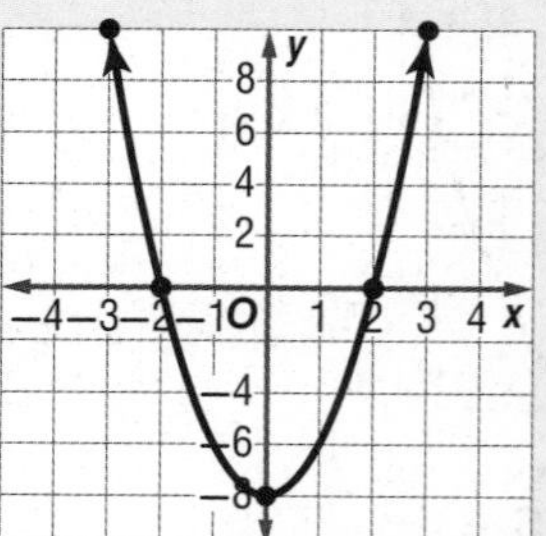

1. Identify any zeros.

The graph crosses the $x$-axis at $x =$ _____ and $x =$ _____. There are _____ zeros.

2. Identify the maximum or minimum.

The graph opens _____________.

The _____________ value of _____ occurs at $x =$ _____.

## Guided Practice

**Use a table to graph each quadratic function.**

1 $y = x^2 - 9$

| x | −4 | −3 | 0 | 3 | 4 |
|---|---|---|---|---|---|
| f(x) | | | | | |

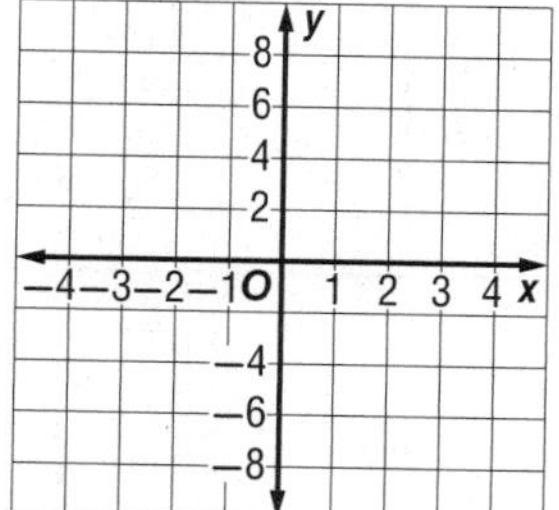

There are 2 solutions, or zeros,

at $x =$ ______ and $x =$ ______.

2 $y = -x^2 + 4$

| x | −3 | −2 | 0 | 2 | 3 |
|---|---|---|---|---|---|
| f(x) | | | | | |

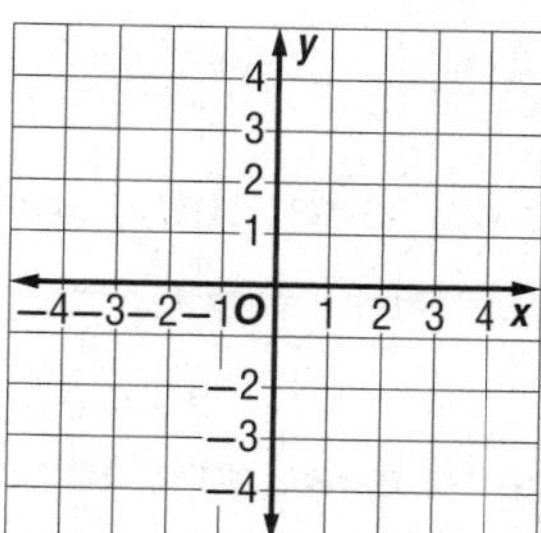

There are 2 solutions, or zeros,

at $x =$ ______ and $x =$ ______.

**Identify any zeros, minimums, or maximums in each graph.**

3

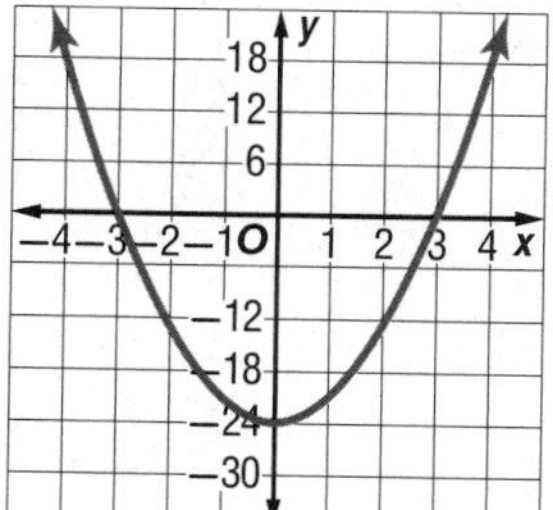

There are 2 solutions, or zeros,

at $x =$ ______ and $x =$ ______.

The ____________ value of

______ occurs at $x =$ ______.

4

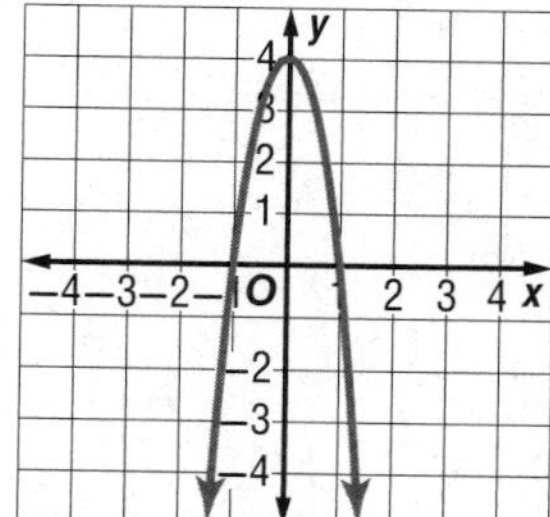

There are 2 solutions, or zeros,

at $x =$ ______ and $x =$ ______.

The ____________ value of

______ occurs at $x =$ ______.



GO ON

## Step by Step Practice

**5** Use a table to graph the quadratic function $y = -2x^2 - 4$.

**Step 1** Make a table.

| $x$ | −2 | −1 | 0 | 1 | 2 |
|---|---|---|---|---|---|
| $f(x)$ | | | | | |

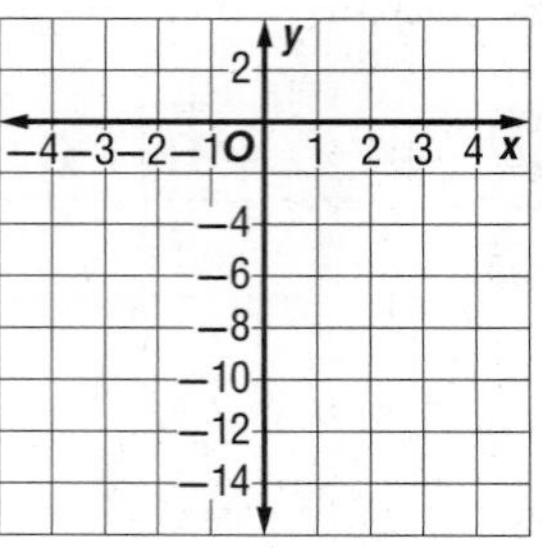


**Step 2** Plot the points.

**Step 3** Are there zeros for this function? ___________

**Use a table to graph each quadratic function.**

**6** $y = -4x^2$

| $x$ | −2 | −1 | 0 | 1 | 2 |
|---|---|---|---|---|---|
| $f(x)$ | | | | | |

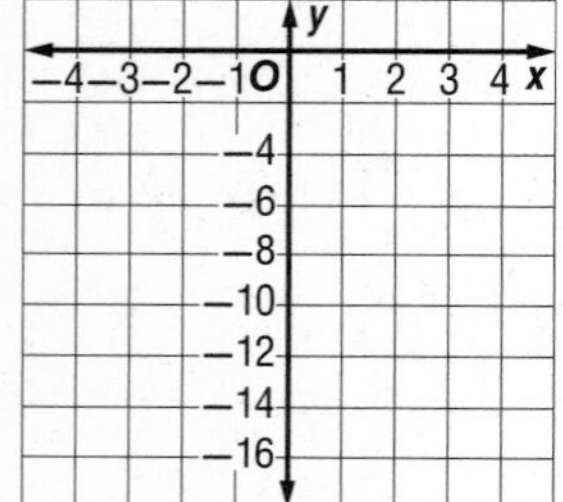


**7** 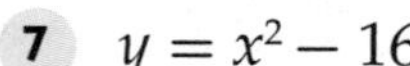 $y = x^2 - 16$

| $x$ | −5 | −4 | 0 | 4 | 5 |
|---|---|---|---|---|---|
| $f(x)$ | | | | | |

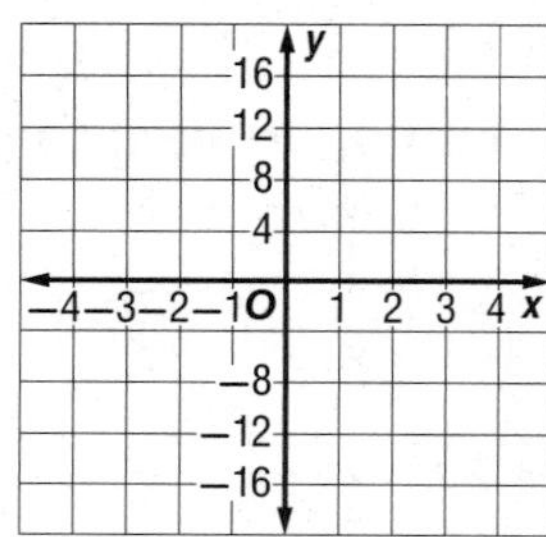


**8** $y = 2x^2 + 3$

| $x$ | −2 | −1 | 0 | 1 | 2 |
|---|---|---|---|---|---|
| $f(x)$ | | | | | |

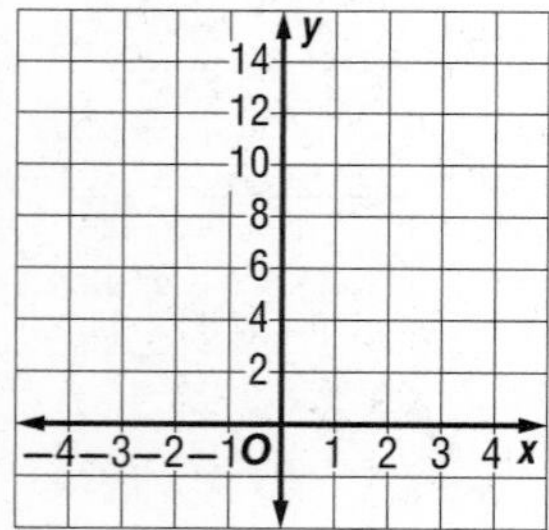


**9** $y = -\frac{1}{2}x^2 - 6$

| $x$ | −4 | −2 | 0 | 2 | 4 |
|---|---|---|---|---|---|
| $f(x)$ | | | | | |

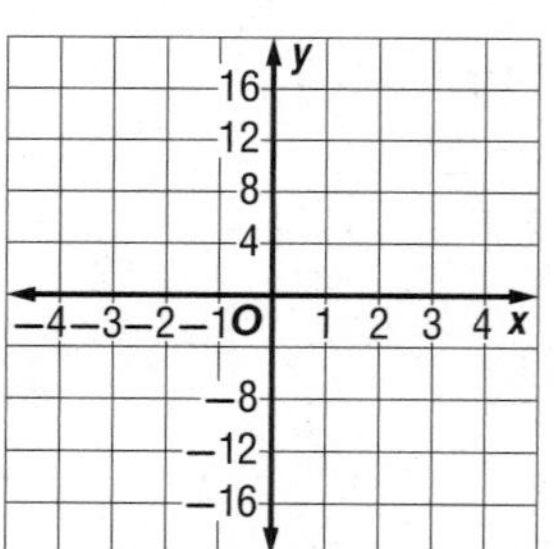

## Step by Step Problem-Solving Practice

**Solve.**

**10** An object is dropped from the window of a building. The height $h$ (in feet) of the object at time $t$ (in seconds) is modeled by the function $h = -9t^2 + 36$. Graph this function.

| x | −3 | −2 | 0 | 2 | 3 |
|---|---|---|---|---|---|
| f(x) | | | | | |

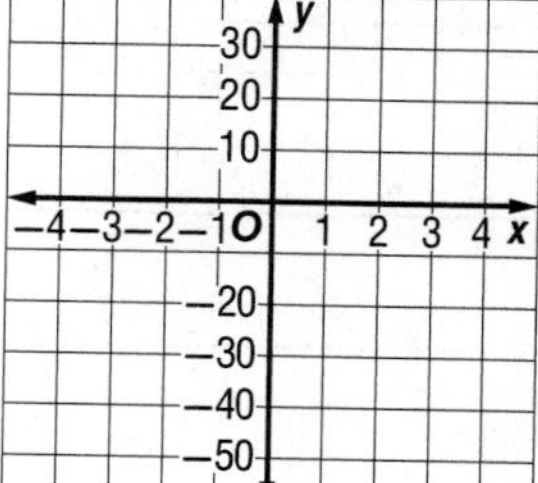

Check off each step.

_____ **Understand: I underlined key words.**

_____ **Plan: To solve the problem, I will** ______________________________.

_____ **Solve: The answer is shown in the graph.**

_____ **Check: I checked my answer by** ______________________________

______________________________.

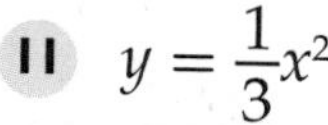

## Skills, Concepts, and Problem Solving

**Use a table to graph each quadratic function.**

**11** $y = \frac{1}{3}x^2$

| x | −6 | −3 | 0 | 3 | 6 |
|---|---|---|---|---|---|
| f(x) | | | | | |

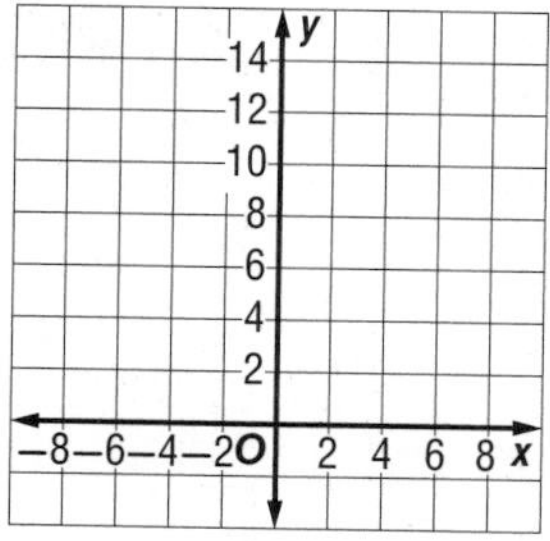

**12** $y = -3x^2 + 25$

| x | −3 | −2 | 0 | 2 | 3 |
|---|---|---|---|---|---|
| f(x) | | | | | |

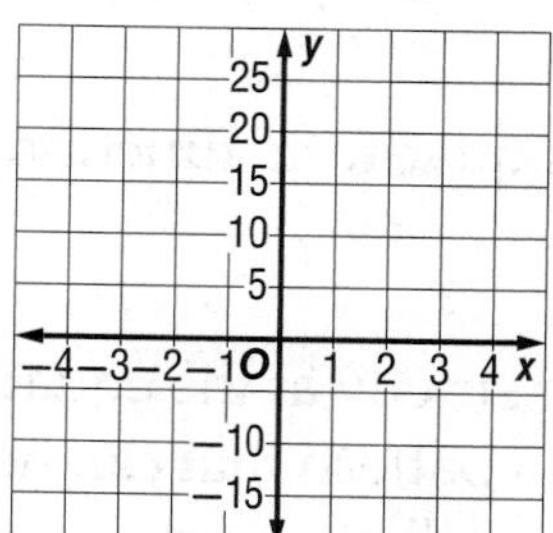



GO ON

**Identify any zeros, minimums, or maximums in each graph.**

**13**

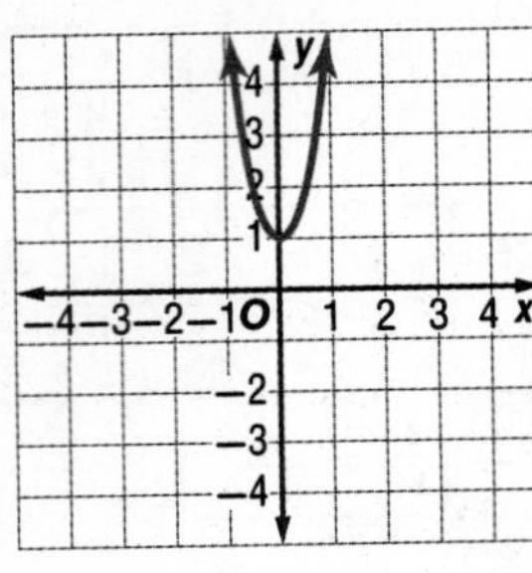


_______________

_______________

**14**

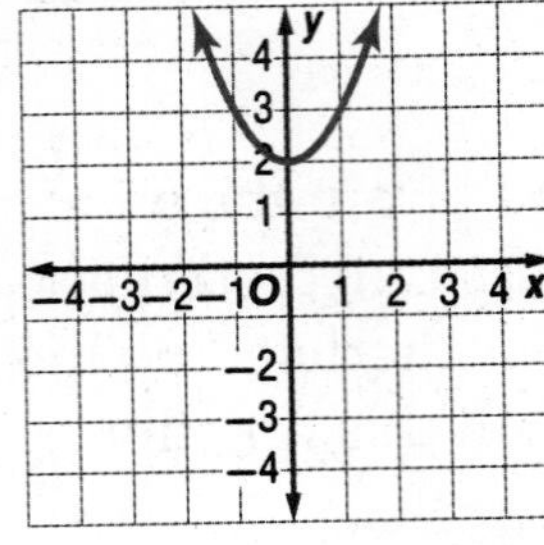


_______________

_______________

**Solve.**

**15** **CONSTRUCTION** Homeowners have hired a fence builder to fence in an area of their lot. The function $y = -x^2 + 600$ models the area of their lot that will not be fenced in. Graph this function.

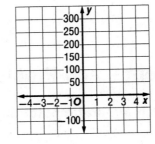

**16** **GRAVITY** A stunt person attached to a bungee jumps from a platform. The function $h = -16t^2 + 320$ can be used to determine the height $h$ (in feet) of the stunt person after $t$ seconds. Graph this function.

**Vocabulary Check** **Write the vocabulary word that completes each sentence.**

**17** A(n) _______________ is a function of the form $y = ax^2 + bx + c$ where $a \neq 0$.

**18** The solutions of a quadratic function are the _______________ of the function.

**19** **Reflect** Looking strictly at the equations throughout this lesson, can you see a pattern that can help you determine whether the parabola will open up or down?

_______________

_______________

STOP

# Lesson 3-6 Families of Graphs

## KEY Concept

Listed below are rules that allow you to graph a function related to a parent function.

### Translation of Linear Equation

The parent function of a linear equation is $y = x$. The table describes the translations of all related linear graphs.

| | |
|---|---|
| $y = x + c$ | Shifts the graph $c$ units up. |
| $y = x - c$ | Shifts the graph $c$ units down. |

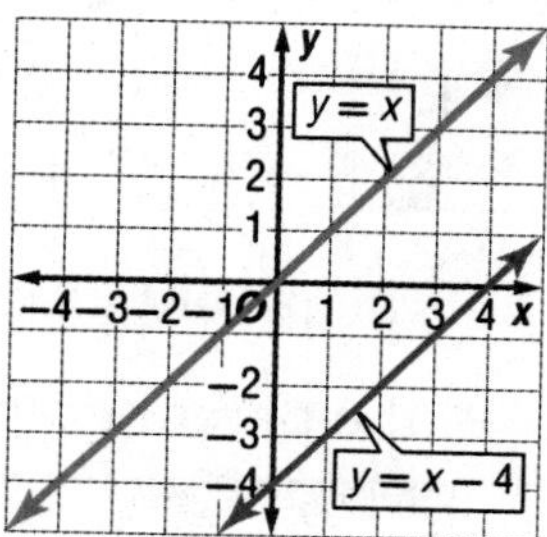


### Translation of Quadratic Functions

The parent function of a quadratic function is $y = x^2$. The table describes the translations of quadratic functions.

| | |
|---|---|
| $y = x^2 + c$ | Shifts the graph $c$ units up. |
| $y = x^2 - c$ | Shifts the graph $c$ units down. |
| $y = (x + c)^2$ | Shifts the graph $c$ units left. |
| $y = (x - c)^2$ | Shifts the graph $c$ units right. |

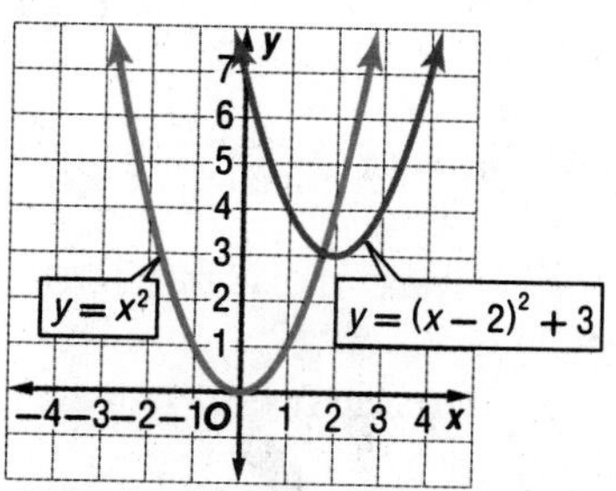


**VOCABULARY**

**family of graphs** graphs of functions with similar features

**linear equation** an equation for which the graph is a straight line

**parent graph** the simplest of the graphs or the anchor graph in a family of graphs

**quadratic function** a function that can be written in the form $y = ax^2 + bx + c$, where $a \neq 0$

**translation** a horizontal or vertical shift of a graph

Because equations can model functions, equations are often written in function notation, where $f(x)$ replaces $y$.

$$y = x + 1 \text{ and } f(x) = x + 1$$

GO ON

## Example 1

**Identify the parent function of $y = x + 3$. Describe the translation.**

1. The parent function of a linear equation is $y = x$.
2. Shift the parent graph 3 units up.

**YOUR TURN!**

**Identify the parent function of $y = x - 5$. Describe the translation.**

1. The parent function of a ______ is $y =$ ______.
2. Shift the parent graph ______ units ______.

## Example 2

**Identify the parent function of $y = (x + 4)^2 - 1$. Describe the translation.**

1. The parent function of a quadratic function is $y = x^2$.
2. Shift the parent graph 4 units left.
3. Shift the parent graph 1 unit down.

**YOUR TURN!**

**Identify the parent function of $y = (x - 3)^2 - 4$. Describe the translation.**

1. The parent function of a ______ ______ is $y =$ ______.
2. Shift the parent graph ______ units ______.
3. Shift the parent graph ______ units ______.

## Guided Practice

**Identify the parent function of each function. Describe each translation.**

**1** $y = x + 4$

The parent function of a ______ equation is $y =$ ______.

Shift the parent graph ______ units up.

.

**2** $y = x + 1$

The parent function of a ______ equation is $y =$ ______.

Shift the parent graph ______ unit ______

**3** $y = x^2 - 3$

The parent function of a ______ function is $y =$ ______.

Shift the parent graph ______ units ______.

**4** $y = (x + 3)^2 + 6$

The parent function of a ______ function is $y =$ ______.

Shift the parent graph ______ units ______ and ______ units ______.

## Step by Step Practice

5 The graph at the right shows a parent function and a related function. Describe the translation.

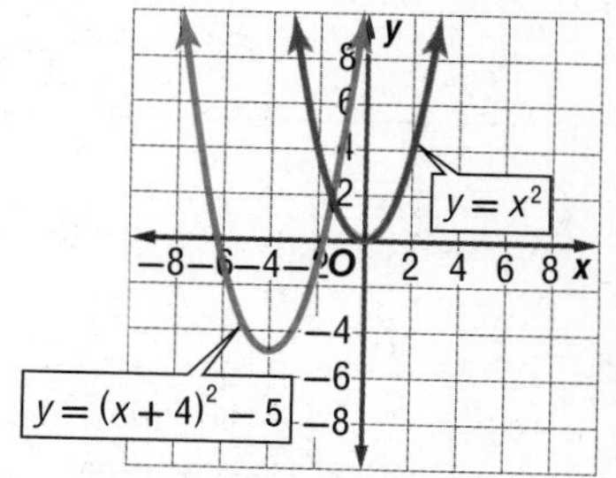


**Step 1** Shift the parent graph _______ units _______.

**Step 2** Shift the parent graph _______ units _______.

**The parent graph and translated function are given. Describe each translation.**

6 $y = x^2$ $y = (x - 1)^2 + 1$

shift: _______ unit(s) _______

shift: _______ unit(s) _______

7 $y = x$ $y = x + \frac{1}{2}$

shift: _______ unit(s) left or right

shift: _______ unit(s) _______

## Step by Step Problem-Solving Practice

**Solve.**

8 The area of a square with side $x$ is modeled by the function $y = x^2$. If you add 10 square feet to the area, the new function of the area is modeled by $y = x^2 + 10$. Describe the translation.

Check off each step.

_______ **Understand: I underlined key words.**

_______ **Plan: To solve the problem, I will** _______________________________________________.

_______ **Solve: The answer is** _______________________________________________.

_______ **Check: I checked my answer by** _______________________________________________.

## Skills, Concepts, and Problem Solving

**Identify the parent function of each function. Describe each translation.**

9 $y = (x + 5)^2 - 2$

_______________________

_______________________

10 $y = x - 9$

_______________________

_______________________

GO ON

**Identify the parent function of each function. Describe each translation.**

11 $y = x + 2.5$

______________________

______________________

12 $y = (x - 7)^2$

______________________

______________________

13 $y = x^2 - 6$

______________________

______________________

14 $y = -3 + x$

______________________

______________________

**Solve.**

15 **WRITING FUNCTIONS** Alma graphs the parent function $y = x^2$. She then graphs a function that is translated 2 units to the left and 3 units up. What is the equation of the translated function?

______________________________________________

16 **MODEL BUILDING** Tia is building a model train set. She models the base for the length of track using the equation $y = x$, where $x$ is the length of track measured in feet and $y$ is the length of the base for the track in feet. If she adds 5 feet to the track, the new function is $y = x + 5$. Describe the translation.

______________________________________________

**Vocabulary Check** **Write the vocabulary word that completes each sentence.**

17 A horizontal or vertical shift of a graph is a(n) ______________.

18 The function $y = x$ is an example of a(n) ______________,

and the function $y = x^2$ is an example of a(n) ______________.

19 Functions with similar features, such as $y = x$ and $y = x - 12$,

are ______________.

20 **Reflect** Consider the function $y = 2x + 3$. Are you able to graph this function using only a vertical shift from the parent function $y = x$? Explain.

______________________________________________

______________________________________________

# Lesson 3-7 Graph Absolute Value Functions

## KEY Concept

To graph a function that is part of a **family of graphs**, it is helpful to graph the **parent graph**, or the simplest of the graphs in a family of graphs, and then **translate**, or shift the parent graph.

The graph below shows the absolute value parent graph $y = |x|$ and the translated graph $y = |x + 1| - 2$.

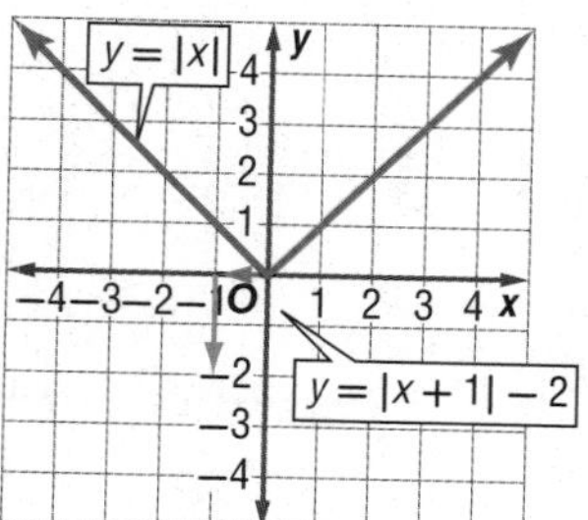


Shift the parent graph 2 units down.

Shift the parent graph 1 unit left.

### VOCABULARY

**absolute value function**
a function that has a V-shaped graph that points upward or downward

**family of graphs**
graphs of functions with similar features

**parent graph**
the simplest of the graphs or the anchor graph in a family of graphs

**translation**
a horizontal or vertical shift of a graph

Adding or subtracting numbers inside the absolute value bars shifts the parent graph left or right. Adding or subtracting numbers outside the absolute value bars shifts the parent graph up or down.

## Example 1

**Graph the absolute value function $y = |x - 3|$.**

1. Graph the parent function.
2. Shift the parent function 3 units right.

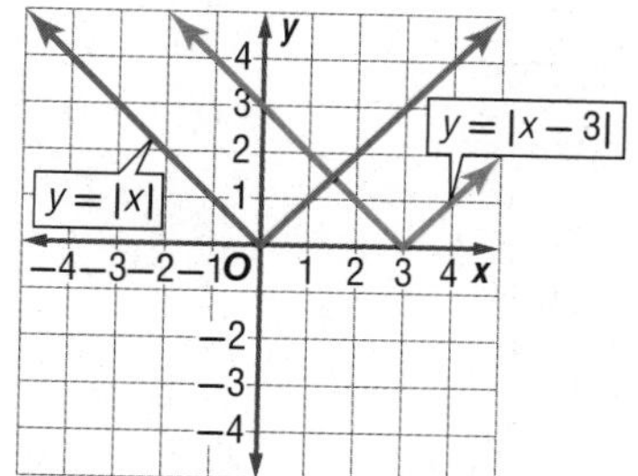


### YOUR TURN!

**Graph the absolute value function $y = |x| + 1$.**

1. Graph the parent function.
2. Shift the parent function _____ unit _____.

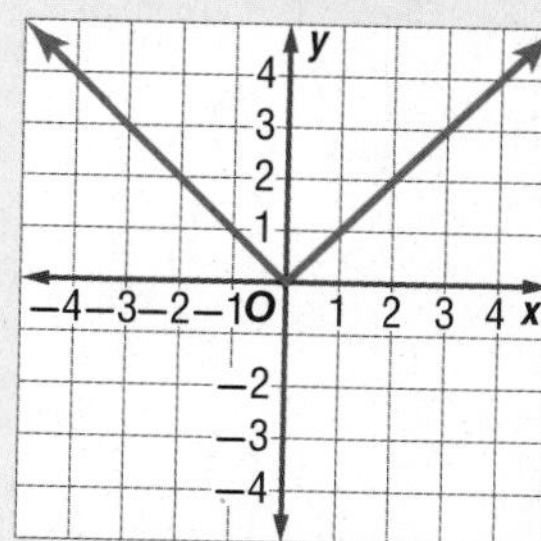

GO ON

## Example 2

**Graph the absolute value function $y = |x + 2| - 4$.**

1. Shift the parent function 2 units left.
2. Shift the graph 4 units down.

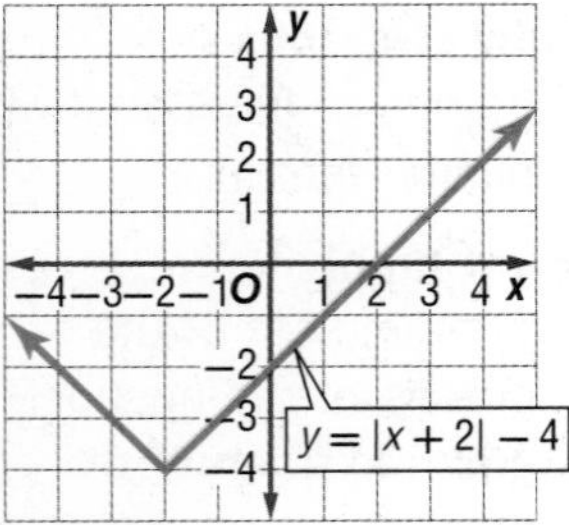


**YOUR TURN!**

**Graph the absolute value function $y = |x - 1| - 2$.**

1. Shift the parent function ____ unit ______.
2. Shift the graph ____ units ______.

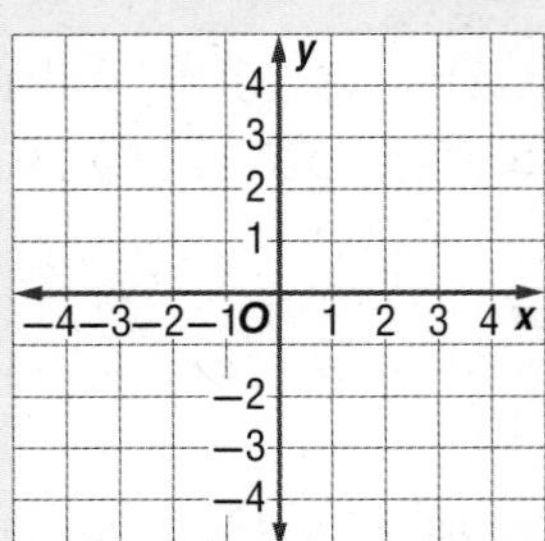

## Guided Practice

**Graph each absolute value function.**

**1** $y = |x - 5|$

Graph the parent function.

Shift the parent graph

______ units ______.

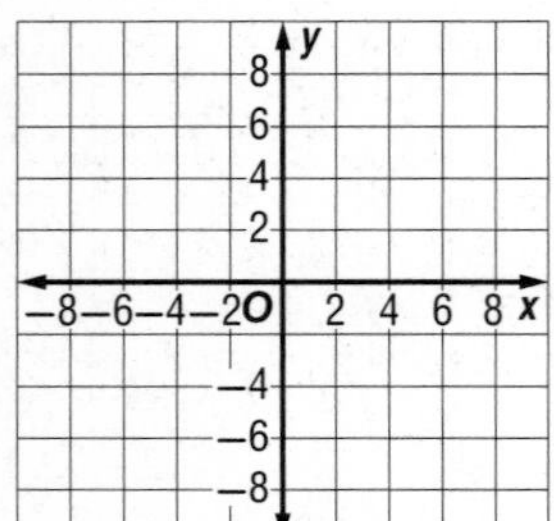

**2** $y = |x| + 2$

Graph the parent function.

Shift the parent graph

______ units ______.

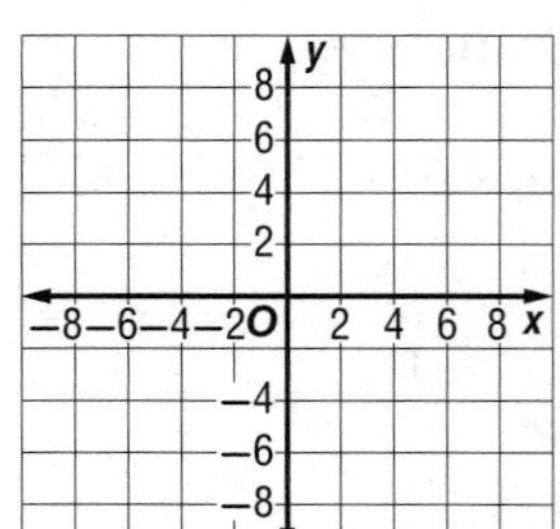

## Step by Step Practice

5 Graph the absolute value function $y = |x - 6| - 1$.

**Step 1** Graph the parent function.

**Step 2** Shift the parent graph ________ units ________.

**Step 3** Shift the graph ________ unit ________.

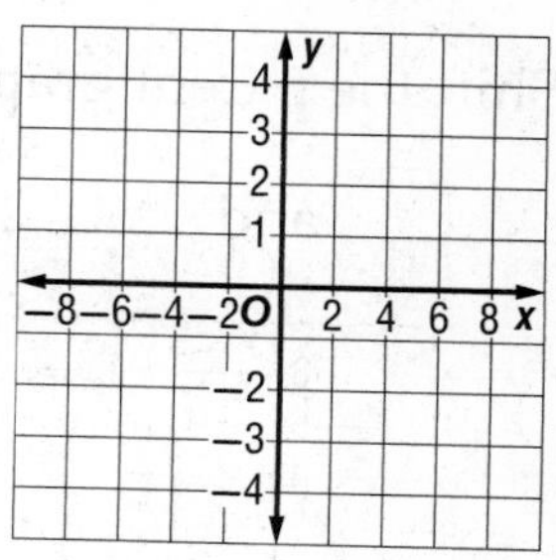

**Graph each absolute value function.**

6 $y = |x - 2| - 2$

Graph the parent function.

Shift the parent graph ________ units ________ and ________ units ________.

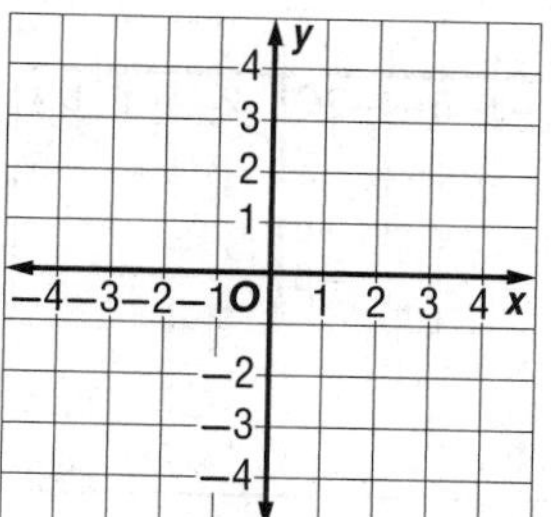

7 $y = |x + 1|$

Graph the parent function.

Shift the parent graph ________ unit ________.

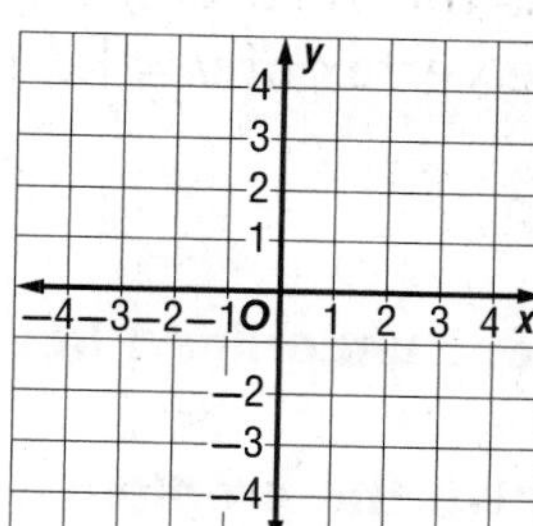

8 $y = |x + 1| + 3$

Graph the parent function.

Shift the parent graph ________ unit ________ and ____ units ________.

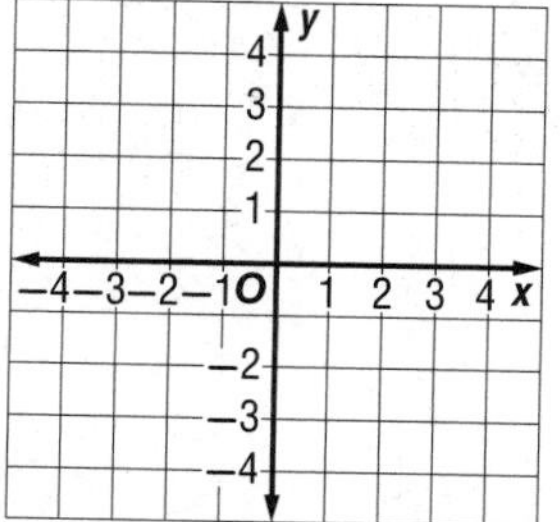

9 $y = |x| - 2$

Graph the parent function.

Shift the parent graph ________ units ________.

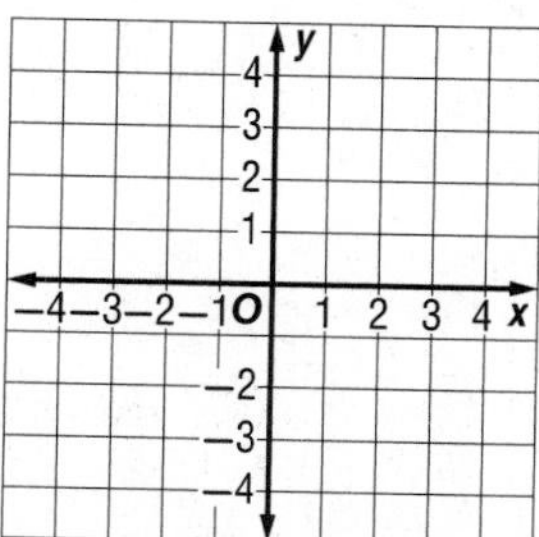

GO ON

**Graph each absolute value function.**

**10** $y = |x - 6| + 1$

Shift the parent graph ______ units ______ and ______ unit ______.

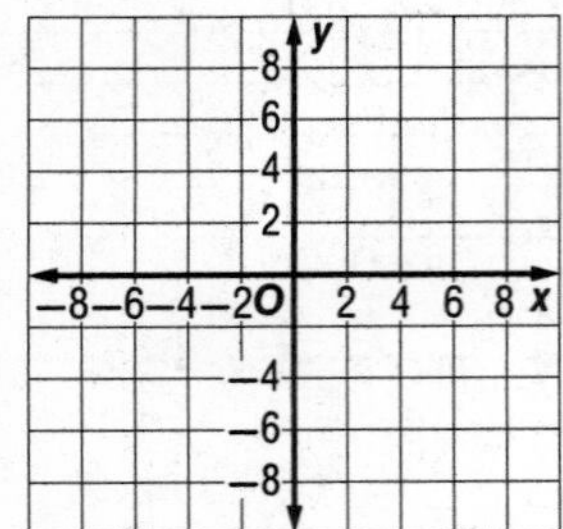

**11** $y = |x + 8|$

Shift the parent graph ______ units ______.

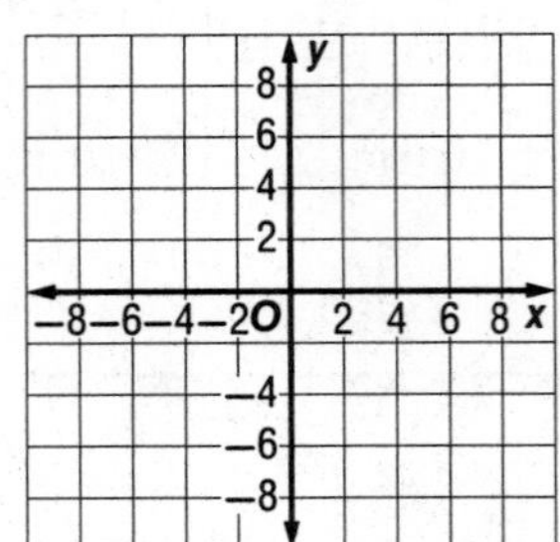

## Step by Step Problem-Solving Practice

**Solve.**

**12** The point at which the graph of an absolute value function changes direction is the vertex. Graph the absolute value function to find the vertex of $y = |x + 3| - 6$.

Check off each step.

______ **Understand: I underlined key words.**

______ **Plan: To solve the problem, I will** ______________________________.

______ **Solve: The answer is** ______________________________.

______ **Check: I checked my answer by** ______________________________

______________________________.

## Skills, Concepts, and Problem Solving

**Graph each absolute value function.**

**13** $y = |x - 7| + 3$

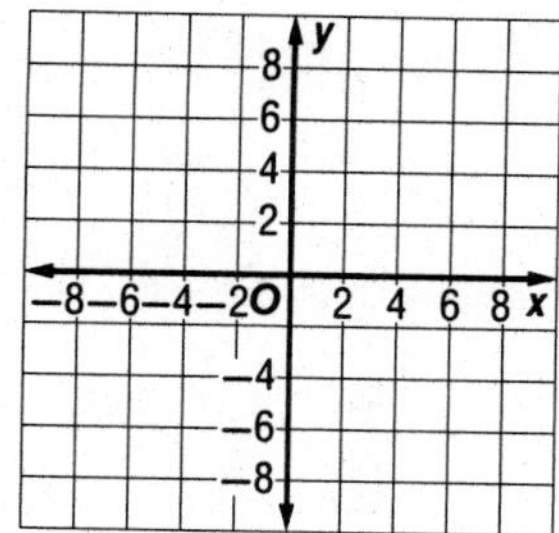

**14** $y = |x| + 9$

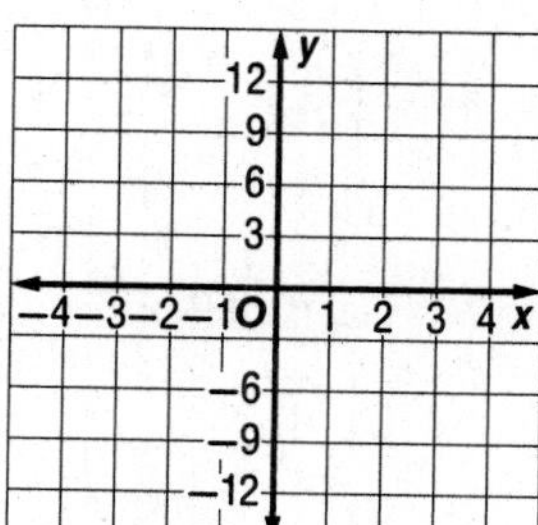

**15** $y = |x - 3|$

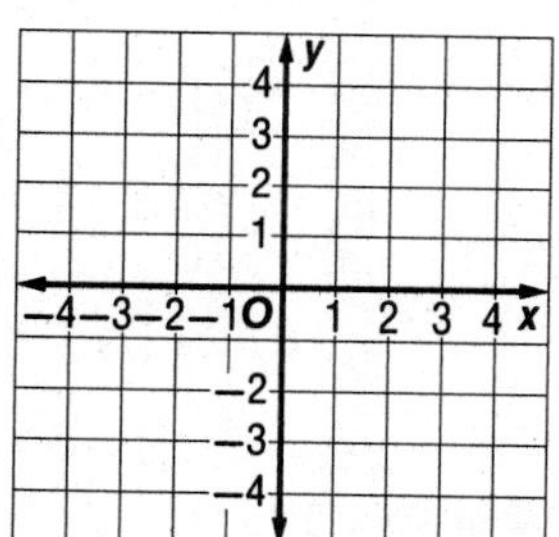

**16** $y = |x + 6| - 1$

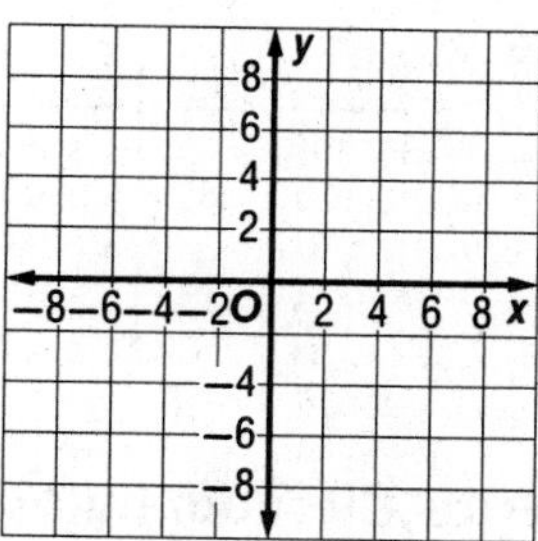

**17** $y = |x + 5| - 5$

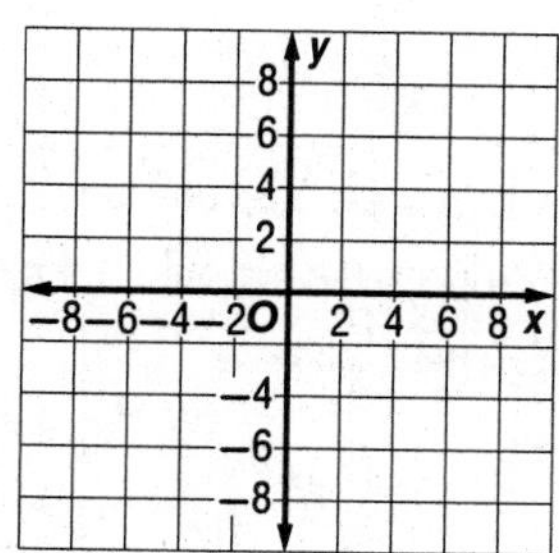

**18** $y = |x - 4| + 2$

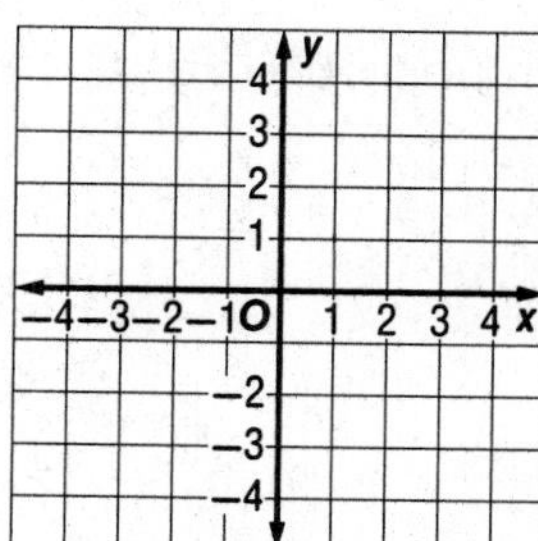



GO ON

**Solve.**

19 **GRAPHING** What is the vertex of the absolute value function $y = |x - 2|$?

_______________

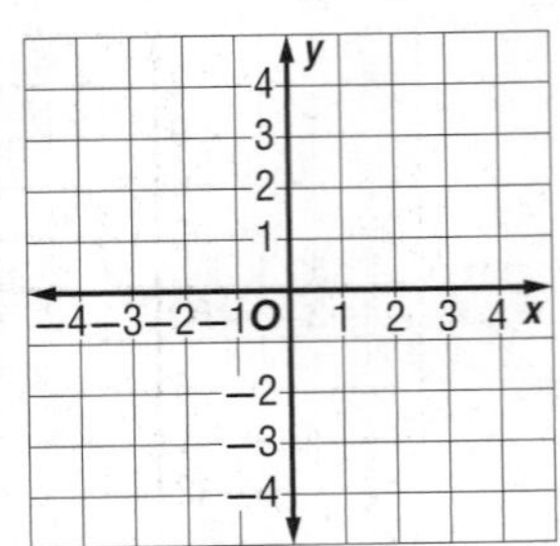

20 **GRAPHIC DESIGN** A company designed part of their logo using the absolute value function $y = |x - 1| + 2$. Show the absolute value part of the logo on a graph.

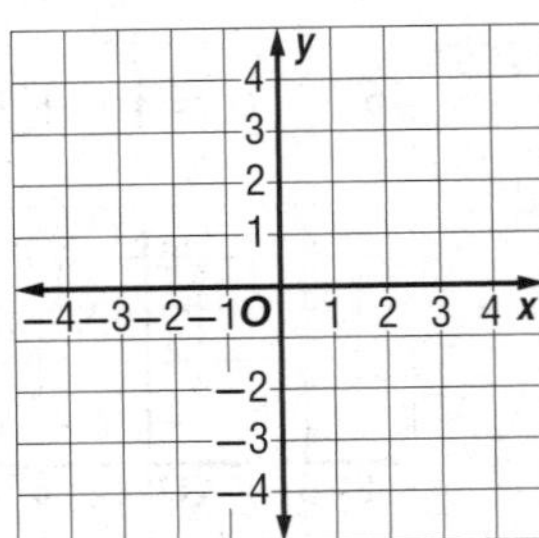

**Vocabulary Check** **Write the vocabulary word that completes each sentence.**

21 A parent graph moves two units up and three units to the left.

A shift in a graph like this is a(n) _______________.

22 A(n) _______________ has a V-shaped graph that points either up or down.

23 **Reflect** Create an absolute value function for each of the following:

where there is only a horizontal translation:

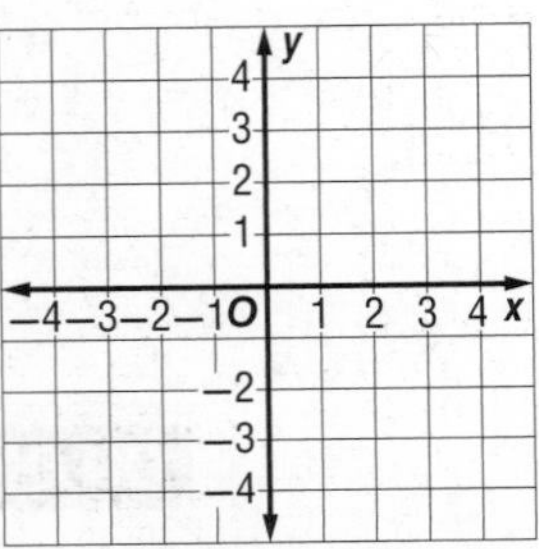

where there is only a vertical translation:

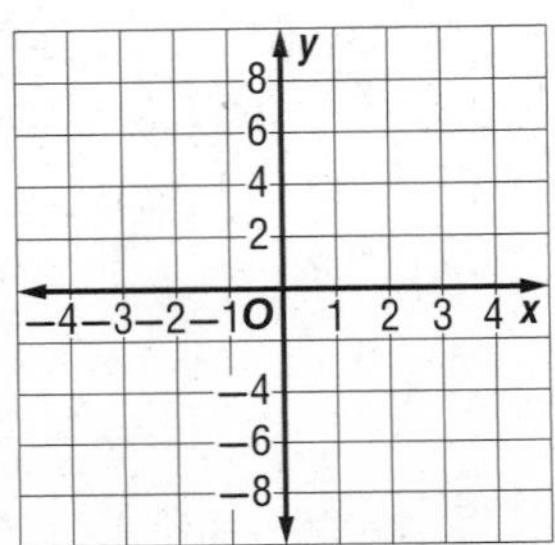

where there is both a horizontal and vertical translation:

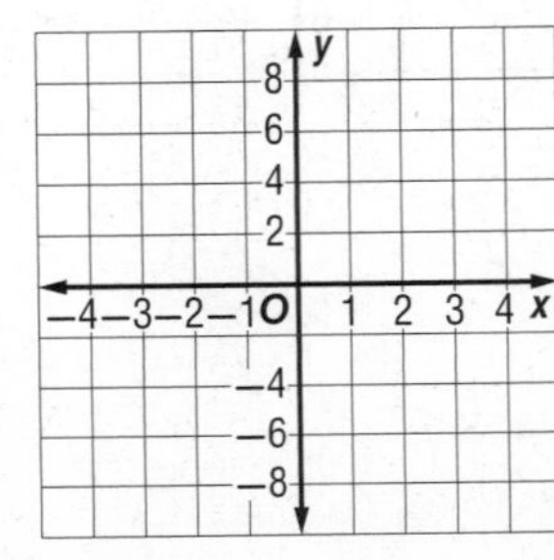

Chapter 3

# Progress Check 3 (Lessons 3-5, 3-6, and 3-7)

**Graph each absolute value function.**

**1** $y = |x + 4| - 2$

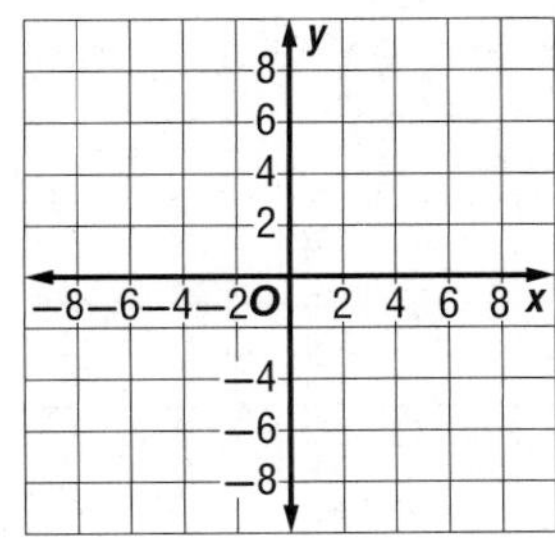

**2** $y = |x| - 3$

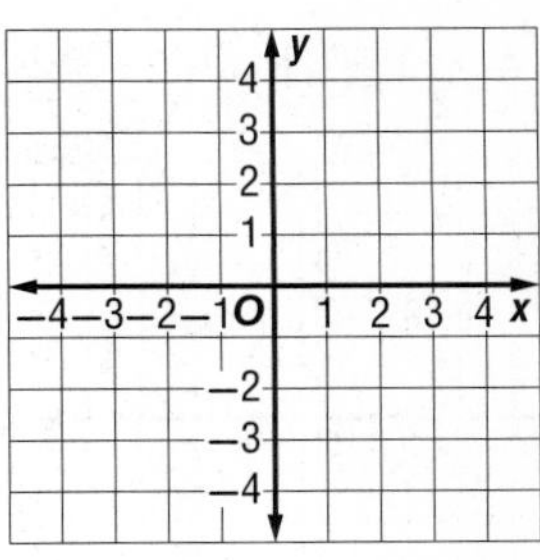

**Identify any zeros, minimums, or maximums in each graph.**

**3** $y = \frac{1}{2}x^2$

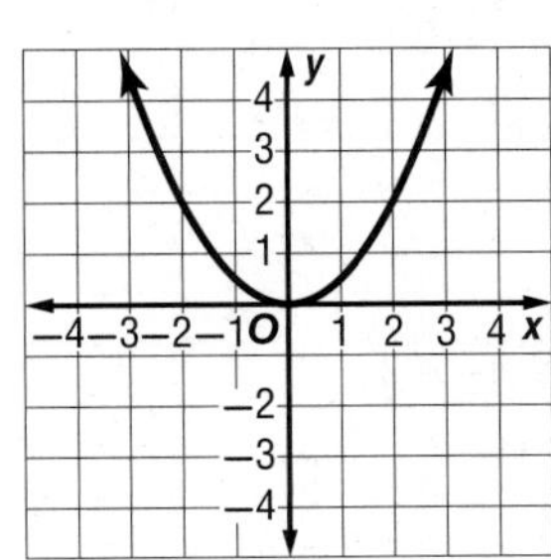

zeros: ______________

minimum or maximum:

______________

**4** $y = -3x^2 + 12$

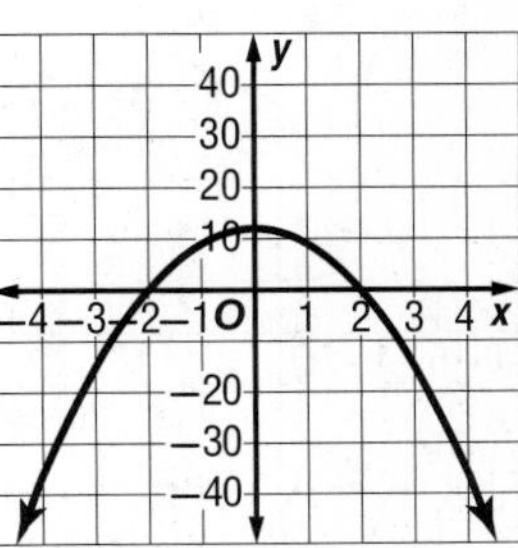

zeros: ______________

minimum or maximum:

______________

**Describe each translation.**

**5** $y = x^2$ $\quad y = (x - 3)^2 + 4$

______________

______________

**6** $y = x$ $\quad y = x + 5$

______________

______________

**Solve.**

**7** **GRAPHING** What is the vertex of the absolute value function $y = |x + 3|$? Graph the equation.

______________

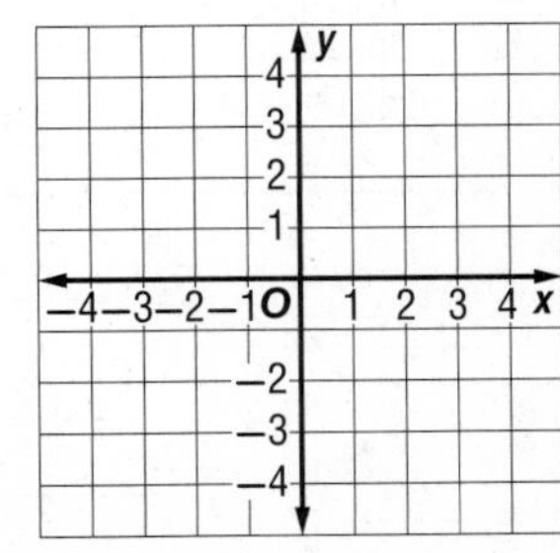

# Chapter 3 Chapter Test

**Graph each equation or inequality.**

**1** $y = \frac{1}{2}x + 1$

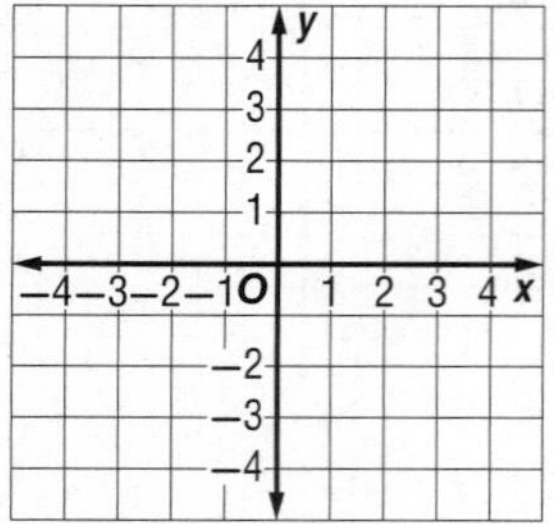

**2** $y \geq \frac{2}{3}x - 3$

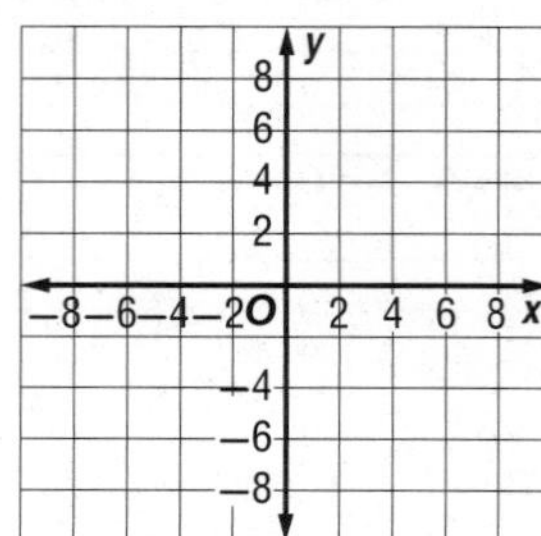

**3** $y = |x - 5| + 2$

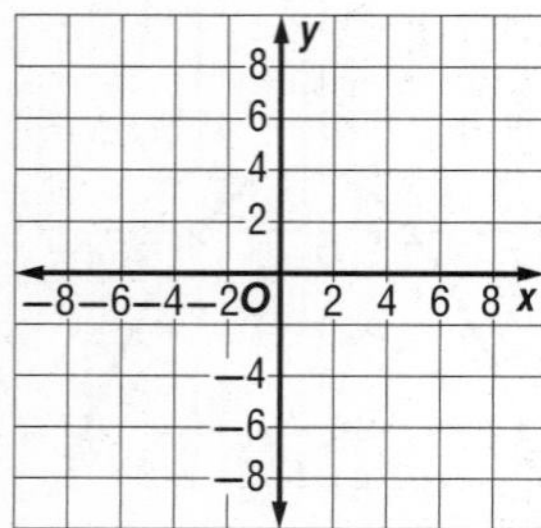

**4** $y = \frac{2}{5}x^2$

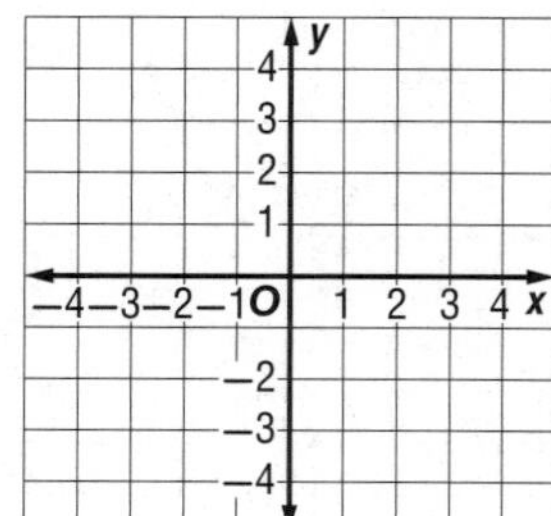

**Solve each system of linear equations or inequalities by graphing.**

**5** $y = -2x + 2$

$y = -\frac{1}{2}x + 5$

__________

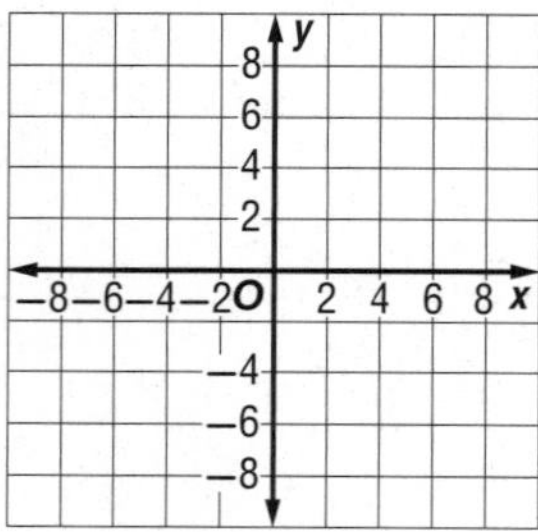

**6** $y > -3$

$y \leq 3x + 5$

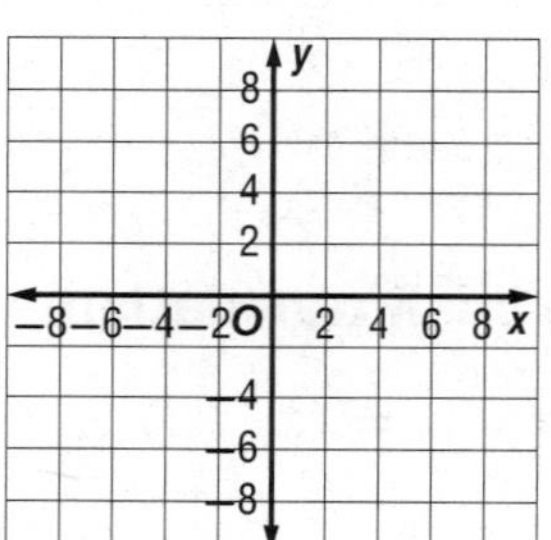

**Identify the parent function of each function. Describe each translation.**

**7** $y = x - 7$

______________________________

______________________________

**8** $y = (x + 2)^2 + 3$

______________________________

______________________________

**Graph using a table.**

9 $y = 3x + 3$

| $x$ | −2 | 0 | 2 |
|---|---|---|---|
| $y$ | | | |

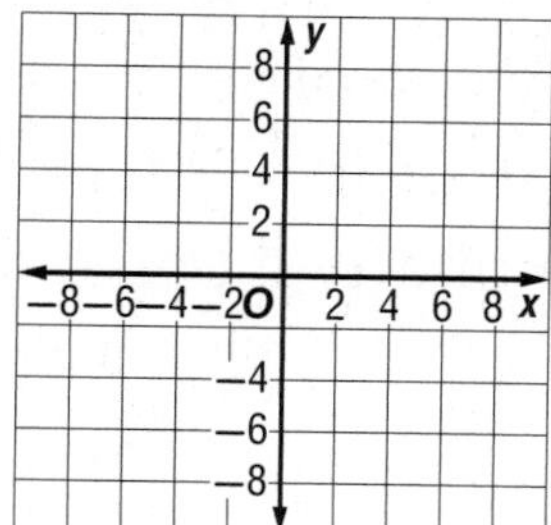

10 $y = x^2 - 2$

| $x$ | −2 | −1 | 0 | 1 | 2 |
|---|---|---|---|---|---|
| $f(x)$ | | | | | |

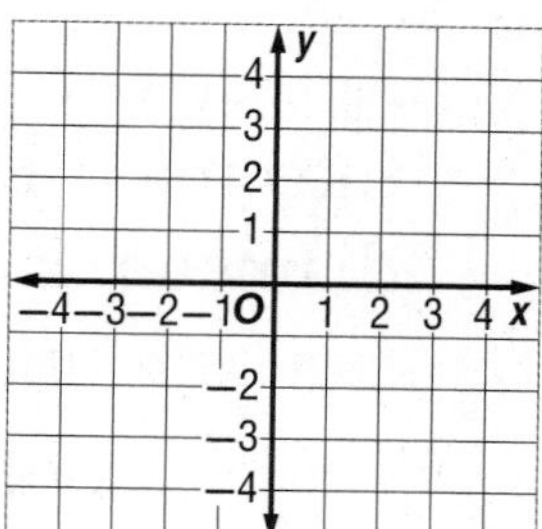

**Solve.**

11 **RECREATION** To ensure swimmer safety, the local pool can have no more than 15 swimmers per lifeguard on duty. If $x$ represents the number of lifeguards and $y$ represents the total number of swimmers, write and graph a linear inequality modeling this regulation.

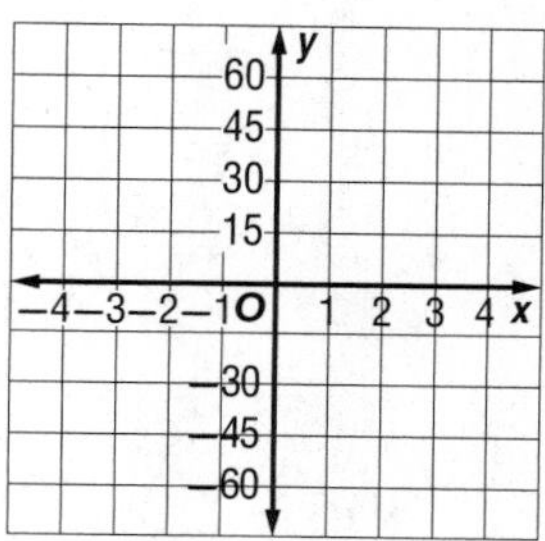

12 **COLLEGE** Jeff and Bart are both working this summer to save money for college. Jeff is selling student handbooks door-to-door for $140 per week plus 20% commission on his sales. Bart is working for a landscaper for a salary of $200 per week. Draw a graph modeling the amount $y$ each person earns in a week.

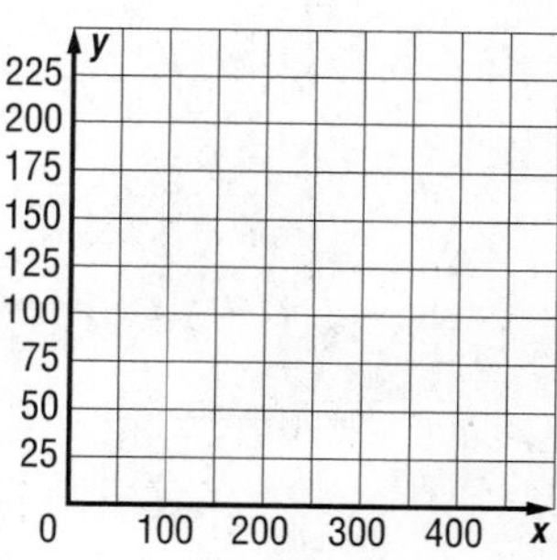

**Correct the mistake.**

13 Tonya graphed the equation $y = (x - 2)^2 + 3$. Her result was the parent graph of $y = x^2$ shifted 2 units left and 3 units up. Did she translate the function correctly? If not, what is the correct answer?

STOP

Chapter 4

# Polynomials

## *What size should the label be?*

A soup company wants to make sure that any can of soup they make has a height that is 2 inches more than its diameter. If they know the diameter, the area of any label can be found by multiplying polynomials.

$$\text{Area} = \ell \cdot h$$

$$\begin{aligned}\text{Area of label} &= (\text{can circumference}) \cdot (\text{can height}) \\ &= (\pi d) \cdot (d + 2) \\ &= \pi d^2 + 2\pi d\end{aligned}$$

**STEP 1 Chapter Pretest** Are you ready for Chapter 4? Take the Chapter 4 Pretest to find out.

**STEP 2 Preview** Get ready for Chapter 4. Review these skills and compare them with what you will learn in this chapter.

| What You Know | What You Will Learn |
|---|---|
| You know that a polygon is a figure with many sides. 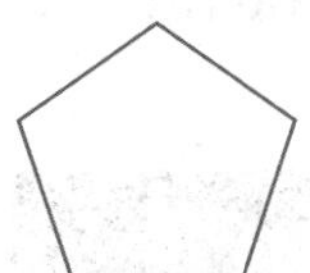 Monotone means one color. A bicycle has two wheels. A triangle has three sides. 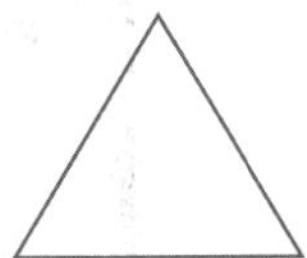 | *Lesson 4-1* A **polynomial** is an expression that contains one or more terms. One term are a **monomial**. **Example:** $2n^3$ Two terms are a **binomial**. **Example:** $5x^2 + 1$ Three terms are a **trinomial**. **Example:** $4m^2 + 3m - 4$ |
| You know that the expressions $3x$ and $5x$ both use the variable $x$. **TRY IT!** Identify the variable. **1** $2n - 1$ and $4n$ ______ **2** $5y$ and $8y + 12$ ______ | *Lesson 4-2* **Like terms** have the same variables and the same exponents. **Examples:** $6x$ and $8x$; $7y^2$ and $-3y^2$; $4a^2b$ and $7a^2b$ |
| You can use the Distributive Property to simplify multiplication expressions. **Example:** $5 \cdot 32 = 5(30 + 2) = 5(30) + 5(2) = 150 + 10 = 160$ **TRY IT!** **3** $3 \cdot 43 =$ ______ **4** $8 \cdot 55 =$ ______ | *Lesson 4-3* Multiplying a monomial by a polynomial is the same as using the Distributive Property. $4(5d + 3d^2) = 4(5d) + 4(3d^2) = 20d + 12d^2$ |

Lesson

# 4-1 Monomials and Polynomials

## KEY Concept

A polynomial is an expression made up of one or more terms. **Terms** of a **polynomial** are separated by addition or subtraction.
Some polynomials have special names based on the number of terms.

| Name | Number of Terms | Example |
|---|---|---|
| monomial | 1 | $4b^5$ |
| binomial | 2 | $7h^2 + 1$ |
| trinomial | 3 | $8x^2 + 5x - 4$ |
| polynomial | 4 or more | $3s^2w - sw + 7w - 1$ |

The **degree of a monomial** is the sum of the exponents of its variables. The **degree of a polynomial** takes on the degree of the term with the greatest degree.

| Polynomial | Degree |
|---|---|
| $10g$ | 1 |
| $4b^3$ | 3 |
| $12a^4b^5$ | $4 + 5 = 9$ |
| 6 | 0 |
| $m^3 + n^2$ | 3 |
| $3s^2 - 6s + 7$ | 2 |
| $x^4 + x^2y^3 - y^3$ | $2 + 3 = 5$ |

Remember $10g = 10g^1$ and 6 is the same as $6x^0$.

### VOCABULARY

**binomial**
a polynomial with two terms

**degree of a monomial**
the sum of the exponents of its variables

**degree of a polynomial**
the degree of the term with the greatest degree

**monomial**
a polynomial with one term

**polynomial**
an expression that contains one or more terms

**term**
a number, a variable, or a product or quotient of numbers and variables

**trinomial**
a polynomial with three terms

You must find the degree of each term in a polynomial before you can determine the degree of the polynomial.

## Example 1

**Classify each polynomial.**

**$6w$ $9a^3 + b$**

1. The polynomial $6w$ has 1 term.
2. This is a monomial.
3. The polynomial $9a^3 + b$ has 2 terms.
4. This is a binomial.

**YOUR TURN!**

**Classify each polynomial.**

**$7x^2y^2 - x^3y^8$ $-m^7n - 9m^4n^5 + 14mn^6$**

1. The polynomial $7x^2y^2 - x^3y^8$ has ____ term(s).
2. This is a ____________.
3. The polynomial $-m^7n - 9m^4n^5 + 14mn^6$ has ____ term(s).
4. This is a ____________.

## Example 2

**Find the degree of $5x^3y^6z$.**

1. There are 3 variables with exponents.
2. Find the sum of the exponents.

   $3 + 6 + 1 = 10$
3. The degree is 10.

**YOUR TURN!**

**Find the degree of $-7hk^4$.**

1. There are ____ variables with exponents.
2. Find the sum of the exponents.

   ____ + ____ = ____
3. The degree is ____.

## Example 3

**Find the degree of the polynomial $9c^2d^4 - 4c^3 + 2c$.**

1. There are 3 terms.
2. Find the degree of each term.

| | |
|---|---|
| $9c^2d^4$ | $2 + 4 = 6$ |
| $-4c^3$ | 3 |
| $2c$ | 1 |

3. The degree of the polynomial is the greatest degree of its terms. The degree of the polynomial is 6.

**YOUR TURN!**

**Find the degree of the polynomial $16s^7 + 12s^5t^4 - 8t^8$.**

1. There are ____ terms.
2. Find the degree of each term.

| | |
|---|---|
| $16s^7$ | 7 |
| $12s^5t^4$ | $5 + 4 = 9$ |
| $-8t^8$ | 8 |

3. The degree of the polynomial is the greatest degree of its terms. The degree of the polynomial is ____.

## Guided Practice

**Classify each polynomial.**

1 $a^5b^2 + 3b^5$

There are ____ terms.

This is a ____________.

2 $10x^{11}$

There is ____ term.

This is a ____________.

3 $5x - y^3 + 2$

There are ____ terms.

This is a ____________.

4 $-4cd^2 - 3b^2$

There are ____ terms.

This is a ____________.

5 $y + 150$

There are ____ terms.

This is a ____________.

6 $12mn + 5n^2 - 15$

There are ____ terms.

This is a ____________.

**Find the degree of each monomial.**

7 $12x^3y^7$

There are ____ variables with exponents.

The sum of the exponents

is ____ + ____ = ____.

The degree is ____.

8 $-11a^6bc$

There are ____ variables with exponents.

The sum of the exponents

is ____ + ____ + ____ = ____.

The degree is ____.

**Find the degree of each polynomial.**

9 $9xy - x$

There are ____ terms.

The degree of $9xy$ is ____ + ____ = ____.

The degree of $x$ is ____.

The degree of the polynomial is ____.

10 $-3p^3 - 2pq + 5q$

There are ____ terms.

The degree of $-3p^3$ is ____.

The degree of $2pq$ is ____ + ____ = ____.

The degree of $5q$ is ____.

The degree of the polynomial is ____.

## Step by Step Practice

**11** Classify the polynomial $22st^7 - 5s^6t^3$. Find its degree.

**Step 1** Count the terms. There are ____ terms. It is a __________.

**Step 2** Find the degree of each term.

The degree of $22st^7$ is ____ + ____ = ____.

The degree of $5s^6t^3$ is ____ + ____ = ____.

**Step 3** The degree of the polynomial is the greatest degree of its terms.

The degree of the polynomial is ____.

**Classify each polynomial. Find its degree.**

**12** $-3x^7$

This is a __________.

The degree is ____.

**13** $rs^{11}t^2 + 6s^5t^4 + r^{14}$

This is a __________.

The degree is ____ + ____ + ____ = ____.

**14** $2p^2 + 5p - 10$

This is a __________.

The degree is ____.

**15** $-x - 1$

This is a __________.

The degree is ____.

## Step by Step Problem-Solving Practice

**Solve.**

**16** **GEOMETRY** In Crystal's design, the area of each square is modeled by the equation $A = 3x^2 - 2x + 15$. Classify this polynomial. What is its degree?

Check off each step.

____ **Understand: I underlined key words.**

____ **Plan: To solve the problem, I will** ______________________________

______________________________.

____ **Solve: The answer is** ______________________________.

____ **Check: I checked my answer by** ______________________________.

GO ON

## Skills, Concepts, and Problem Solving

**Classify each polynomial. Find its degree.**

**17** $rt^4 - rs^3t - 8$

____________________

**18** $-5xy$

____________________

**19** $a^2 - b^3$

____________________

**20** $-12mn^{11}p^9 + p^{12} - 6np^3$

____________________

**21** $t - 4$

____________________

**22** $2xy - x - y$

____________________

**23** $a - b - c$

____________________

**24** $-km^5pq^5$

____________________

**25** $-b^3 - abc^2 + 10$

____________________

**Solve.**

**26** **GEOMETRY** In Bianca's design, the area of each square is modeled by the polynomial $x^2 + 10x + 25$. Classify this polynomial and find its degree.

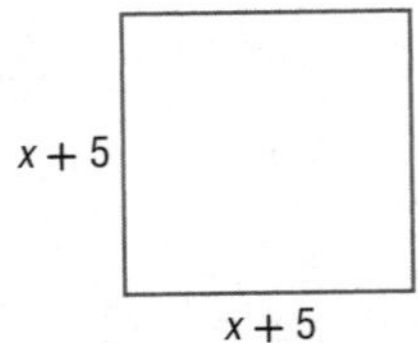


__________________________________

**27** **POPULATION** The annual growth of the population of a certain town is modeled by the polynomial $12g^3 + 153$, where $g$ is a determined growth factor. Classify this polynomial and find its degree.

__________________________________

**Vocabulary Check** **Write the vocabulary word that completes each sentence.**

**28** To find the __________________________, look at the degree of the term with the highest degree.

**29** The __________________________ is the sum of the exponents of its variables.

**30** An expression that contains one or more monomials is a(n)

__________________________.

**31** **Reflect** Create an example of each of the following: a trinomial of the 12th degree, a binomial of the 1st degree, and a monomial with three variables and a degree of 4.

____________________________________________________________

____________________________________________________________

STOP

Lesson 4-2

# Add and Subtract Polynomials

## KEY Concept

You need to identify like terms to add and subtract polynomials. **Like terms** have the same variables and the same exponents.

| Examples that are like terms | Examples that are not like terms |
|---|---|
| $3x$ and $12x$ | $2a$ and $7b$ |
| $9y^2$ and $-2y^2$ | $-4x$ and $5x^2$ |
| $8x^2y$ and $14x^2y$ | $7c^2d$ and $9cd^2$ |

To add like terms, add only the coefficients and keep the variables and exponents the same.

$$3x + 12x = (3 + 12)x = 15x$$

$$9y^2 + -2y^2 = (9 + (-2))y^2 = 7y^2$$

$$8x^2y + 14x^2y = (8 + 14)x^2y = 22x^2y$$

To add polynomials, add like terms.

$$(x^2 - 2x + 2) + (-x^2 + x + 1)$$

$$= (x^2 + (-2x) + 2) + (-x^2 + x + 1)$$

$$= (x^2 + (-x^2)) + (-2x + x) + (2 + 1)$$

$$= 0x^2 + (-x) + 3$$

$$= -x + 3$$

To subtract polynomials, subtract each like term of the polynomial being subtracted.

$$(4y^3 + 7y^2 + 1) - (-2y^3 + 3y^2 + 5)$$

$$= (4y^3 - (-2y^3)) + (7y^2 - 3y^2) + (1 - 5)$$

$$= 6y^3 + 4y^2 - 4$$

### VOCABULARY

**Associative Property of Addition**
the grouping of the addends does not change the sum

**Commutative Property of Addition**
the order in which two numbers are added does not change the sum

**like terms**
terms that have the same variables to the same power

**polynomial**
an expression that contains one or more terms

Simplifying polynomials includes adding like terms. Write the terms of a simplified polynomial in descending order starting with the term with the greatest exponent.



GO ON

## Example 1

**Find the sum of $(8g^3 + g^2)$ and $(7g^3 - 5g^2 + 2)$.**

1. Identify like terms.

   $(8g^3 + g^2) + (7g^3 - 5g^2 + 2)$

2. Group like terms.

   $(8g^3 + 7g^3) + (g^2 - 5g^2) + 2$

3. Add like terms.

   $15g^3 - 4g^2 + 2$

### YOUR TURN!

**Find the sum of $(2w^3 + 4w^2)$ and $(-3w^3 + w^2 + 11)$.**

1. Identify like terms.

   $(2w^3 + 4w^2) + (-3w^3 + w^2 + 11)$

2. Group like terms.

   (____ − ____) + (____ + ____) + ____

3. Add like terms.

   ____ + ____ + ____

## Example 2

**Find the difference of $(8n^3 - n^2 + 3n) - (-2n^3 + 4n^2 + 5n)$.**

1. Identify like terms.

   $(8n^3 - n^2 + 3n) - (-2n^3 + 4n^2 + 5n)$

2. Group like terms.

   $(8n^3 - (-2n^3)) + (-n^2 - 4n^2) + (3n - 5n)$

3. Subtract like terms.

   $10n^3 - 5n^2 - 2n$

### YOUR TURN!

**Find the difference of $(2x^4y^5 + 2xy^5) - (x^4y^5 + 3xy^5 + 2x^4)$.**

1. Identify like terms.

   $(2x^4y^5 + 2xy^5) - (x^4y^5 + 3xy^5 + 2x^4)$

2. Group like terms.

   (____ − ____) + (____ − 3____) − ____

3. Subtract like terms.

   ________

## Guided Practice

**Find each sum.**

**1** $(2a^2 - 4) + (a^2 + 9)$

Group like terms.

$(2a^2 +$ ____$) + (-4 +$ ____$)$

Add like terms.

____ + ____

**2** $(x - 6) + (-5x^2 + 9x - 5)$

Group like terms.

$-5x^2 + (x +$ ____$) + (-6 +$ ____$)$

Add like terms.

____ + ____ − ____

## Step by Step Practice

**3** Find $(-7x^2y^2 + 2xy - y^2) - (8x^2y^2 + 5xy - 6y^2)$.

**Step 1** Identify like terms.

$(-7x^2y^2 + 2xy - y^2) - (8x^2y^2 + 5xy - 6y^2)$

**Step 2** Group like terms. Subtract each term of the second polynomial.

(____________) + (____________) + (____________)

**Step 3** Subtract like terms.

__________ − __________ + __________

**Find each difference.**

**4** $(-5b^3 + 2b) - (7b^3 + 6b)$

Group like terms.

$(-5b^3 -$ _____) + $(2b -$ _____)

Subtract like terms.

______ − ______

**5** $(8x - y + 1) - (-4x + 2y - 4)$

Group like terms.

(_____ $- (-4x)) + (-y -$ _____) + (_____ $- (-4))$

Subtract like terms.

______ − ______ + ______

**6** $(p^6q^3 - p^2q + p) - (3p^6q^3 - 2p^2q + 5p)$

$(p^6q^3 -$ ______) + (______ $- (-2p^2q)) + (p -$ ______)

______ + ______ − ______

**7** $(-9g^6 + 6f^2g^2 + 3) - (3g^6 + 6f^2g^2 - 9)$

$(-9g^6 -$ ______) + $(6f^2g^2 -$ ______) + (______ $- (-9))$

______ + ______



GO ON

Find each sum or difference.

**8** $(x^3 - x^2 + 4) - (5x^3 - 2x^2 + 12)$

(________) + (____________) + (________)

________________________

**9** $(-2z^3 - 3z^2 + 6) + (5z^2 - 10)$

________ + (____________) + (________)

________________________

**10** $(8a^2b + 2b^2) + (-2a^2b - b^2)$

(____________) + (____________)

________________________

**11** $(-14g - 11) - (-5g - 15)$

(____________) + (____________)

________________________

**12** $(w^3 + 2w^2 - 3w) - (w^3 + 4w^2 + 6w)$

(________) + (__________) + (__________)

________________________

**13** $(-8b^2 + b + 6) + (b^2 - 20)$

(__________) + ____ + (__________)

________________________

## Step by Step Problem-Solving Practice

**Solve.**

**14** **GEOMETRY** What is the perimeter of the figure at the right? Write the answer in simplest form.

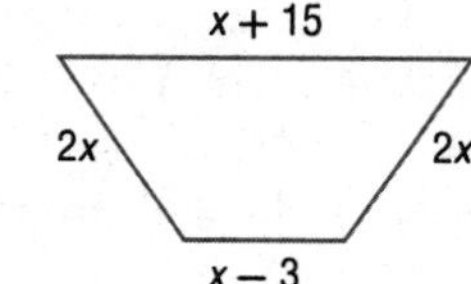


Perimeter is the sum of the sides.

$\text{side}_1 + \text{side}_2 + \text{side}_3 + \text{side}_4$

________ + ________ + ________ + ________

= ____________ + ____________

= ____________

Check off each step.

______ **Understand: I underlined key words.**

______ **Plan: To solve the problem, I will** ________________________.

______ **Solve: The answer is** ________________________.

______ **Check: I checked my answer by** ________________________

________________________________________.

## Skills, Concepts, and Problem Solving

**Find each sum or difference.**

**15** $(-3x^2 + x) - (-x^2 + 4) =$ ____________________

**16** $(-h^5 + 5h^4 + h^3) + (7h^5 - 2h^4 + 6h^2) =$ ____________________

**17** $(5a^2b^2 - 6b^2 + 3b) + (-a^2b^2 - b^2 + b) =$ ____________________

**18** $(xy - x - 9) - (10 + x) =$ ____________________

**19** $(z^4 - 8z^2 - 2z) - (-4z^4 + 3z^2) =$ ____________________

**20** $(-12p^2q^2 - 5p^2q) - (-9p^2q^2 - 6p^2q) =$ ____________________

**21** $(10g - 8) + (12g^2 - 5g - 1) =$ ____________________

**22** $(2w^5 - w^3 + 5wx) - (2w^4 - w^3 + wx) =$ ____________________

**Solve.**

**23** **GARDENING** Sam is planting a garden that is in the shape of a rectangle with area $2x^2 + 13x + 15$. He will also fence a square area within the garden to plant a patch of carrots. The square will have an area of $x^2 + 4x + 4$. What is the area of the garden that is left to plant other vegetables?

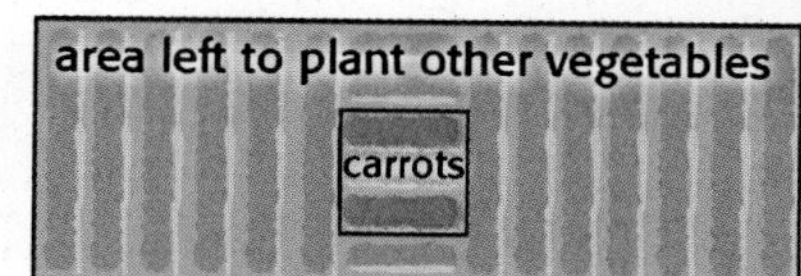


____________________

**24** **GEOMETRY** Find a simplified expression for the perimeter of the figure at the right.

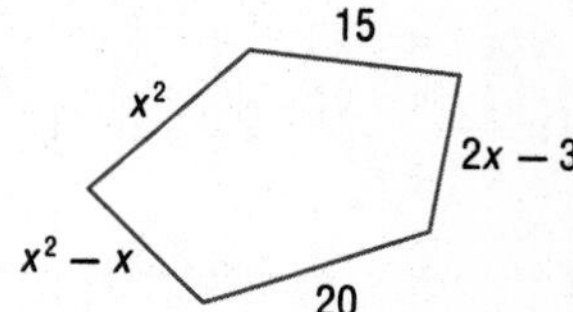


____________________

**Vocabulary Check** **Write the vocabulary word that completes the sentence.**

**25** Terms that have the same variables to the same power, such as $xy^2$ and $2xy^2$, are called ____________________.

**26** **Reflect** When two polynomials with like terms are added together, does the simplified result have to be a polynomial? Explain. Give an example.

____________________

____________________

Chapter 4

# Progress Check 1 (Lessons 4-1 and 4-2)

**Classify each polynomial. Find its degree.**

1 $r^2s^3t - 8s$

2 $-3x^2y$

3 $c^2d^2 - d^3$

4 $-4m^6n^9 + m^{10} - 18m^3n^5$

5 $3x^3 + 5x - 9$

6 $x - 9$

7 $a^9b^3c^2$

8 $-pq^5r + p^3q^3 + r^8$

9 $-a^3 - bc^2 + 10c^4$

10 $16s^7 + 5rs - r^3$

11 $xy^3z^8$

12 $r^7s^8t + rs^5t^{10}v$

**Find each sum or difference.**

13 $(-2x^2 + 3x) - (-x^2 + 4x - 3) =$

14 $(k^3 + 5k^2 + 8k) + (4k^3 - 6k^2 + 9) =$

15 $(rs^2 - 6s^2) + (-7rs^2 + 12s^2) =$

16 $(xy + 3x - 14) - (2 + x) =$

17 $(y^4 - 7y^3 - 2y^2 - 4y - 1) - (5y^4 + 17y^2) =$

18 $(2a^2b^2 - 5a^2b) - (-7a^2b^2 - 11a^2b) =$

19 $(24h - 8) + (12h^2 - 12h - 31) =$

20 $(8m^5 + 4m^3 + mn) + (m^4 + 3m^2 + mn) =$

**Solve.**

21 **EDUCATION** At time $t$, the number of boys enrolled at Legion High School is modeled by the expression $20t + 281$, and the number of girls enrolled is modeled by the expression $22.8t + 300$. Create a simplified expression to show the total enrollment at Legion High School at time $t$.

Lesson 4-3

# Multiply Polynomials

## KEY Concept

You can use a model to show how to multiply a monomial by a polynomial or a polynomial by a polynomial.

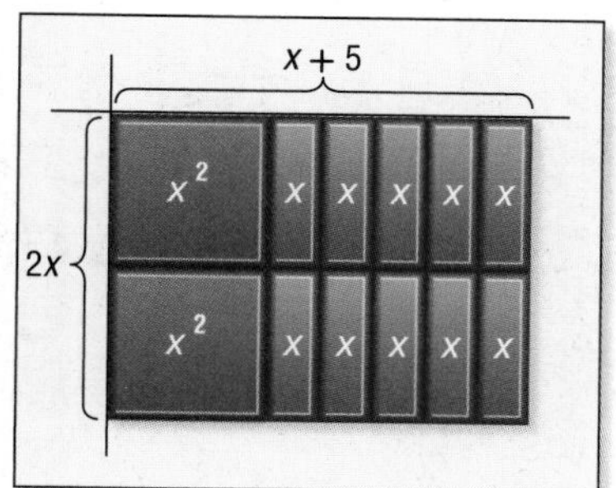


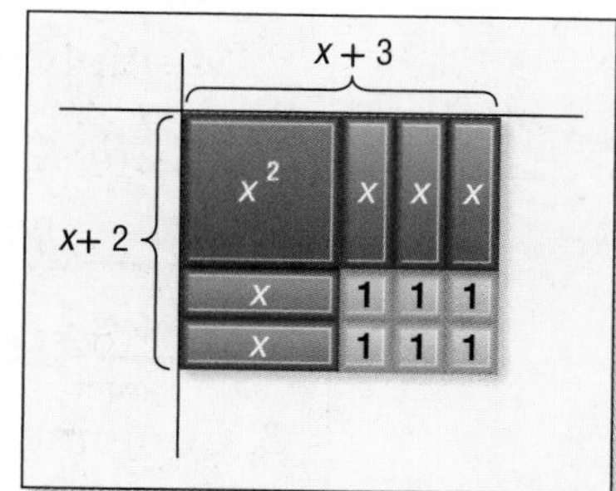


$2x(x + 5) = 2x(x) + 2x(5)$

$= 2x^2 + 10x$

$(x + 2)(x + 3) = x(x + 3) + 2(x + 3)$

$= x(x) + x(3) + 2(x) + 2(3)$

$= x^2 + 3x + 2x + 6$

$= x^2 + 5x + 6$

To multiply polynomials, use the Product of Powers Property.

$$x^2 \bullet x^3 = x^{2+3} = x^5$$

### VOCABULARY

**Distributive Property**
to multiply a sum by a number, you can multiply each addend by the number and add the products

**monomial**
a polynomial with one term

**polynomial**
an expression that contains one or more terms

Multiplying a monomial by a polynomial is the same as using the Distributive Property. Multiplying a binomial by a polynomial is the same as using the Distributive Property two times and then adding like terms.

## Example 1

**Find the product of $7(3s + 5s^2)$.**

1. Use the Distributive Property.

   $7(3s + 5s^2) = 7(3s) + 7(5s^2)$

2. Multiply. Write in simplest form.

   $7(3s) + 7(5s^2) = 21s + 35s^2$

### YOUR TURN!

**Find the product of $5x(3x^3 + 12x)$.**

1. Use the Distributive Property.

   $5x(3x^3 + 12x) = 5x(\_\_\_\_) + 5x(\_\_\_\_)$

2. Multiply. Write in simplest form.

   $5x(\_\_\_\_) + 5x(\_\_\_\_) = \_\_\_\_ + \_\_\_\_$

GO ON

## Example 2

**Find the product $(a + 1)(a + 2)$.**

1. Use the Distributive Property two times with the first binomial.

$(a + 1)(a + 2) = a(a + 2) + 1(a + 2)$

2. Multiply.

$a(a + 2) + 1(a + 2)$

$= a(a) + a(2) + 1(a) + 1(2)$

$= a^2 + 2a + a + 2$

3. Combine like terms.

$a^2 + 3a + 2$

**YOUR TURN!**

**Find the product $(b + 4)(b + 5)$.**

1. Use the Distributive Property two times with the first binomial.

$(b + 4)(b + 5) =$ ____$(b + 5) +$ ____$(b + 5)$

2. Multiply.

____$(b + 5) +$ ____$(b + 5)$

$=$ ____$(b) +$ ____(____) $+$ ____$(b) +$ ____$(5)$

$=$ ____ $+$ ____ $+$ ____ $+$ ____

3. Combine like terms.

____ $+$ ____ $+$ ____

## Guided Practice

**Find each product.**

**1** $2x(5x^2 + 7)$

$= 2x($______$) + 2x($______$)$

$=$ ______ $+$ ______

**2** $-4(9g - 3g^2)$

$= -4($______$) - 4($______$)$

$=$ ______ $+$ ______

**3** $3y(-y^2 + 4y - 12)$

$= 3y($______$) + 3y(4y) + 3y(-12)$

$=$ ______ $+$ ______ $-$ ______

**4** $-a^3(5a^3 + 6a - 3a^2)$

$= -a^3($______$) - a^3($______$) - a^3($______$)$

$=$ ______ $-$ ______ $+$ ______

**5** $2xy(4x^4 + 7x^3 + 8y)$

$= 2xy($______$) + 2xy($______$) + 2xy($______$)$

$=$ ______ $+$ ______ $+$ ______

**6** $-5ab(3a^2b^2 - ab + 5b^2)$

$= -5ab($______$) - 5ab($______$) - 5ab($______$)$

$=$ ______ $+$ ______ $-$ ______

## Step by Step Practice

7 Find the product of $(7x - 3)(7x + 3)$.

**Step 1** Use the Distributive Property two times with the first binomial.

$(7x + 3)(7x - 3) = \_\_\_\_(7x - 3) + \_\_\_\_(7x - 3)$

**Step 2** Multiply.

$\_\_\_\_(7x - 3) + \_\_\_\_(7x - 3) = \_\_\_\_(7x) + \_\_\_\_(-3) + \_\_\_\_(7x) + \_\_\_\_(-3)$

$= \_\_\_\_ + (\_\_\_\_) + \_\_\_\_ + (\_\_\_\_)$

**Step 3** Add like terms. $= \_\_\_\_ - \_\_\_\_$

**Find each product.**

8 $(9x + 1)(5x + 4) =$ ________

9 $(6 - y)(1 - 4y) =$ ________

10 $(3a + 2)(2a - 2) =$ ________

11 $(-y + 8)(6y - 1) =$ ________

12 $(4p - 3q)(9p + q) =$ ________

13 $(x - 1)(x + 1) =$ ________

## Step by Step Problem-Solving Practice

**Solve.**

14 **VOLUME** The formula for the volume of a cylinder is $V = \pi r^2 h$, where $r$ is the radius of the cylinder and $h$ is the height of the cylinder. Rewrite the formula if the height of the cylinder is twice the radius plus 5. Write this new formula in simplest form.

$V = \pi r^2 h$

$V =$ ________

Check off each step.

____ **Understand: I underlined key words.**

____ **Plan: To solve the problem, I will** ________.

____ **Solve: The answer is** ________.

____ **Check: I checked my answer by** ________.

GO ON

## Skills, Concepts, and Problem Solving

**Find each product.**

**15** $(5p - 10)(5p + 10) =$ ______

**16** $4x(2x^2 - 3x) =$ ______

**17** $(8y - z)(7y - z) =$ ______

**18** $2pq(5p^2q^4 + 3pq - 6p) =$ ______

**19** $9y(9xy^3 - 3) =$ ______

**20** $(-10y - 4)(3y - 5) =$ ______

**21** $(4x + 2)(5x - 8) =$ ______

**22** $-g(11g^4 - 10g) =$ ______

**Solve.**

**23** **GEOMETRY** The number of diagonals that can be drawn from one vertex of a polygon is modeled by the expression $\frac{n}{2}(n - 3)$, where $n$ is the number of sides in the polygon. Find this product.

______

**24** **ARCHITECTURE** An architect is creating a base for a model he plans to build. The base is in the shape of the rectangle shown at the right. What is the area of the base?

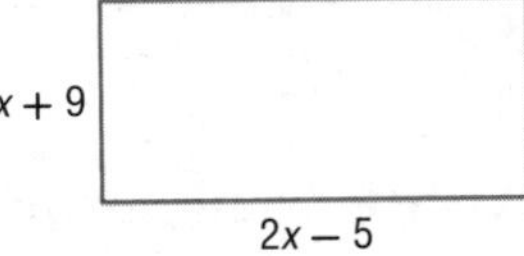


______

**Vocabulary Check** **Write the vocabulary word that completes each sentence.**

**25** Use the ______ to multiply a sum by a number; this allows you to multiply each addend by the number and add the products.

**26** A(n) ______ is an expression that contains more than one term.

**27** **Reflect** Let $x$ and $x + 1$ represent two consecutive integers. Find the product of these two integers. Will the product be even or odd? Explain.

______

______

STOP

Lesson 4-4

# Factor Polynomials

## KEY Concept

You can factor a **polynomial** by writing it as a product of the greatest common factor of its terms and another polynomial.

First find the **greatest common factor (GCF)** of the terms of $2x^2 + 8x + 4x^3$.

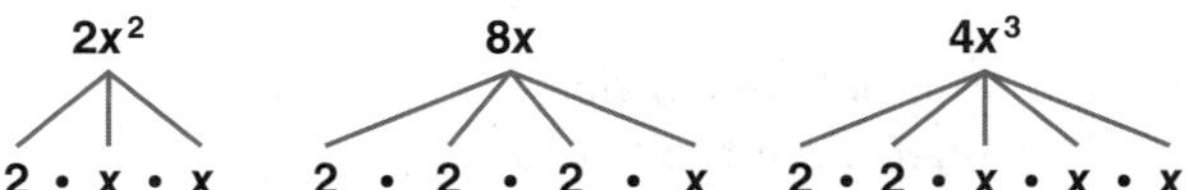


The GCF is $2 \cdot x$ or $2x$.

Write each term in the polynomial as a product of a different term and the GCF. Factor out the GCF.

$$2x^2 + 8x + 4x^3 = (2x \cdot x) + (2x \cdot 2 \cdot 2) + (2x \cdot 2 \cdot x \cdot x)$$
$$= 2x(x + 4 + 2x^2)$$

### VOCABULARY

**greatest common factor (GCF)**
the product of the prime factors common to two or more integers

**polynomial**
an expression that contains one or more term

You can check that you have factored correctly by using the Distributive Property.

## Example 1

**Find the greatest common factor (GCF) of the terms of $10y^3 - 15y^2 + 5y^5$.**

1. Factor each term.

$10y^3 = 5 \cdot 2 \cdot y \cdot y \cdot y$

$-15y^2 = -1 \cdot 5 \cdot 3 \cdot y \cdot y$

$5y^5 = 5 \cdot y \cdot y \cdot y \cdot y \cdot y$

2. Name the common factors.

$5, y, y$

3. Find the product of the common factors.

$\text{GCF} = 5 \cdot y \cdot y = 5y^2$

### YOUR TURN!

**Find the greatest common factor (GCF) of the terms of $18ab + 6a^3b + 12a^2b^3$.**

1. Factor each term.

$18ab =$ ____________

$6a^3b =$ ____________

$12a^2b^3 =$ ____________

2. Name the common factors.

____________

3. Find the product of the common factors.

GCF = ____________ = ______

GO ON

## Example 2

**Factor $4h^2 + 6h^4$.**

1. Factor each term.

   $4h^2 = 2 \cdot 2 \cdot h \cdot h$ $\quad$ $6h^4 = 3 \cdot 2 \cdot h \cdot h \cdot h \cdot h$

2. Name the common factors.

   $2, h, h$

3. Find the product of the common factors.

   $\text{GCF} = 2 \cdot h \cdot h = 2h^2$

4. Write the polynomial as a product of the GCF and another polynomial.

   $4h^2 + 6h^4 = 2h^2(2 + 3h^2)$

The only factor not used from the list for $4h^2$ is 2. The factors not used from the list for $6h^4$ are $3 \cdot h \cdot h$.

**YOUR TURN!**

**Factor $3st^2 + 9s^3t^2$.**

1. Factor each term.

   $3st^2 =$ ________________ $\quad$ $9s^3t^2 =$ ________________________

2. Name the common factors.

   ____________

3. Find the product of the common factors.

   GCF = ________________ = ______

4. Write the polynomial as a product of the GCF and another polynomial.

   $3st^2 + 9s^3t^2 =$ _____(_____ + _____)

## Guided Practice

**Find the GCF of the terms in each polynomial.**

**1** $a^2b^3 + ab^4$

$a^2b^3 =$ ______________________________

$ab^4 =$ ______________________________

GCF = ______________________________

**2** $24x^3 + 6x$

$24x^3 =$ ______________________________

$6x =$ ______________________________

GCF = ______________________________

## Step by Step Practice

**3** Factor $10g^2 + 12g^4 - 8g$.

**Step 1** Factor each term.

$10g^2 =$ ______________

$12g^4 =$ ______________

$8g =$ ______________

**Step 2** Name the common factors. ______________

**Step 3** Find the product of their common factors.

GCF = ________ = ______

**Step 4** Write the polynomial as a product of the GCF and another polynomial.

$10g^2 + 12g^4 - 8g =$ ______(______ + ______ − ______)

**Find the GCF. Then factor each polynomial.**

**4** $3g^2 + 12g^4$

$3g^2 =$ ______________

$12g^4 =$ ______________

GCF = ______________

$3g^2 + 12g^4 =$ ______________

**5** $-20s^2t^2 - 5s^3t^2$

$-20s^2t^2 =$ ______________

$-5s^3t^2 =$ ______________

GCF = ______________

$-20s^2t^2 - 5s^3t^2 =$ ______________

**6** $16x^3y^2 + 4y^5$

$16x^3y^2 =$ ______________

$4y^5 =$ ______________

GCF = ______________

$16x^3y^2 + 4y^5 =$ ______________

**7** $6m^2n^4 - 18m^2n^2 + 15mn$

$6m^2n^4 =$ ______________

$-18m^2n^2 =$ ______________

$15mn =$ ______________

GCF = ______________

$6m^2n^4 - 18m^2n^2 + 15mn =$

______________

## Step by Step Problem-Solving Practice

**Solve.**

**8** **AREA** The area of a rectangle is modeled by the expression $2x^5 - 30x^3$. The width is the GCF. Factor this expression to find the width.

Area = $2x^5 - 30x^3$

factors of $2x^5$: ____________________

factors of $30x^3$: ____________________

common factors: ____________________

Check off each step.

_____ **Understand: I underlined key words.**

_____ **Plan: To solve the problem, I will** ____________________.

_____ **Solve: The answer is** ____________________.

_____ **Check: I checked my answer by using the** ____________________.

## Skills, Concepts, and Problem Solving

**Factor each polynomial.**

**9** $-28x^3y - 42x$

GCF: ____________________

____________________

**10** $8p^9 - 24p^7$

GCF: ____________________

____________________

**11** $u^6v^4 + 9u^4v^6$

GCF: ____________________

____________________

**12** $-6b^3 + 9b^2 - 15$

GCF: ____________________

____________________

**13** $16a^4b - 10a^3b^7 - 6ab$

GCF: ____________________

____________________

**14** $-3w^3 + 15w^2 - 27w$

GCF: ____________________

____________________

**Factor each polynomial.**

15 $-4mn - 2mn^2 - 12m^2n^2 =$ ____________________

16 $45x^5y^2 + 18x^3y^2 =$ ____________________

17 $-5x^5 - 10x^2 - 20x =$ ____________________

18 $-21h^5 + 27h^3 =$ ____________________

19 $16a^6 - 4a^2 + 12a =$ ____________________

20 $20g^8 - 4g^4 + 16g^2 =$ ____________________

**Solve.**

21 **GEOMETRY** The area of the rectangle at the right is modeled by the polynomial $15x^4 - 9x^3$. What expression models the missing width?

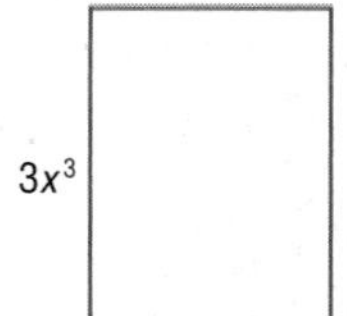


____________________

22 **MANUFACTURING** The diagram at the right is a mold for a cube-shaped manufacturing part. The volume of the liquid copper needed for the mold is modeled by the expression $20x^2 - 80x + 64$. Factor the expression.

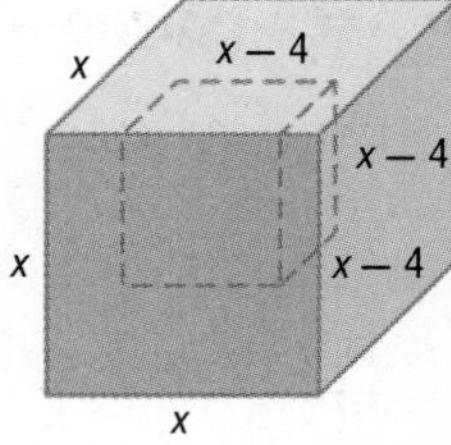


____________________

**Vocabulary Check** **Write the vocabulary word that completes each sentence.**

23 An expression such as $2x^5y - 4xy$, which contains more than one term, is a(n) ____________________.

24 The product of the prime factors common to two or more integers is the ____________________.

25 **Reflect** If the area of a rectangle is modeled by the expression $x^3 + 4x^2$, what are two expressions that could possibly model the length and the width of the rectangle? Explain your answer.

____________________

____________________

____________________

STOP

Chapter 4

# Progress Check 2 (Lessons 4-3 and 4-4)

**Find each product.**

1 $(3p - 7)(3p + 7) =$

2 $-3x(4x^2 - x) =$

3 $4g(8g^3 + 11g) =$

4 $(7x + 4)(2x - 6) =$

5 $5x(10xy^3 - 2y) =$

6 $(-6y - 9)(7y - 5) =$

7 $pq(3p^2q^4 + 5p^2q - 8p) =$

8 $(y^2 - 8z)(y - 11z) =$

**Factor each polynomial.**

9 $-8x^5y - 4xy^2 =$

10 $45k^3 + 63k =$

11 $-12p^4 + 6pq^2 =$

12 $15r + 21r^2 - 12 =$

13 $-6x^4y^4 - 12x^2y^2 - 24xy =$

14 $10mn^2 - 4mn^3 + 14mn =$

15 $24x^6y^3 + 8x^5 =$

16 $12rst - 4r^2s^2 + 3r^4s^2 =$

**Solve.**

17 **LANDSCAPING** Robert is clearing a spot in his yard to put a swing set. The clearing is in the shape of the rectangle shown below. What polynomial models the area of the clearing?

$3x + 5$

$x - 2$

Lesson 4-5

# Factor Trinomials: $x^2 + bx + c$

## KEY Concept

Just as numbers have **factors**, polynomials also have factors. When you find the factors of a polynomial, you find two binomials whose product is the given **trinomial**. Factoring trinomials is the inverse of multiplying binomials.

### Multiplying Binomials

You can use the Distributive Property to multiply binomials.

$$(x + 4)(x + 2) = x(x) + x(2) + 4(x) + 4(2)$$
$$= x^2 + 6x + 8$$

### Factoring a Trinomial

To factor the trinomial $x^2 + bx + c$, find two numbers that have a sum of $b$ and a product of $c$.

$$x^2 + 6x + 8 = (x + 4)(x + 2)$$

In the trinomial above, the coefficient of the middle term is 6, which is the sum of 2 and 4. The constant term is 8, which is the product of 2 and 4.

**VOCABULARY**

**factors**
the quantities being multiplied to get a product

**trinomial**
a polynomial with three terms

When $c$ is positive, the factors will either both be positive or both be negative. When $c$ is negative, one factor will be positive and one will be negative. You may not have to list all factors before you find the pair that sums to the middle term.

## Example 1

**Factor $x^2 + 8x + 12$.**

1. Write two binomials using factors of $x^2$.

   $(x + \square)(x + \square)$

2. List all pairs of factors of 12. Show their sums.

3. Use the pair of factors with a sum of 8 in the binomials. $(x + 2)(x + 6)$

4. Check. $x^2 + 8x + 12 \stackrel{?}{=} (x + 2)(x + 6)$

   $= x^2 + 6x + 2x + 12$

   $= x^2 + 8x + 12$ ✓

| Factors of 12 | Sum of Factors |
|---|---|
| 1 and 12 | 13 |
| 2 and 6 | 8 |
| 3 and 4 | 7 |

**YOUR TURN!**

**Factor $x^2 + 9x + 20$.**

1. Write two binomials using factors of $x^2$. $(x + \square)(x + \square)$
2. List all pairs of factors of _____. Show their sums.
3. Use the pair of factors with a sum of _____ in the binomials. $(x +$ _____$)(x +$ _____$)$
4. Check. $x^2 + 9x + 20 \stackrel{?}{=} (x +$ _____$)(x +$ _____$)$

   $= x^2 +$ _____$x +$ _____$x +$ _____

   $= x^2 +$ _____$x +$ _____

| Factors of 20 | Sum of Factors |
|---|---|
| _____ and _____ | _____ |
| _____ and _____ | _____ |
| _____ and _____ | _____ |

## Example 2

**Factor $x^2 - 3x - 18$.**

1. Write two binomials using factors of $x^2$.

   $(x + \square)(x + \square)$

2. List all pairs of factors of −18. Show their sums.

| Factors of −18 | Sum of Factors |
|---|---|
| −1 and 18 | 17 |
| 1 and −18 | −17 |
| −2 and 9 | 7 |
| 2 and −9 | −7 |
| −3 and 6 | 3 |
| 3 and −6 | −3 |

3. Use the pair of factors with a sum of −3 in the binomials.

   $(x + 3)(x + (-6))$

4. Simplify by rewriting the binomials using only one sign in each.

   $(x + 3)(x - 6)$

5. Check: $x^2 - 3x - 18 \stackrel{?}{=} (x + 3)(x - 6)$

   $= x^2 - 6x + 3x - 18$

   $= x^2 - 3x - 18$ ✓

**YOUR TURN!**

**Factor $x^2 + x - 6$.**

1. Write two binomials using factors of $x^2$.

   $(x + \square)(x + \square)$

2. List all pairs of factors of _____. Show their sums.

| Factors of −6 | Sum of Factors |
|---|---|
| _____ and _____ | _____ |
| _____ and _____ | _____ |
| _____ and _____ | _____ |
| _____ and _____ | _____ |

3. Use the pair of factors with a sum of _____ in the binomials.

   $(x +$ _____$)(x + ($_____$))$

4. Simplify by rewriting the binomials using only one sign in each.

   $(x -$ _____$)(x +$ _____$)$

5. Check: $x^2 + x - 6 \stackrel{?}{=} (x -$ _____$)(x +$ _____$)$

   $= x^2 +$ _____$x -$ _____$x -$ _____

   $= x^2 +$ _____ $-$ _____

## Guided Practice

**Factor each trinomial.**

1 $x^2 + 5x + 6$

| Factors of 6 | Sum of Factors |
|---|---|
| ____ and ____ | ____ |
| ____ and ____ | ____ |
| ____ and ____ | ____ |
| ____ and ____ | ____ |

Use the pair of factors with a sum

of _____ in the binomials.

$x^2 + 5x + 6 = (x$ _____$)(x$ _____$)$

2 $x^2 - 7x + 10$

| Factors of 10 | Sum of Factors |
|---|---|
| ____ and ____ | ____ |
| ____ and ____ | ____ |
| ____ and ____ | ____ |
| ____ and ____ | ____ |

Use the pair of factors with a sum

of _____ in the binomials.

$x^2 - 7x + 10 = (x$ _____$)(x$ _____$)$

## Step by Step Practice

3 Factor $x^2 - 8x + 16$.

**Step 1** Write two binomials using factors of $x^2$. $(x + \square)(x + \square)$

**Step 2** List all pairs of factors of _____.
Show their sums.

| Factors of 16 | Sum of Factors |
|---|---|
| ____ and ____ | ____ |
| ____ and ____ | ____ |
| ____ and ____ | ____ |
| ____ and ____ | ____ |
| ____ and ____ | ____ |
| ____ and ____ | ____ |

**Step 3** Use the pair of factors with a sum of _____ in the binomials.

(_____)(_____)

**Step 4** Check: $x^2 - 8x + 16 \stackrel{?}{=} (x$ _____$)(x$ _____$)$

$\stackrel{?}{=} x^2$ _____ $x$ _____ $x$ _____

$= x^2 -$ _____ $+$ _____

GO ON

**Factor each trinomial.**

4 $x^2 + 2x - 15$

| Factors of −15 | Sum of Factors |
|---|---|
| _____ and _____ | _____ |
| _____ and _____ | _____ |
| _____ and _____ | _____ |
| _____ and _____ | _____ |

The pair that has a sum of _____ is

_____ and _____.

$x^2 + 2x - 15 = (x$ _____$)(x$ _____$)$

5 $x^2 + 2x - 35$

| Factors of −35 | Sum of Factors |
|---|---|
| _____ and _____ | _____ |
| _____ and _____ | _____ |
| _____ and _____ | _____ |
| _____ and _____ | _____ |

The pair that has a sum of _____ is

_____ and _____.

$x^2 - 2x - 35 = (x$ _____$)(x$ _____$)$

6 $x^2 + 5x + 6$

| Factors of 6 | Sum of Factors |
|---|---|
| _____ and _____ | _____ |
| _____ and _____ | _____ |
| _____ and _____ | _____ |
| _____ and _____ | _____ |

The pair that has a sum of _____ is

_____ and _____.

$x^2 + 5x + 6 =$ (_______)(_______)

7 $b^2 - 4b + 3$

| Factors of 3 | Sum of Factors |
|---|---|
| _____ and _____ | _____ |
| _____ and _____ | _____ |

The pair that has a sum of _____ is

_____ and _____.

$b^2 - 4b + 3 =$ (_______)(_______)

8 $y^2 - 6y - 27$

| Factors of −27 | Sum of Factors |
|---|---|
| _____ and _____ | _____ |
| _____ and _____ | _____ |
| _____ and _____ | _____ |
| _____ and _____ | _____ |

The pair that has a sum of _____ is

_____ and _____.

$y^2 - 6y - 27 =$ _______________

9 $g^2 + 9g + 14$

| Factors of 14 | Sum of Factors |
|---|---|
| _____ and _____ | _____ |
| _____ and _____ | _____ |
| _____ and _____ | _____ |
| _____ and _____ | _____ |

The pair that has a sum of _____ is

_____ and _____.

$g^2 + 9g + 14 =$ _______________

## Step by Step Problem-Solving Practice

**Solve.**

**10** NUMBER SENSE Find a value of $b$ that is greater than 30 and less than 100 that makes the polynomial $x^2 + bx - 105$ factorable. Factor the trinomial into two binomials.

| Factors of −105 | Sum of Factors |
|---|---|
| ______ and ______ | ______ |
| ______ and ______ | ______ |
| ______ and ______ | ______ |
| ______ and ______ | ______ |
| ______ and ______ | ______ |
| ______ and ______ | ______ |

________________________________

Check off each step.

______ Understand: I underlined key words.

______ Plan: To solve the problem, I will ________________________________.

______ Solve: The answer is ________________________________.

______ Check: I checked my answer by ________________________________.

## Skills, Concepts, and Problem Solving

**Factor each trinomial.**

**11** $t^2 + 12t + 35 =$ ______________

**12** $y^2 + 12y - 64 =$ ______________

**13** $w^2 - 14w + 49 =$ ______________

**14** $a^2 - a - 72 =$ ______________

**15** $h^2 - 38h + 72 =$ ______________

**16** $q^2 - 6q - 16 =$ ______________

GO ON

**Factor each trinomial.**

17 $k^2 - 19k + 84 =$ ______

18 $t^2 - 21t - 100 =$ ______

19 $x^2 + 7x - 30 =$ ______

**Solve.**

20 **PARKING LOT DESIGN** If the polynomial $x^2 + 12x + 27$ models the area of the parking lot shown below, what expression models the width?

______

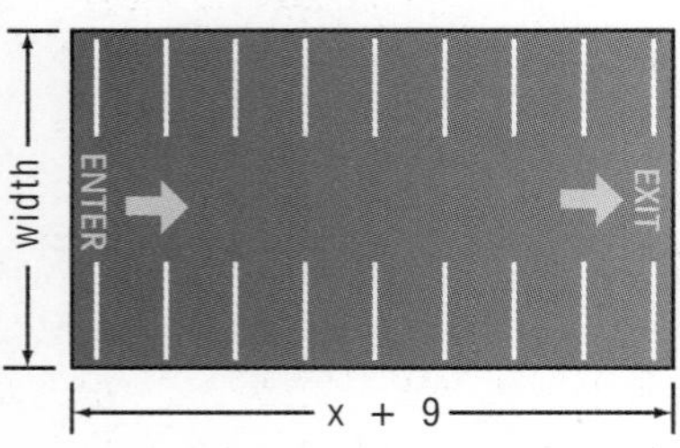


21 **NUMBER SENSE** Find a value of $b$ that is between 0 and $-7$ that makes the polynomial $x^2 + bx - 21$ factorable. Then factor the polynomial into two binomials.

______

**Vocabulary Check** **Write the vocabulary word that completes each sentence.**

22 A polynomial with three terms, such as $x^2 + 5x + 6$, is a(n) ______.

23 ______ are quantities that are multiplied to get a product.

24 **Reflect** The trinomial $x^2 + bx + c$ can be factored into two binomials. If $c > 0$, what can you assume about the factors of $c$?

______

______

______

STOP

Lesson 4-6

# Factor Trinomials: $ax^2 + bx + c$

## KEY Concept

When the squared term of a **trinomial** has a coefficient other than 1, there are more **factors** and combinations to consider when finding its two binomial factors.

There are three steps to factor a trinomial in $ax^2 + bx + c$ form.

1. Consider factors of $ax^2$.
2. Consider factors of $c$.
3. Find the right combination that has a sum of $bx$.

$$6x^2 + 14x + 4$$

1. Factors of $6x^2$: **$2x$ and $3x$** or **$6x$ and $x$**
2. Factors of 4: **2 and 2** or **4 and 1**
3. Use trial and error to find which factor pairs have a sum of $14x$.

Multiply one factor from $6x^2$ and one factor from 4. Multiply different combinations of the other factors. Then add to find a sum of $14x$.

$$2x \cdot 2 + 3x \cdot 2 = 4x + 6x = 10x$$

$$3x \cdot 1 + 2x \cdot 4 = 3x + 8x = 11x$$

$$2x \cdot 1 + 3x \cdot 4 = 2x + 12x = 14x \checkmark$$

Use the $x$-terms as the first terms in the binomials and fill in the other factors based on FOIL.

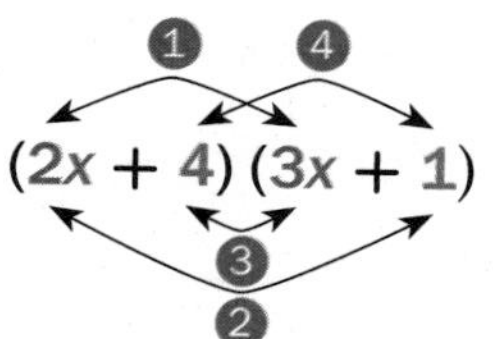


1 First terms
2 Outside terms
3 Inside terms
4 Last terms

So, $6x^2 + 14x + 4 = (2x + 4)(3x + 1)$.

### VOCABULARY

**factors**
the quantities being multiplied to get a product

**trinomial**
a polynomial with three terms

You should check your factoring by multiplying the binomials to verify that the product is the same as the original trinomial.



GO ON

## Example 1

**Factor $3x^2 + 11x + 10$.**

1. List the factors and check combinations.

| Factors of $3x^2$ | Factors of 10 | Sum of Factors |
|---|---|---|
| $x$ and $3x$ | 1 and 10 | $x \cdot 1 + 3x \cdot 10 = 31x$ |
| | | $x \cdot 10 + 3x \cdot 1 = 13x$ |
| $x$ and $3x$ | 2 and 5 | $x \cdot 2 + 3x \cdot 5 = 17x$ |
| | | $x \cdot 5 + 3x \cdot 2 = 11x$ ✓ |

2. Use the correct factors to write the binomials based on FOIL.

$3x^2 + 11x + 10 = (x + 2)(3x + 5)$

3. Check.

$3x^2 + 11x + 10 \stackrel{?}{=} (x + 2)(3x + 5)$

$= 3x^2 + 5x + 6x + 10$

$= 3x^2 + 11x + 10$ ✓

**YOUR TURN!**

**Factor $5x^2 + 12x + 4$.**

1. List the factors and check combinations.

| Factors of $5x^2$ | Factors of 4 | Sum of Factors |
|---|---|---|
| _____ and _____ | _____ and _____ | |
| | | |
| _____ and _____ | _____ and _____ | |

2. Use the correct factors to write the binomials based on FOIL.

$5x^2 + 12x + 4 =$ (__________)(__________)

3. Check.

$5x^2 + 12x + 4 \stackrel{?}{=}$ (__________)(__________)

$= 5x^2 +$ _____ $+$ _____ $+$ _____

$= 5x^2 +$ _____ $+$ _____

## Example 2

**Factor $14x^2 + x - 3$.**

1. List the factors and check combinations.

| Factors of $14x^2$ | Factors of $-3$ | Sum of Factors |
|---|---|---|
| $x$ and $14x$ | $-1$ and $3$ | $x \cdot (-1) + 14x \cdot 3 = 41x$ |
| | | $x \cdot 3 + 14x \cdot (-1) = -11x$ |
| $2x$ and $7x$ | $1$ and $-3$ | $2x \cdot 1 + 7x \cdot (-3) = -19x$ |
| | | $2x \cdot (-3) + 7x \cdot 1 = x$ ✓ |

2. Use the correct factors to write the binomials based on FOIL.

$14x^2 + x - 3 = (2x + 1)(7x - 3)$

3. Check.

$14x^2 + x - 3 \stackrel{?}{=} (2x + 1)(7x - 3)$

$= 14x^2 - 6x + 7x - 3$

$= 14x^2 + x - 3$ ✓

**YOUR TURN!**

**Factor $4x^2 - 8x - 5$.**

1. List the factors and check combinations.

| Factors of $4x^2$ | Factors of $-5$ | Sum of Factors |
|---|---|---|
| ____ and ____ | ____ and ____ | |
| | | |
| ____ and ____ | ____ and ____ | |
| | | |

2. Use the correct factors to write the binomials based on FOIL.

$4x^2 - 8x - 5 =$ (__________)(__________)

3. Check.

$4x^2 - 8x - 5 \stackrel{?}{=}$ (__________)(__________)

$= 4x^2 +$ _____ $-$ _____ $-$ _____

$= 4x^2 -$ _____ $-$ _____

GO ON

## Guided Practice

**Factor each trinomial.**

1. $2x^2 + 5x + 3$

| Factors of $2x^2$ | Factors of 3 | Sum of Factors |
|---|---|---|
| ____ and ____ | ____ and ____ | $x \cdot$ ____ $+ 2x \cdot$ ____ $=$ ____ |
| | | ____ $\cdot 3 +$ ____ $\cdot 1 =$ ____ |

$2x^2 + 5x + 3 = (2x +$ ____$)($____ $+$ ____$)$

## Step by Step Practice

2. Factor $6x^2 + 13x - 28$.

**Step 1** List factors and check combinations.

| Factors of $6x^2$ | Factors of $-28$ | Sum of Factors |
|---|---|---|
| ____ and ____ | ____ and ____ | |
| | | |
| ____ and ____ | ____ and ____ | |

**Step 2** Use the correct factors to write the binomials based on FOIL.

$6x^2 + 13x - 28 = ($________$)($________$)$

**Step 3** Check.

$6x^2 + 13x - 28 \stackrel{?}{=} ($________$)($________$)$

$= 6x^2 -$ ____ $x +$ ____ $x -$ ____

$= 6x^2 +$ ____ $x -$ ____

Identify the factors of ______ and ______.

Using the FOIL, which products have a sum of 13?

______ and ______; ______ and ______.

Therefore, $6x^2 + 13x - 28 = ($________$)($________$)$

**Factor each trinomial.**

3 $10x^2 + 57x + 11$

| Factors of $10x^2$ | Factors of 11 | Sum of Factors |
|---|---|---|
| ____ and ____ | ____ and ____ | $x$ • ____ + $10x$ • ____ = ____ |
| | | ____ • 11 + ____ • 1 = ____ |
| ____ and ____ | ____ and ____ | ____ • 1 + ____ • 11 = ____ |

$10x^2 + 57x + 11 =$ (____ + ____)(____ + ____)

4 $9x^2 + 18x - 7$

| Factors of $9x^2$ | Factors of $-7$ | Sum of Factors |
|---|---|---|
| ____ and ____ | ____ and ____ | |
| | | |
| ____ and ____ | ____ and ____ | |
| | | |
| ____ and ____ | ____ and ____ | |

$9x^2 + 18x - 7 =$ (________)(________)

5 $2x^2 - x - 10$

| Factors of $2x^2$ | Factors of $-10$ | Sum of Factors |
|---|---|---|
| ____ and ____ | ____ and ____ | |
| | | |
| ____ and ____ | ____ and ____ | |
| | | |
| ____ and ____ | ____ and ____ | |
| | | |
| ____ and ____ | ____ and ____ | |
| | | |

$2x^2 - x - 10 =$ (________)(________)



GO ON

## Step by Step Problem-Solving Practice

**Solve.**

**6** **BUILDING DESIGN** Find an expression that models the length of the floor plan shown if the area is modeled by the polynomial $2x^2 + 9x - 5$. Factor the trinomial into two binomials.

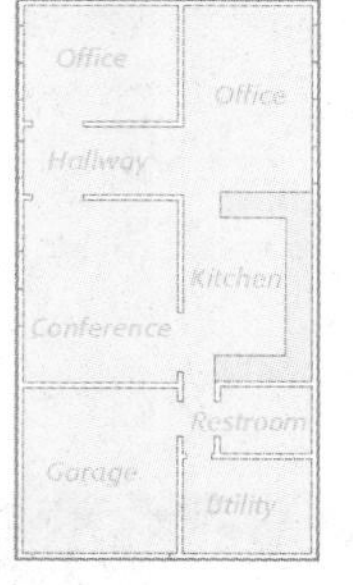


Area = length • width

$2x^2 + 9x - 5 = (\square + \square)(x + 5)$

| Factors of $2x^2$ | Factors of $-5$ | Sum of Factors |
|---|---|---|
| | | |
| | | |
| | | |
| | | |

$2x^2 + 9x - 5 =$ ____________________

Check off each step.

______ **Understand: I underlined key words.**

______ **Plan: To solve the problem, I will** ______________________________.

______ **Solve: The answer is** ______________________________.

______ **Check: I checked my answer by** ______________________________.

## Skills, Concepts, and Problem Solving

**Factor each trinomial.**

7. $36z^2 + 30z + 6 =$ __________

8. $24h^2 - 27h - 42 =$ __________

9. $2k^2 - 24k + 40 =$ __________

10. $11a^2 + 52a - 15 =$ __________

11. $4x^2 + 2x - 20 =$ __________

12. $40y^2 + 42y + 2 =$ __________

13. $6q^2 - 29q + 35 =$ __________

14. $14b^2 + 23b - 15 =$ __________

15. $18x^2 - 45x + 18 =$ __________



**Solve.**

16. **AREA** The area of the rectangle pictured to the right is modeled by the expression $10x^2 + 11x - 8$. What binomial models the length of the rectangle?

$2x - 1$

__________

**Vocabulary Check** **Write the vocabulary word that completes each sentence.**

17. Two __________ of 8 are 2 and 4.

18. A polynomial with three terms is a(n) __________.

19. **Reflect** What is the maximum number of combinations of factors you would have to test to factor for the trinomial $11x^2 - 12x + 1$? Explain your answer.

__________

__________

__________

STOP

Chapter 4

# Progress Check 3 (Lessons 4-5 and 4-6)

**Factor each trinomial.**

1 $x^2 + 2x - 15$

| Factors of −15 | Sum of Factors |
|---|---|
| _____ and _____ | _____ |
| _____ and _____ | _____ |
| _____ and _____ | _____ |
| _____ and _____ | _____ |

$x^2 + 2x - 15 = (x$ ________ $)(x$ ________ $)$

2 $x^2 - 3x - 28$

| Factors of 28 | Sum of Factors |
|---|---|
| _____ and _____ | _____ |
| _____ and _____ | _____ |
| _____ and _____ | _____ |
| _____ and _____ | _____ |
| _____ and _____ | _____ |
| _____ and _____ | _____ |

$x^2 - 3x - 28 = (x$ ________ $)(x$ ________ $)$

3 $x^2 + 5x + 6 =$ ____________________

4 $b^2 - 4b + 3 =$ ____________________

5 $y^2 + 2y - 63 =$ ____________________

6 $g^2 - 13g + 40 =$ ____________________

**Factor each trinomial.**

7 $2x^2 + 11x + 5 =$ ____________________

8 $10x^2 - 37x + 21 =$ ____________________

9 $12x^2 - 56x + 60 =$ ____________________

10 $15x^2 + 33x - 36 =$ ____________________

11 $10x^2 + 46x + 24 =$ ____________________

12 $7x^2 + 52x - 32 =$ ____________________

**Solve.**

13 **BANNER** The area of a rectangular banner for a parade is modeled by the polynomial $12x^2 + 32x - 35$. What binomials represent the length and width of the banner?

____________________________________

Lesson 4-7

# Factor Difference of Perfect Squares

## KEY Concept

A difference of perfect squares can be factored into two binomials. A binomial that is a difference of **perfect squares** is a binomial of the form $a^2 - b^2$.

For all real numbers $a$ and $b$: $a^2 - b^2 = (a + b)(a - b)$

You can work backwards to check.

$$(a + b)(a - b) = a^2 - ab + ab - b^2$$
$$= a^2 - b^2 \checkmark$$

Here is an example. $x^2 - 64 = (x + 8)(x - 8)$

Work backwards to check.

$$(x + 8)(x - 8) = x^2 - 8x + 8x - 64$$
$$= x^2 - 64 \checkmark$$

$x^2$ is the perfect square of $x$.
64 is the perfect square of 8.

So, $x^2 - 64$ is the difference of perfect squares.

**VOCABULARY**

**perfect square**
a number or expression that is the square of another number or expression

For a binomial to be a difference of squares, the two perfect squares must be separated by a subtraction sign. Some binomials may need to be simplified by factoring out the GCF first.

### Example 1

**Factor $x^2 - 9$.**

1. Verify that each term is a perfect square.

   $\sqrt{x^2} = x$ $\qquad$ $\sqrt{9} = 3$

2. Substitute the factors into the binomial factors of a difference of squares.

   $a^2 - b^2 = (a + b)(a - b)$

   $x^2 - 9 = (x + 3)(x - 3)$

3. Check.

   $x^2 - 9 \stackrel{?}{=} (x + 3)(x - 3)$

   $\stackrel{?}{=} x^2 - 3x + 3x - 9$

   $= x^2 - 9 \checkmark$

### YOUR TURN!

**Factor $y^2 - 36$.**

1. Verify that each term is a perfect square.

   $\sqrt{y^2} =$ _____ $\qquad$ $\sqrt{36} =$ _____

2. Substitute the factors into the binomial factors of a difference of squares.

   $a^2 - b^2 = (a + b)(a - b)$

   $y^2 - 36 =$ (_____ + _____)(_____ − _____)

3. Check.

   $y^2 - 36 \stackrel{?}{=} (y +$ _____$)(y -$ _____$)$

   $\stackrel{?}{=}$ _____ − _____ + _____ − _____

   $=$ _____ − _____

GO ON

## Example 2

**Factor $4x^2 - 81$.**

1. Verify that each term is a perfect square.

$\sqrt{4x^2} = 2x$ $\sqrt{81} = 9$

2. Substitute the factors into the binomial factors of a difference of squares.

$4x^2 - 81 = (2x + 9)(2x - 9)$

3. Check.

$$4x^2 - 81 \stackrel{?}{=} (2x + 9)(2x - 9)$$
$$\stackrel{?}{=} 4x^2 - 18x + 18x - 81$$
$$= 4x^2 - 81 \checkmark$$

**YOUR TURN!**

**Factor $25x^2 - 49$.**

1. Verify that each term is a perfect square.

$\sqrt{25x^2} =$ ____ $\sqrt{49} =$ ____

2. Substitute the factors into the binomial factors of a difference of squares.

$25x^2 - 49 =$ (____ + ____)(____ − ____)

3. Check.

$25x^2 - 49 \stackrel{?}{=}$ (____ + ____)(____ − ____)

$\stackrel{?}{=}$ ____ − ____$x$ + ____$x$ − ____

= ____ − ____

## Example 3

**Factor $2x^2 - 242$.**

1. Factor out the GFC. $2(x^2 - 121)$
2. Verify that each term is a perfect square.

$\sqrt{x^2} = x$ $\sqrt{121} = 11$

3. Substitute the factors into the binomial factors of a difference of squares.

$x^2 - 121 = (x + 11)(x - 11)$

4. Write the product of all factors.

$2(x + 11)(x - 11)$

5. Check.

$$2(x + 11)(x - 11)$$
$$\stackrel{?}{=} 2(x^2 - 121)$$
$$= 2x^2 - 242 \checkmark$$

**YOUR TURN!**

**Factor $27x^3 - 48x$.**

1. Factor out the GFC. ____________
2. Verify that each term is a perfect square.

$\sqrt{9x^2} =$ ____ $\sqrt{16} =$ ____

3. Substitute the factors into the binomial factors of a difference of squares.

$9x^2 - 16 =$ (____ + ____)(____ − ____)

4. Write the product of all factors.

____(____ + ____)(____ − ____)

5. Check.

$3x$ (____ + ____)(____ − ____)

$\stackrel{?}{=} 3x($____ − ____)

= ____ − ____



## Guided Practice

**Factor each difference of squares.**

**1** $x^2 - 25$

$x^2 - 25 =$ (____ + ____)(____ − ____)

**2** $y^2 - 121$

$y^2 - 121 =$ (____ + ____)(____ − ____)

## Step by Step Practice

**3** Factor $3a^2 - 192$.

**Step 1** Is there a GCF to factor out? ______ Factor out a GFC. ______

**Step 2** Verify that each term is a perfect square. $\sqrt{a^2}$ = ______ $\sqrt{64}$ = ______

**Step 3** Substitute the factors into the binomial factors of a difference of squares. $a^2 - 64$ = (____ + ____)( − ____)

**Step 4** Write the product of all factors. ____(____ + ____)(____ − ____)

**Step 5** Check: $3a^2 - 192 \stackrel{?}{=}$ ____ (____ + ____)(____ − ____)

$\stackrel{?}{=}$ 3(____ − ____ + ____ − ____)

$\stackrel{?}{=}$ 3(____ − ____)

= ____ − ____

**Factor each difference of squares.**

**4** $36x^2 - 4$

4(____ − ____) = 4(____ + ____)(____ − ____)

**5** $169y^2 - 49$

(____ + ____)(____ − ____)

## Step by Step Problem-Solving Practice

**Solve.**

**6** **CONSTRUCTION** Samir needs a piece of wood that looks like the diagram at the right. The smaller square will be removed from the larger square. Write an expression to model the area of the remaining wood. Write in factored form.

The area of the larger square is ______.

The area of the smaller square is ______.

Check off each step.

______ **Understand: I underlined key words.**

______ **Plan: To solve the problem, I will** ______.

______ **Solve: The answer is** ______.

______ **Check: I checked my answer by** ______.

GO ON

## Skills, Concepts, and Problem Solving

**Factor each difference of squares.**

7 $x^2 - 81 =$ ____________________

8 $16b^2 - 49 =$ ____________________

9 $y^2 - 169 =$ ____________________

10 $7x^2 - 28 =$ ____________________

11 $4k^2 - 121 =$ ____________________

12 $2h^2 - 72 =$ ____________________

13 $363x^2 - 108 =$ ____________________

14 $p^2 - 64 =$ ____________________

15 $9m^2 - 4 =$ ____________________

16 $27f^2 - 27 =$ ____________________

17 $12x^2 - 75 =$ ____________________

18 $64d^2 - 100 =$ ____________________

**Solve.**

19 **GARDENING** Margo's backyard is in the shape of a square. She is planning a 9 ft by 9 ft square garden for one corner of her yard as shown in the diagram below. Write an expression that will show the area of the remaining portion of her backyard. Then factor this expression.

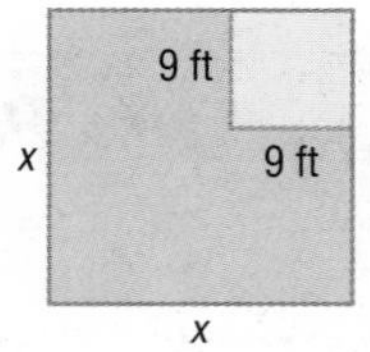


____________________________________________

____________________________________________

**Vocabulary Check** **Write the vocabulary word that completes the sentence.**

20 The terms $4v^2$ and 144 are examples of ____________________.

21 **Reflect** If the terms of a binomial are separated by a subtraction sign, but they are not perfect squares, what should you look for to determine if the binomial can be factored?

____________________________________________

____________________________________________

STOP

Lesson 4-8

# Factor Perfect Square Trinomials

## KEY Concept

A **perfect square trinomial** is a trinomial of the form $a^2 + 2ab + b^2$ or $a^2 - 2ab + b^2$. These trinomials are equal to the sum or the difference of a perfect square.

For all real numbers $a$ and $b$ where $a$ is the square root of the first term and $b$ is the square root of that last term:

$$a^2 + 2ab + b^2 = (a + b)(a + b) = (a + b)^2$$

$$a^2 - 2ab + b^2 = (a - b)(a - b) = (a - b)^2$$

Here are two examples.

$$x^2 + 10x + 25 = (x + 5)(x + 5) = (x + 5)^2$$

$$z^2 - 14z + 49 = (z - 7)(z - 7) = (z - 7)^2$$

**VOCABULARY**

**perfect square**
a number that is the square of another number

**trinomial**
a polynomial with three terms

You should check your factoring by multiplying the binomials to verify that the product is the same as the original trinomial.

## Example 1

**Factor $x^2 + 18x + 81$.**

1. Rewrite the first and last terms.

   $x^2 + 18x + 81$
   $= x \cdot x + 18x + 9 \cdot 9$

2. Verify that the middle term equals $2ab$.

   $18x = 2(x \cdot 9)$

3. Substitute the factors into the binomial factors of the trinomial.

   $x^2 + 18x + 81 = (x + 9)(x + 9)$
   $= (x + 9)^2$

4. Check.

   $(x + 9)^2 = (x + 9)(x + 9)$
   $= x^2 + 9x + 9x + 81$
   $= x^2 + 18x + 81$ ✓

**YOUR TURN!**

**Factor $x^2 + 14x + 49$.**

1. Rewrite the first and last terms.

   $x^2 + 14x + 49$
   $=$ ____ $\cdot$ ____ $+ 14x +$ ____ $\cdot$ ____

2. Verify that the middle term equals $2ab$.

   ____ $= 2($____ $\cdot$ ____$)$

3. Substitute the factors into the binomial factors of the trinomial.

   $x^2 + 14x + 49 = ($____ $+$ ____$)($____ $+$ ____$)$
   $= ($____ $+$ ____$)^2$

4. Check.

   $($____ $+$ ____$)^2 = ($____ $+$ ____$)($____ $+$ ____$)$
   $=$ ____ $+$ ____ $+$ ____ $+$ ____
   $=$ ____ $+$ ____ $+$ ____

## Example 2

**Factor $9x^2 - 12x + 4$.**

1. Rewrite the first and last terms.

$$9x^2 - 12x + 4 = 3x \cdot 3x - 12x + 2 \cdot 2$$

2. Verify that the middle term equals $2ab$.

$$-12x = -2(3x \cdot 2)$$

3. Substitute the factors into the binomial factors of the trinomial.

$$9x^2 - 12x + 4 = (3x - 2)(3x - 2)$$
$$= (3x - 2)^2$$

4. Check.

$$(3x - 2)^2 = (3x - 2)(3x - 2)$$
$$= 9x^2 - 6x - 6x + 4$$
$$= 9x^2 - 12x + 4 \checkmark$$

**YOUR TURN!**

**Factor $25x^2 - 40x + 16$.**

1. Rewrite the first and last terms.

$25x^2 - 40x + 16 =$ ____ • ____ $- 40x +$ ____ • ____

2. Verify that the middle term equals $2ab$.

$25x^2 - 40x + 16$

____ $= -2($____ • ____$)$

3. Substitute the factors into the binomial factors of the trinomial.

$25x^2 - 40x + 16 = ($____ $-$ ____$)($____ $-$ ____$)$

$= ($____ $-$ ____$)^2$

4. Check.

$($____ $-$ ____$)^2 = ($____ $-$ ____$)($____ $-$ ____$)$

$=$ ____ $-$ ____ $-$ ____ $+$ ____

$=$ ____ $-$ ____ $+$ ____

## Guided Practice

**Factor each perfect square trinomial.**

**1** $x^2 + 8x + 16$

Rewrite the first and last terms.

$x^2 + 8x + 16$

$= \_\_\_\_ \cdot \_\_\_\_ + 8x + \_\_\_\_ \cdot \_\_\_\_$

Verify that the middle term equals $2ab$.

$\_\_\_\_ = 2(\_\_\_\_ \cdot \_\_\_\_)$

Substitute the factors.

$x^2 + 8x + 16 = (\_\_\_\_ + \_\_\_\_)(\_\_\_\_ + \_\_\_\_)$

$= (\_\_\_\_ + \_\_\_\_)^2$

**2** $x^2 + 20x + 100$

Rewrite the first and last terms.

$x^2 + 20x + 100$

$= \_\_\_\_ \cdot \_\_\_\_ + 10x + \_\_\_\_ \cdot \_\_\_\_$

Verify that the middle term equals $2ab$.

$\_\_\_\_ = 2(\_\_\_\_ \cdot \_\_\_\_)$

Substitute the factors.

$x^2 + 20x + 100 = (\_\_\_\_ + \_\_\_\_)(\_\_\_\_ + \_\_\_\_)$

$= (\_\_\_\_ + \_\_\_\_)^2$

**3** $x^2 - 22x + 121$

Rewrite the first and last terms.

$x^2 - 22x + 121$

$= \_\_\_\_ \cdot \_\_\_\_ - 22x + \_\_\_\_ \cdot \_\_\_\_$

Verify that the middle term equals $2ab$.

$\_\_\_\_ = -2(\_\_\_\_ \cdot \_\_\_\_)$

Substitute the factors.

$x^2 - 22x + 121 = (\_\_\_\_ - \_\_\_\_)(\_\_\_\_ - \_\_\_\_)$

$= (\_\_\_\_ - \_\_\_\_)^2$

**4** $x^2 - 6x + 9$

Rewrite the first and last terms.

$x^2 - 6x + 9$

$= \_\_\_\_ \cdot \_\_\_\_ - 6x + \_\_\_\_ \cdot \_\_\_\_$

Verify that the middle term equals $2ab$.

$\_\_\_\_ = -2(\_\_\_\_ \cdot \_\_\_\_)$

Substitute the factors.

$x^2 - 6x + 9 = (\_\_\_\_ - \_\_\_\_)(\_\_\_\_ - \_\_\_\_)$

$= (\_\_\_\_ - \_\_\_\_)^2$



GO ON

## Step by Step Practice

**5** Factor $16x^2 + 56x + 49$.

**Step 1** Rewrite the first and last terms.

$16x^2 + 56x + 49 =$ _____ • _____ $+ 56x +$ _____ • _____

**Step 2** Verify that the middle term equals $2ab$.

_____ = 2(_____ • _____)

**Step 3** Substitute the factors into the binomial factors of the trinomial.

$16x^2 + 56x + 49 =$ (_____ + _____)(_____ + _____)

= (_____ + _____)$^2$

**Step 4** Check.

(_____ + _____)$^2$ = (_____ + _____)(_____ + _____)

= _____ + _____ + _____ + _____

$= 16x^2 + 56x + 49$

**Factor each perfect square trinomial.**

**6** $4x^2 - 36x + 81$

$4x^2 - 36x + 81 =$ _____ • _____ − 2(_____ • _____) + _____ • _____

$4x^2 - 36x + 81 =$ (__________)(__________) = (__________)$^2$

**7** $49x^2 + 42x + 9$

$49x^2 + 42x + 9 =$ _____ • _____ + 2(_____ • _____) + _____ • _____

$49x^2 + 42x + 9 =$ (__________)(__________) = (__________)$^2$

**Factor the perfect square trinomial.**

8 $16x^2 + 48x + 36$

$16x^2 + 48x + 36 =$ ______ • ______ + 2(______ • ______) + ______ • ______

$16x^2 + 48x + 36 = 4($__________$)($__________$) = 4($__________$)^2$

## Step by Step Problem-Solving Practice

**Solve.**

9 **BUILDING DESIGN** The area of a square base of a building is $4x^2 + 20x + 25$. Write an expression to model the length of a side.

______________________________________________

______________________________________________

______________________________________________

Check off each step.

______ **Understand: I underlined key words.**

______ **Plan: To solve the problem, I will** ______________________________.

______ **Solve: The answer is** ______________________________.

______ **Check: I checked my answer by** ______________________________.

## Skills, Concepts, and Problem Solving

**Factor each perfect square trinomial.**

10 $x^2 + 12x + 36 =$ ______________

11 $9b^2 - 30b + 25 =$ ______________

12 $k^2 + 30k + 225 =$ ______________

GO ON

**Factor the perfect square trinomial.**

13 $100q^2 - 160q + 64 =$ ________________

14 $y^2 - 14y + 49 =$ ________________

15 $81x^2 + 18x + 1 =$ ________________

16 $16p^2 - 88p + 121 =$ ________________

17 $4y^2 + 48y + 144 =$ ________________

18 $z^2 - 40z + 400 =$ ________________

**Solve.**

19 **AREA** The area of the square shown at the right is represented by the polynomial $49y^2 - 84y + 36$. Find an expression that represents the length of a side.

$49y^2 - 84y + 36$

________________________________

20 **GEOMETRY** The area of the figure shown at the right is $16x^2 + 64x + 64$. What is the sum of $a$ and $b$?

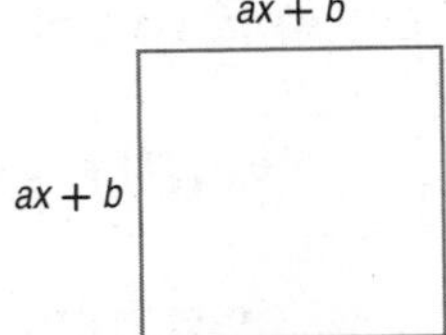


________________________________

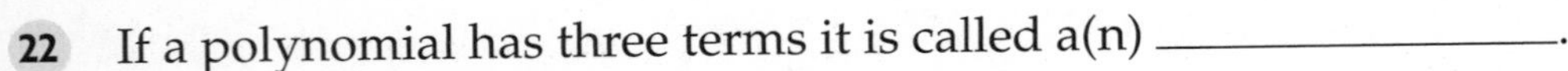

**Vocabulary Check** **Write the vocabulary word that completes each sentence.**

21 A number, such as 25, that is the square of another number is a(n) ________________.

22 If a polynomial has three terms it is called a(n) ________________.

23 **Reflect** Create a perfect square trinomial and write an explanation on how you would factor it for someone who has never learned how to factor a perfect square trinomial.

________________________________

________________________________

STOP

# Chapter 4 Progress Check 4 (Lessons 4-7 and 4-8)

**Factor each difference of squares.**

1 $4x^2 - 9 =$ __________

2 $9y^2 - 100 =$ __________

3 $x^2 - 144 =$ __________

4 $y^2 - 16 =$ __________

5 $27x^2 - 147 =$ __________

6 $4y^2 - 324 =$ __________

**Factor each perfect square trinomial.**

7 $25m^2 - 50m + 25 =$ __________

8 $36x^2 + 120x + 100 =$ __________

9 $4x^2 - 48x + 144 =$ __________

10 $64x^2 + 144x + 81 =$ __________

11 $4c^2 + 32c + 64 =$ __________

12 $80b^2 - 280b + 245 =$ __________

**Solve.**

13 **PICTURE FRAME** Write an expression that models the area of the frame around the picture as shown below. Factor this expression.

__________

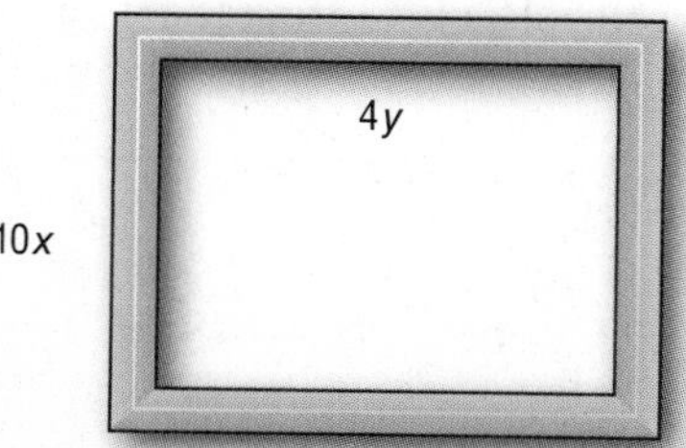

Chapter 4

# Chapter Test

**Classify each polynomial. Find its degree.**

**1** $4g^2h^3$ ____________

**2** $9 - x^8$ ____________

**3** $f^4d^2 - d^7$ ____________

**4** $3a^2b + a^2 - 8b$ ____________

**5** $5x$ ____________

**6** $x^3 - 9$ ____________

**Find each sum or difference.**

**7** $(12x^2 - 5x) - (6x^2 + 2x - 10) =$ ____________

**8** $(4n^3 + n^2 + 3n) + (4n^3 - 3n^2 + 2) =$ ____________

**Find each product.**

**9** $(s - 7)(3s + 2) =$ ____________

**10** $-2x(x^2 - 9) =$ ____________

**11** $(g + 1)(g - 6) =$ ____________

**12** $(5x - 4)(2x - 3) =$ ____________

**Factor each polynomial.**

**13** $25p^3 + 40p^2 =$ ____________

**14** $3w + 21w^2 - 30 =$ ____________

**Factor each trinomial.**

**15** $y^2 + 15y + 54 =$ ____________

**16** $w^2 - 6w - 55 =$ ____________

**17** $a^2 - 16a + 28 =$ ____________

**18** $6q^2 - 29q + 35 =$ ____________

**19** $14b^2 + 23b - 15 =$ ____________

**20** $18x^2 - 45x + 18 =$ ____________

**21** $12v^2 + 64v + 20 =$ ____________

**22** $4x^2 - 4x + 1 =$ ____________

**23** $2g^2 - 4g - 48 =$ ____________

**Factor each difference of squares.**

24 $25x^2 - 400 =$ __________

25 $10y^2 - 360 =$ __________

26 $4x^2 - 9 =$ __________

27 $9y^2 - 1 =$ __________

**Factor each perfect square trinomial.**

28 $x^2 - 24x + 144 =$ __________

29 $x^2 + 4x + 4 =$ __________

30 $4m^2 + 32m + 64 =$ __________

31 $36x^2 - 60x + 25 =$ __________

**Solve.**

32 **SPORTS** The number of boys in the community soccer league is modeled by the expression $14t + 156$, where $t$ is the number of teams in the league. The number of girls is modeled by one-half the number of boys plus 45. Write a simplified expression to show the total children in the community soccer league.

__________

__________

**Correct the mistake.**

33 George was given the problem $(n + 1)(n - 1) = 224$ and was told to use logic to determine the value of $n$. His response is given below. What did George forget to do?

> The expression $(n + 1)(n - 1)$ is the difference of the two perfect squares, $n^2$ and 1. So, $(n + 1)(n - 1) = n^2 - 1 = 224$. Therefore, $n^2 = 224 + 1 = 225$, so, $n = 225$.

__________

__________

STOP

# Chapter 5 Probability

## *Who will win the race?*

In a political race with 6 candidates, you can use permutations to calculate how many ways the candidates can finish in first, second, and third place.

**STEP 1 Chapter Pretest** Are you ready for Chapter 5? Take the Chapter 5 Pretest to find out.

**STEP 2 Preview** Get ready for Chapter 5. Review these skills and compare them with what you will learn in this chapter.

| What You Know | What You Will Learn |
|---|---|
| You know how to multiply.<br>**Example:** $4 \cdot 3 \cdot 2 \cdot 1 = 24$<br>**TRY IT!**<br>1 $6 \cdot 5 \cdot 4 \cdot 3 \cdot 2 \cdot 1 =$ ______<br>2 $7 \cdot 6 \cdot 5 \cdot 4 \cdot 3 \cdot 2 \cdot 1 =$ ______ | *Lesson 5-1*<br>A factorial is a product of all counting numbers starting with $n$ and counting backward to 1.<br>**Example:** $5! = 5 \cdot 4 \cdot 3 \cdot 2 \cdot 1$<br>$= 120$ |
| You know that if a coin is flipped, the probability of getting heads is $\frac{1}{2}$.<br>  | *Lesson 5-3*<br>Suppose a coin is flipped twice. What is the probability of getting two heads?<br>If $A$ and $B$ are independent events,<br>$P(A \text{ and } B) = P(A) \cdot P(B) = \frac{1}{2} \cdot \frac{1}{2} = \frac{1}{4}$. |
| You know how to find area.<br>**Example:**<br>Find the area of the rectangle.<br>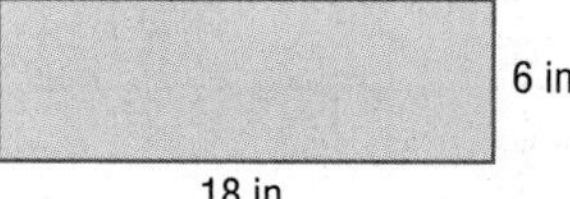<br>$A = \ell w$<br>$A = 18 \cdot 6$<br>$A = 108$<br>The area is 108 square inches. | *Lesson 5-5*<br>You can use the area of figures to find geometric probability.<br><br>If a point in region $A$ is chosen at random, then $P(B)$ that the point is in region $B$ is<br>$P(B) = \frac{\text{area of region } B}{\text{area of region } A}$. |

Lesson 5-1

# Permutations

## KEY Concept

In situations that involve arrangements for which order is important, use the factorial products and the permutation formula. One example might be picking an order for a baseball lineup.

An exclamation mark is used for factorials. It indicates a special multiplication operation.

$$n! = n(n-1)(n-2)\dots 1$$

$$4! = 4(4-1)(4-2)(4-3)$$

$$= 4 \cdot 3 \cdot 2 \cdot 1$$

$$= 24$$

When counting the number of permutations possible for a situation, use the formula below.

**Permutation Formula**

$${}_nP_r = \frac{n!}{(n-r)!}$$

$${}_5P_2$$

- 2: number of factors or arrangements
- 5: number of objects from which to choose

### VOCABULARY

**factor**
a number that is multiplied by another number

**factorial**
a product of all counting numbers starting with $n$ and counting backward to 1

**permutation**
an arrangement of objects in which order is important

## Example 1

**Simplify 4!.**

1. Rewrite 4! as a factorial product.

   $4! = 4 \cdot 3 \cdot 2 \cdot 1$

2. Simplify.

   $4! = 4 \cdot 3 \cdot 2 \cdot 1$

   $= 24$

### YOUR TURN!

**Simplify (7 – 2)!.**

1. Rewrite (7 – 2)! as a factorial product.

   $(7-2)! = 5! =$ ____________

2. Simplify.

   $5! =$ ____________

   $=$ ____________

## Example 2

**Find $_5P_3$.**

1. Write $n$ and $r$ values. $n = 5$ $r = 3$
2. Substitute $n$ and $r$ values into the formula.

$$_5P_3 = \frac{5!}{(5-3)!}$$
$$= \frac{5 \cdot 4 \cdot 3 \cdot \cancel{2 \cdot 1}}{\cancel{2 \cdot 1}}$$
$$= 5 \cdot 4 \cdot 3$$

3. Simplify.

$5 \cdot 4 \cdot 3 = 60$

$_5P_3 = 60$

**YOUR TURN!**

**Find $_8P_4$.**

1. Write $n$ and $r$ values. $n =$ ____ $r =$ ____
2. Substitute $n$ and $r$ values into the formula.

$$_{\square}P_{\square} = \frac{\square!}{(\square - \square)!}$$
$$= \frac{\square}{\square}$$
$$= \_\_\_ \cdot \_\_\_ \cdot \_\_\_ \cdot \_\_\_$$

3. Simplify.

____ • ____ • ____ • ____ = ____

$_8P_4 =$ ____

## Example 3

**How many ways can six members of a track team be chosen for a four-member relay race?**

1. Write $n$ and $r$ values. $n = 6$ $r = 4$
2. Substitute $n$ and $r$ values into the formula.

$$_6P_4 = \frac{6!}{(6-4)!}$$
$$= \frac{6 \cdot 5 \cdot 4 \cdot 3 \cdot \cancel{2 \cdot 1}}{\cancel{2 \cdot 1}}$$
$$= 6 \cdot 5 \cdot 4 \cdot 3$$

3. Simplify. $6 \cdot 5 \cdot 4 \cdot 3 = 360$
4. There are 360 ways to choose 4 members for a four-member relay.

**YOUR TURN!**

**How many three-digit locker combinations can be made from the digits 1, 3, 4, 5, 7, 8, and 9?**

1. Write $n$ and $r$ values. $n =$ ____ $r =$ ____
2. Substitute $n$ and $r$ values into the formula.

$$_{\square}P_{\square} = \frac{\square!}{(\square - \square)!}$$
$$= \frac{\square}{\square}$$
$$= \_\_\_ \cdot \_\_\_ \cdot \_\_\_$$

3. Simplify. ____ • ____ • ____ = ____
4. There are ______ ways to choose a 3-digit locker combination.

## Guided Practice

**Simplify each factorial.**

**1** 9! = ____ • ____ • ____ • ____ • ____ • ____ • ____ • ____ • ____

= ______

**2** (29 − 26)! = ____!

= ____ • ____ • ____

= ______

**Simplify each factorial.**

3 $7! =$ ______________

$=$ ________

4 $(18 - 14)! =$ ____!

$=$ ____ $\cdot$ ____ $\cdot$ ____ $\cdot$ ____

$=$ ________

## Step by Step Practice

5 Find ${}_{10}P_7$.

**Step 1** Write $n$ and $r$ values. $n =$ ____ $r =$ ____

**Step 2** Substitute $n$ and $r$ values into the formula.

$${}_{\square}P_{\square} = \frac{\square!}{(\square - \square)!}$$

$$= \frac{\square}{\square}$$

$=$ ____ $\cdot$ ____ $\cdot$ ____

**Step 3** Simplify. ____ $\cdot$ ____ $\cdot$ ____ $=$ ____________

**Step 4** ${}_{10}P_3 =$ ____________

**Find each permutation.**

6 ${}_{8}P_{4}$

$n =$ ____ $r =$ ____

${}_{8}P_{4} = \frac{\square!}{(\square - \square)!} =$ ______________________ $=$ ____ $\cdot$ ____ $\cdot$ ____ $\cdot$ ____ $=$ ________

7 ${}_{12}P_{2}$

$n =$ ____ $r =$ ____

${}_{12}P_{2} = \frac{\square!}{(\square - \square)!} =$ ______________________________

$=$ ____ $\cdot$ ____ $=$ ________

## Step by Step Problem-Solving Practice

**Solve.**

**8** **STUDENT ELECTIONS** There are 20 students on the student council. Four students will be chosen from the council to hold the offices of President, Vice-President, Secretary, and Treasurer. Assuming no one person may hold more than one office, how many different possibilities of officers are there?

$_{\square}P_{\square} = \dfrac{\square!}{(\square - \square)!} = $ ______________________ = __________

Check off each step.

_____ **Understand: I underlined key words.**

_____ **Plan: To solve the problem, I will** ______________________.

_____ **Solve: The answer is** ______________________.

_____ **Check: I checked my answer by** ______________________.

## Skills, Concepts, and Problem Solving

**Simplify each factorial.**

**9** 2! = __________

**10** (9 − 3)! = __________

**11** 8! = __________

**12** (18 − 15)! = __________

**13** (11 − 10)! = __________

**14** 4! = __________

**Find each permutation.**

**15** $_{8}P_{4}$ = __________

**16** $_{6}P_{4}$ = __________

**17** $_{7}P_{6}$ = __________

**18** $_{5}P_{3}$ = __________

**19** $_{15}P_{5}$ = __________

**20** $_{100}P_{2}$ = __________

GO ON

**Solve.**

21 **CODES** Andrew is creating a 4-digit garage code from the ten digits 0 through 9. How many different combinations are possible if he does not repeat any digits in the code?

_______________

22 **PUBLIC SPEAKING** The twelve students in Mrs. Franco's speech class are giving informative speeches, beginning on Friday. Only five of the students will have time to speak on Friday. The remaining students in the class will speak on Monday. They are choosing the order in which they will speak by a drawing names. How many different orders of speakers are possible on Friday?

_______________

23 **CHILD DEVELOPMENT** A preschooler is playing with the blocks below. The child lines the blocks up in a row. In how many different ways can the child line up the blocks?

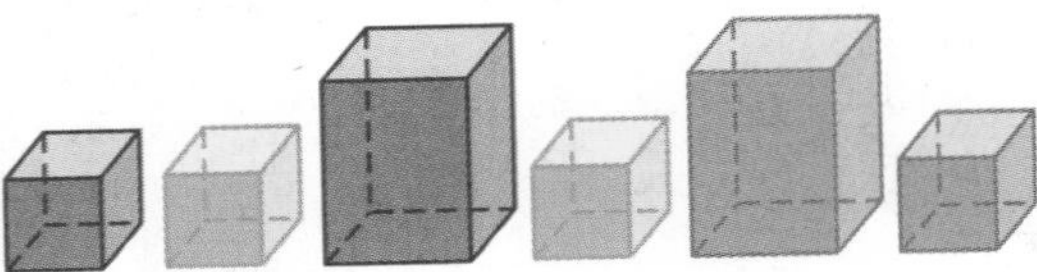

_______________

**Vocabulary Check** **Write the vocabulary word that completes each sentence.**

24 A(n) _______________ is a number that is multiplied by another number.

25 An arrangement of objects in which the order is important is a(n) _______________.

26 The symbol ! is shown to represent a(n) _______________, which is a product of all counting numbers starting with *n* counting backward to 1.

27 **Reflect** Which is greater, 6! or ${}_6P_5$? Explain.

_______________

_______________

STOP

Lesson 5-2

# Combinations

## KEY Concept

In situations that involve arrangements for which order is **not** important, use the factorial products and the combination formula. One example might be picking a committee at a meeting.

When counting the number of combinations possible for a situation, use the formula below.

**Combination Formula**

$$_nC_r = \frac{n!}{(n-r)!r!}$$

$_4C_2$

number of factors or arrangements

number of objects from which to choose

### VOCABULARY

**combination**
an arrangement of objects in which order is not important

**factor**
a number that is multiplied by another number

**factorial**
a product of all counting numbers starting with $n$ counting backward to 1

**permutation**
an arrangement of objects in which order is important

## Example 1

**Find $_7C_4$.**

1. Write $n$ and $r$ values.
   $n = 7 \quad r = 4$
2. Substitute $n$ and $r$ values into the formula.
   $$_7C_4 = \frac{7!}{(7-4)!4!}$$
   $$= \frac{7 \cdot 6 \cdot 5 \cdot \cancel{4} \cdot \cancel{3} \cdot \cancel{2} \cdot \cancel{1}}{3 \cdot 2 \cdot 1 \cdot \cancel{4} \cdot \cancel{3} \cdot \cancel{2} \cdot \cancel{1}}$$
3. Simplify.
   $$\frac{7 \cdot \cancel{6} \cdot 5}{\cancel{3} \cdot \cancel{2} \cdot 1} = \frac{7 \cdot 5}{1} = 35$$

### YOUR TURN!

**Find $_9C_6$.**

1. Write $n$ and $r$ values.
   $n =$ _____ $r =$ _____
2. Substitute $n$ and $r$ values into the formula.
   $$_{\square}C_{\square} = \frac{\square!}{(\square - \square)!\square!}$$
   $$= \frac{\square}{\square}$$
3. Simplify.
   $$\frac{\square}{\square} = \frac{\square}{\square} = \_\_\_\_$$

GO ON

## Example 2

**There are 10 players on a basketball team. Without regard to positions, how many ways can five players be chosen for a starting line up?**

1. Write $n$ and $r$ values. $n = 10$ $r = 5$
2. Substitute $n$ and $r$ values into the formula.

$$_{10}C_5 = \frac{10!}{(10-5)!5!}$$

$$= \frac{10 \cdot 9 \cdot 8 \cdot 7 \cdot 6 \cdot \cancel{5} \cdot \cancel{4} \cdot \cancel{3} \cdot \cancel{2} \cdot \cancel{1}}{5 \cdot 4 \cdot 3 \cdot 2 \cdot 1 \cdot \cancel{5} \cdot \cancel{4} \cdot \cancel{3} \cdot \cancel{2} \cdot \cancel{1}}$$

3. Simplify. $\frac{\overset{2}{\cancel{10}} \cdot \overset{3}{\cancel{9}} \cdot \cancel{8} \cdot 7 \cdot 6}{\cancel{5} \cdot \cancel{4} \cdot \cancel{3} \cdot \cancel{2} \cdot 1} = \frac{2 \cdot 3 \cdot 7 \cdot 6}{1} = 252$
4. There are 252 ways to choose 5 players for a starting line up.

**YOUR TURN!**

**There are 12 members of a committee. Without regard for the office, how many ways can a three-member subcommittee be selected?**

1. Write $n$ and $r$ values. $n =$ ______ $r =$ ______
2. Substitute $n$ and $r$ values into the formula.

$$_{\square}C_{\square} = \frac{\square!}{(\square - \square)!\square!} = \frac{\square \cdot 11 \cdot 10 \cdot \cancel{9} \cdot \cancel{8} \cdot \cancel{7} \cdot \cancel{6} \cdot \cancel{5} \cdot \cancel{4} \cdot \cancel{3} \cdot \cancel{2} \cdot \cancel{1}}{\cancel{9} \cdot \cancel{8} \cdot \cancel{7} \cdot \cancel{6} \cdot \cancel{5} \cdot \cancel{4} \cdot \cancel{3} \cdot \cancel{2} \cdot \cancel{1} \cdot \square \cdot 2 \cdot 1}$$

3. Simplify. $\frac{\square}{\square} = \frac{\square}{\square} =$ ______
4. There are ______ ways to choose three members.

## Guided Practice

**Set up the formula for each combination.**

**1** $_5C_3$

$$= \frac{\square!}{(\square - \square)!\square!}$$

= ____________________

**2** $_9C_4$

$$= \frac{\square!}{(\square - \square)!\square!}$$

= ____________________

## Step by Step Practice

**3** Find ${}_8C_3$.

**Step 1** Write the values for $n$ and $r$. $n =$ ______ $r =$ ______

**Step 2** Substitute. ${}_8C_3 = \dfrac{\square!}{(\square - \square)!\square!} =$ ______

**Step 3** Simplify. $\dfrac{\square}{\square} = \dfrac{\square}{\square} =$ ______

**Find each combination.**

**4** ${}_8C_6 = \dfrac{\square!}{(\square - \square)!\square!}$

$= \dfrac{\square}{\square} =$ ______

**5** ${}_6C_2 = \dfrac{\square!}{(\square - \square)!\square!}$

$= \dfrac{\square}{\square} =$ ______

**6** $\dfrac{6!}{(6-4)!4!} =$ ______

**7** $\dfrac{9!}{(9-3)!3!} =$ ______



## Step by Step Problem-Solving Practice

**Solve.**

**8** **BASKETBALL** Of the twelve players on the basketball team, only five can play at a time. In how many ways can the different players be chosen?

${}_nC_r = {}_{12}C_5 = \dfrac{12!}{(12-5)!5!} = \dfrac{\square}{\square}$

$= \dfrac{\square}{\square} =$ ______

Check off each step.

______ **Understand: I underlined key words.**

______ **Plan: To solve the problem, I will** ______________________________.

______ **Solve: The answer is** ______________________________.

______ **Check: I checked my answer by** ______________________________.

## Skills, Concepts, and Problem Solving

**Find each combination.**

9 $_{8}C_{3}$ = ____________

10 $_{12}C_{4}$ = ____________

11 $_{20}C_{2}$ = ____________

12 $_{5}C_{3}$ = ____________

13 $_{10}C_{8}$ = ____________

14 $_{18}C_{5}$ = ____________

**Solve.**

15 **GARDENS** A landscaper can choose from 9 varieties of rose bushes. In how many different ways can she choose 4 bushes?

____________________________

16 **FOOD** How many different five-topping pizzas can be created from the toppings listed below?

| | | | | |
|---|---|---|---|---|
| feta | sausage | mushrooms | pineapple | bacon |
| pepperoni | onions | green peppers | ham | olives |
| anchovies | extra cheese | spinach | roasted garlic | banana peppers |

____________________________

17 **SPORTS** Three girls are selected from the 18 members of the varsity softball team to be captains. How many different groups of captains could be selected?

____________________________

**Vocabulary Check** **Write the vocabulary word that completes each sentence.**

18 A(n) ____________ is an arrangement in which the order of the objects is important.

19 If the order of the objects in an arrangement is not important, then it is a(n) ____________.

20 **Reflect** A teacher has 15 stickers, 12 erasers, and 7 tokens in a prize box. Each student will pick 3 prizes to keep. Is this a permutation or a combination?

____________________________

____________________________

STOP

Chapter 5

# Progress Check 1 (Lessons 5-1 and 5-2)

**Simplify each factorial.**

1 $(13 - 9)! =$ ____________  2 $(8 - 2)! =$ ____________  3 $7! =$ ____________

**Find each permutation.**

4 ${}_6P_3 =$ ____________  5 ${}_9P_7 =$ ____________  6 ${}_7P_4 =$ ____________

7 ${}_8P_2 =$ ____________  8 ${}_5P_4 =$ ____________  9 ${}_9P_5 =$ ____________

**Find each combination.**

10 ${}_9C_3 =$ ____________  11 ${}_{10}C_6 =$ ____________  12 ${}_{20}C_4 =$ ____________

13 ${}_{15}C_4 =$ ____________  14 ${}_4C_1 =$ ____________  15 ${}_{12}C_5 =$ ____________

**Solve.**

16 **PHOTOS** Anita can choose a package of senior photos from 12 poses taken by the photographer. In how many different ways can Anita choose 6 of the photos?

____________________

17 **SPORTS** There are 8 runners competing in the mile run at a track meet. How many different ways can 8 runners finish in first, second, or third place?

____________________

18 **CODES** Mae has to choose a 3-digit personal identification number (PIN) to access her voicemail. She has 10 digits, 0 through 9, from which to choose. How many different combinations are possible if she does not repeat any digits in the PIN?

____________________

Lesson 5-3

# Multiply Probabilities

## KEY Concept

Probability sometimes involves a situation in which two or more events occur. In calculating probability in these types of situations, you need to know if the two events are independent of each other or if the second event is dependent on the first event.

### Probability of Two Independent Events

If $A$ and $B$ are independent events,

$$P(A \text{ and } B) = P(A) \cdot P(B).$$

Suppose a coin is flipped and a number from 1–3 is picked at random. What is the probability of getting heads and a number greater than 1?

There is 1 heads and 2 sides of the coin, so $P(A) = \frac{1}{2}$.

There are 2 numbers greater than one and 3 total numbers, so $P(B) = \frac{2}{3}$.

$P(A \text{ and } B) = \frac{1}{2} \cdot \frac{2}{3} = \frac{2}{6} = \frac{1}{3}$

Multiply the numerators. Multiply the denominators. Simplify if possible.

### Probability of Two Dependent Events

If $A$ and $B$ are dependent events,

$$P(A \text{ then } B) = P(A) \cdot P(B \text{ after } A).$$

Suppose the letters in MATHEMATICS are placed in a box. What is the probability of picking a T, followed by an A?

There are 2 Ts and 11 total letters, so $P(A) = \frac{2}{11}$.

There are 2 As, and only 10 letters in the box, so $P(B \text{ after } A) = \frac{2}{10}$

$P(A \text{ and } B) = \frac{2}{11} \cdot \frac{2}{10} = \frac{4}{110} = \frac{2}{55}$

Notice the denominator of the second probability is reduced by 1.

**VOCABULARY**

**compound events**
events that involve two or more simple events

**dependent events**
events in which the outcome of one event affects the outcome of the other events

**independent events**
events in which the outcome of one event does not affect the outcome of the other events

In the case of independent or dependent events, find the probability of each event and then multiply the fractions.

## Example 1

**One red and one blue number cube both with numbers 1–6 are rolled. What is the probability you will roll a 4 on the red cube and roll an odd number on the blue cube?**

1. Are the events independent or dependent? **independent**

2. Find $P$(red 4). There is 1 four and 6 sides.

   $\frac{1}{6}$

3. Find $P$(blue odd). There are 3 odd numbers and 6 sides.

   $\frac{3}{6}$

4. Find $P$(red 4 and blue odd).

   $\frac{1}{6} \cdot \frac{3}{6} = \frac{3}{36} = \frac{1}{12}$

**YOUR TURN!**

**There are 4 red marbles, 5 blue marbles, and 1 green marble in a bag. What is the probability you draw a red marble, and after replacing the marble, draw a green marble?**

1. Are the events independent or dependent? ________________

2. Find $P$(red). There are ______ red marbles and ______ total marbles.

   ______

3. Find $P$(green). There is ______ green marble and ______ total marbles.

   ______

4. Find $P$(red and green).

   ______ • ______ = ______ = ______



## Example 2

**There are 6 yellow tiles, 4 pink tiles, and 5 blue tiles in a drawer. What is the probability you draw a pink tile, and without replacing the tile, draw a blue tile?**

1. Are the events independent or dependent? **dependent**

2. Find $P$(pink). $\frac{\text{4 pink tiles}}{\text{15 total tiles}}$

3. Find $P$(blue after pink). $\frac{\text{5 blue tiles}}{\text{14 total tiles}}$

4. Find $P$(pink then blue).

   $\frac{4}{15} \cdot \frac{5}{14} = \frac{20}{210} = \frac{2}{21}$

**YOUR TURN!**

**There are 8 purple chips, 3 orange chips, and 9 white chips in a bag. What is the probability you draw a white chip, and without replacing the chip, draw a purple chip?**

1. Are the events independent or dependent?

   ________________

2. Find $P$(white). $\frac{\square \text{ white chips}}{\square \text{ total chips}}$

3. Find $P$(purple after white). $\frac{\square \text{ purple chips}}{\square \text{ total chips}}$

4. Find $P$(white then purple).

   ______ • ______ = ______ = ______

## Guided Practice

**Find each probability.**

1. $P(\text{blue}) = \frac{4}{7}$; $P(\text{green}) = \frac{2}{7}$

   Find $P$(blue and green).

   This is a(n) ______________ probability.

   $P$(blue and green)

   $= P(\text{blue}) \cdot P(\text{green}) = \frac{\square}{\square} \cdot \frac{\square}{\square} = \frac{\square}{\square}$

2. $P(\text{odd}) = \frac{4}{9}$; $P(\text{even after odd}) = \frac{5}{8}$

   Find $P$(odd then even).

   This is a(n) ______________ probability.

   $P$(odd then even)

   $= P(\text{odd}) \cdot P(\text{even after odd})$

   $= \frac{\square}{\square} \cdot \frac{\square}{\square} = \frac{\square}{\square} = \frac{\square}{\square}$

## Step by Step Practice

3. There are 8 markers in a bag: 3 blue, 4 orange, and 1 green. What is the probability that you draw an orange marker, and after replacing the orange marker, draw a green marker?

**Step 1** Are the events independent or dependent? ______________

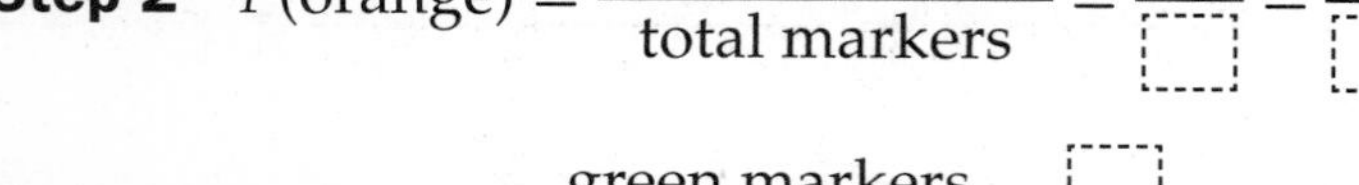

**Step 2** $P(\text{orange}) = \frac{\text{orange markers}}{\text{total markers}} = \frac{\square}{\square} = \frac{\square}{\square}$

**Step 3** $P(\text{green}) = \frac{\text{green markers}}{\text{total markers}} = \frac{\square}{\square}$

**Step 4** $P(\text{orange then green}) = \frac{\square}{\square} \cdot \frac{\square}{\square} = \frac{\square}{\square}$

**Find each probability.**

4. A jar contains 7 purple marbles, 5 white marbles, and 8 black marbles. What is the probability that a black marble will be drawn, and without replacement, another black marble will be drawn?

   The events are ______________.

   $P(\text{black}) =$ ________

   $P(\text{black after black}) =$ ________

   $P(\text{black then black}) =$ ________

5. Each number 1 through 15 is on a spinner. What is the probability that the number 8 will be spun, and then an odd number will be spun?

   The events are ______________.

   $P(8) =$ ________

   $P(\text{odd}) =$ ________

   $P(8 \text{ and odd}) =$ ________

**Find each probability.**

6. A green number cube and a yellow number cube are rolled. What is the probability that the green cube will show an even number, and the yellow cube will show an even number.

   The events are ____________________.

   $P$(green even) = ________

   $P$(yellow even) = ________

   $P$(green even and yellow even) = ________

7. Three pink pencils, 4 red pencils, and 5 purple pencils are in a desk. What is the probability that a pink pencil will be taken out, and without replacement, a purple pencil will be taken out?

   The events are ____________________.

   $P$(pink) = ________

   $P$(purple after pink) = ________

   $P$(pink then purple) = ________

## Step by Step Problem-Solving Practice

**Solve.**

8. **TEAM CAPTAIN** There are 8 girls and 10 boys in the freshman math league at Taylor High School. Each week, two team captains are chosen randomly. Find the probability that both captains will be boys.

   $P$(first captain is boy) = ________ = ________

   $P$(second captain is boy) = ________

   $P$(boy then boy) = ________ = ________

   Check off each step.

   ______ **Understand: I underlined key words.**

   ______ **Plan: To solve the problem, I will** ________________________________.

   ______ **Solve: The answer is** ________________________________.

   ______ **Check: I checked my answer by** ________________________________

   ________________________________________________.

GO ON

## Skills, Concepts, and Problem Solving

**A bag contains 4 red tiles, 7 pink tiles, 3 white tiles, 5 green tiles, and 1 black tile. Find the probability of each of the following.**

9 A green tile is drawn, and without replacing it, a white tile is drawn.

____________________

10 A red tile is drawn, and after replacing it, another red tile is drawn.

____________________

11 A pink tile is drawn, and without replacing it, another pink tile is drawn.

____________________

12 A black tile is drawn, and without replacing it, another black tile is drawn.

____________________

**Solve. Write the answer in simplest form.**

13 **ELECTRONICS** Three of the 24 computers that Forever Electronics has in stock are damaged. If the store sells 2 computers, what is the probability that both were damaged?

____________________

14 **SURVEYS** Marcus is randomly surveying his classmates about their pets' health. In his class, 10 of the 16 girls have pets and 3 of the 10 boys have pets. What is the probability that the first girl and the first boy surveyed do not have pets?

____________________

**Vocabulary Check** **Write the vocabulary word that completes each sentence.**

15 Events in which the outcome of one event affects the outcome of the other events are ____________________.

16 ____________________ are events in which the outcome of one event does not affect the outcome of the other events.

17 **Reflect** A team of soccer players randomly chose from a variety of drinks in a cooler. Would the probability of players taking a certain type of drink be considered independent or dependent events? Why?

________________________________________

________________________________________

# Lesson 5-4 Add Probabilities

## KEY Concept

Probability sometimes involves a situation in which two or more events occur. Sometimes those events can occur at the same time and sometimes they cannot. Events that cannot occur at the same time are **mutually exclusive** events. Events that can occur at the same time are **inclusive** events.

### Probability of Mutually Exclusive Events

If $A$ and $B$ are mutually exclusive events,

$$P(A \text{ or } B) = P(A) + P(B).$$

Suppose Rita chooses a dog or a cat at a pet store. This is a mutually exclusive situation because Rita cannot choose a pet that is both a dog and a cat.

To find the probability, add the probability that she chooses a dog to the probability that she chooses a cat.

$$P(\text{cat or dog}) = P(\text{cat}) + P(\text{dog})$$

### Probability of Inclusive Events

If $A$ and $B$ are inclusive events,

$$P(A \text{ or } B) = P(A) + P(B) - P(A \text{ and } B).$$

Suppose Rita chooses a pet that is both young and white. This is an inclusive event because Rita can choose a young white pet that is either a dog or cat.

To find the probability, add the probability that she chooses a young pet to the probability that she chooses a white pet. Then, subtract the probability that she chooses a young **and** white pet.

$$P(\text{young or white}) = P(\text{young}) + P(\text{white}) - P(\text{young and white})$$

**VOCABULARY**

**compound events**
events that involve two or more simple events

**inclusive**
events that can occur at the same time

**mutually exclusive**
events that cannot occur at the same time

To simplify the probabilities you will need to rewrite the fractions so that they have common denominators. Write answers that are fractions in simplest form.



GO ON

## Example 1

**A jar contains 3 red buttons, 5 blue buttons, and 2 green buttons. If one button is selected from the jar, what is the probability that it is red or blue?**

1. Is this event mutually exclusive or inclusive?

   mutually exclusive

   Can a button be both red and blue?

2. Find $P$(red). $\frac{\text{3 red buttons}}{\text{10 total buttons}}$

3. Find $P$(blue). $\frac{\text{5 blue buttons}}{\text{10 total buttons}}$

   $P(\text{red or blue}) = P(\text{red}) + P(\text{blue})$

   $= \frac{3}{10} + \frac{5}{10} = \frac{8}{10} = \frac{4}{5}$

### YOUR TURN!

**A bag contains 4 white softballs, 5 yellow softballs, and 3 green softballs. If one softball is selected from the bag, what is the probability that it is white or yellow?**

1. Is this event mutually exclusive or inclusive?

   ____________

2. Find $P$(white). $\frac{\square \text{ white softballs}}{\square \text{ total softballs}}$

3. Find $P$(yellow). $\frac{\square \text{ yellow softballs}}{\square \text{ total softballs}}$

   $P(\text{white or yellow}) = P(\text{white}) + P(\text{yellow})$

   = _____ + _____ = ______

## Example 2

**Marcie has 2 white t-shirts, 4 blue t-shirts, 5 white dress shirts, and 4 blue dress shirts. What is the probability Marcie will pick a white shirt or a dress shirt?**

1. Is this a mutually exclusive event or an inclusive event?

   inclusive

   Can a shirt be both white and a dress shirt?

2. Find $P$(white). $\frac{7}{15}$

3. Find $P$(dress). $\frac{9}{15}$

4. Find $P$(white and dress). $\frac{5}{15}$

   $P(\text{white or dress})$

   $= P(\text{white}) + P(\text{dress}) - P(\text{white and dress})$

   $= \frac{7}{15} + \frac{9}{15} - \frac{5}{15} = \frac{11}{15}$

### YOUR TURN!

**Nate has 4 black sweaters, 2 blue sweaters, 3 black shirts, and 1 blue shirt. What is the probability Nate will pick a blue top or a shirt?**

1. Is this a mutually exclusive event or an inclusive event?

   ____________

2. Find $P$(blue). ______

3. Find $P$(shirt). ______

4. Find $P$(blue and shirt). ______

   $P(\text{blue or shirt})$

   $= P(\text{blue}) + P(\text{shirt}) - P(\text{blue and shirt})$

   = _____ + _____ − _____ = _____

## Guided Practice

**Find the probability of each mutually exclusive event.**

**1** $P(\text{green}) = \frac{3}{5}$

$P(\text{yellow}) = \frac{1}{5}$

$P(\text{green or yellow})$

$= \frac{\square}{\square} + \frac{\square}{\square} = \frac{\square}{\square}$

**2** $P(\text{short-sleeved}) = \frac{4}{14}$

$P(\text{long-sleeved}) = \frac{8}{14}$

$P(\text{short-sleeved or long sleeved})$

$= \frac{\square}{\square} + \frac{\square}{\square} = \frac{\square}{\square} = \frac{\square}{\square}$

### Step by Step Practice

**3** A bag contains children's blocks. There are 7 blue striped blocks, 6 green striped blocks, 3 blue spotted blocks, and 5 green spotted blocks. If one block is selected from the bag, what is the probability that it will be blue or striped?

**Step 1** Is this a mutually exclusive event or an inclusive event?

____________________

**Step 2** Find $P(\text{blue})$. ________

**Step 3** Find $P(\text{striped})$. ________

**Step 4** Find $P(\text{blue and striped})$. ________

$P(\text{blue or striped})$ = ________ + ________ − ________ = ________

**Find the probability of each inclusive event.**

**4** $P(\text{pink}) = \frac{8}{17}$

$P(\text{circle}) = \frac{10}{17}$

$P(\text{pink and circle}) = \frac{6}{17}$

$P(\text{pink or circle})$

$= \frac{\square}{\square} + \frac{\square}{\square} - \frac{\square}{\square} = \frac{\square}{\square}$

**5** $P(\text{red}) = \frac{7}{24}$

$P(\text{shorts}) = \frac{8}{24}$

$P(\text{red and shorts}) = \frac{2}{24}$

$P(\text{red or shorts})$

$= \frac{\square}{\square} + \frac{\square}{\square} - \frac{\square}{\square} = \frac{\square}{\square}$

GO ON

**Find each probability.**

**6** The probability of a white marble being pulled from a bag is $\frac{3}{8}$. The probability of a black marble being pulled from a bag is $\frac{2}{8}$. Find the probability of a white or a black marble being pulled from the bag.

Is this event mutually exclusive or inclusive?

____________________

$P$(black) = ______

$P$(white) = ______

$P$(black or white) = ____________

**7** A pencil case contains 3 black pens, 2 black crayons, 4 red pens, and 1 red crayon. If one item is selected from the bag, what is the probability that it will be red or a pen?

Is this event mutually exclusive or inclusive?

____________________

$P$(red) = ______

$P$(pen) = ______

$P$(red and pen) = ______

$P$(red or pen) =

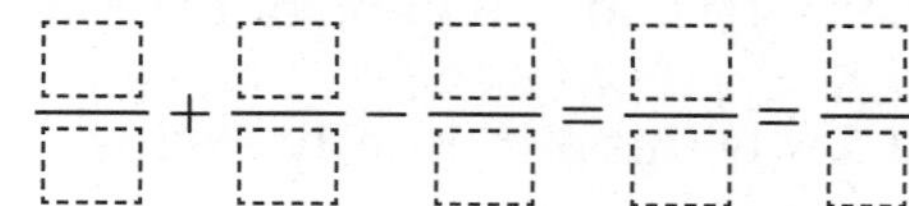

**8** A group of cards with pictures of shapes are placed in a stack. There are 5 cards with circles, 7 cards with squares, 3 cards with hearts, and 10 cards with triangles. If one card is selected from the stack, what is the probability that it will have a heart or circle?

Is this event mutually exclusive or inclusive?

____________________

$P$(heart) = ______

$P$(circle) = ______

$P$(heart or circle) = ____________

**9** A drawer contains 6 white shirts, 4 blue shirts, and 8 black shirts. If one item is picked from the drawer, what is the probability that it will be a black shirt or a white shirt?

Is this event mutually exclusive or inclusive?

____________________

$P$(black shirt) = ______

$P$(white shirt) = ______

$P$(black shirt or white shirt) =

____________________

## Step by Step Problem-Solving Practice

**Solve.**

**10** SURVEY Gloria surveyed 26 students. Eight of the students preferred chocolate ice cream and nine preferred vanilla ice cream. Seven of the students preferred chocolate frozen yogurt and two preferred vanilla frozen yogurt. If one student is chosen at random, what is the probability that the student prefers chocolate or frozen yogurt?

Is this event mutually exclusive or inclusive? ____________________

$P$(chocolate) = ________ $P$(yogurt) = ________

$P$(chocolate and yogurt) = ________ $P$(chocolate or yogurt)

$$= \frac{\square}{\square} + \frac{\square}{\square} - \frac{\square}{\square} = \frac{\square}{\square}$$

Check off each step.

______ **Understand: I underlined key words.**

______ **Plan: To solve the problem, I will** ____________________.

______ **Solve: The answer is** ____________________.

______ **Check: I checked my answer by** ____________________.

## Skills, Concepts, and Problem Solving

**A group of cards are placed in a stack. Four of the cards have red circles, 8 have yellow squares, 6 have green squares, 2 have red triangles, 5 have yellow circles, and 10 have green triangles. One card is picked at random. Find each probability.**

**11** The card has a red or green shape.

____________________

**12** The card has a yellow shape or a circle.

____________________

**13** The card has a red shape or a square.

____________________

**14** The card has a triangle or a red shape.

____________________

GO ON

**Solve. Write the answer in simplest form.**

**15** **ACADEMICS** The table at the right shows the declared majors of 50 college students. Each student has only one major. If one surveyed student is chosen randomly, what is the probability that the student has a major in engineering?

_______________

| Major | Number of Students |
|---|---|
| Chemical Engineer | 8 |
| History | 5 |
| Psychology | 7 |
| Mechanical Engineer | 12 |
| Secondary Education | 10 |
| Other | 8 |

**16** **SCIENCE** Samantha studied ten different tree species. Six of the trees had green foliage and were deciduous. Two trees had green foliage and were conifers. Two trees had red foliage and were deciduous. Samantha will give a speech on one of the trees. What is the probability that the tree she chooses is red or deciduous?

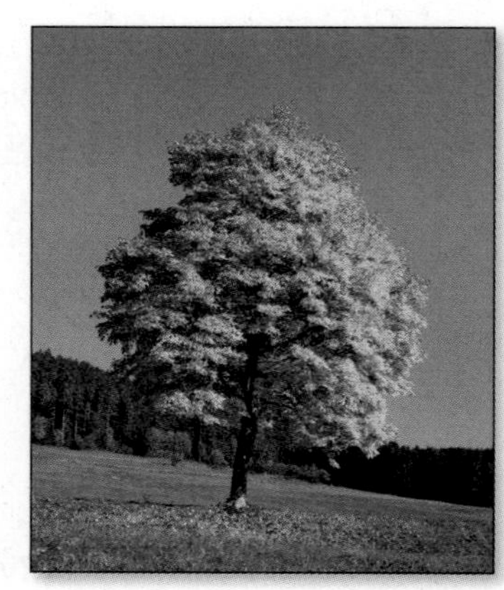

_______________

**17** **MUSIC** Several students in the Partridge School choir can play instruments. Two of the students sing alto and play piano, three sing soprano and play guitar, one sings tenor and plays guitar, and four sing soprano and play piano. If a student is randomly selected, what is the probability that the student sings soprano or plays piano?

_______________

**Vocabulary Check** **Write the vocabulary word that completes each sentence.**

**18** _______________ events cannot occur at the same time.

**19** If two or more events are involved in a situation, this is an example of _______________.

**20** Events that can occur at the same time are _______________ events.

**21** **Reflect** Suppose you chose a random student from your class. Is the probability that the student is a girl or has a letter *e* in his or her name a mutually exclusive or inclusive event? Explain.

_______________

_______________

_______________

STOP

Lesson 5-5

# Geometric Probability

## KEY Concept

When calculating geometric probability, you will use formulas you learned in geometry.

### Geometric Probability with Area

If a point in region $A$ is chosen at random, then the $P(B)$ that the point is in region $B$ is shown below.

$$P(B) = \frac{\text{area of region } B}{\text{area of region } A}$$

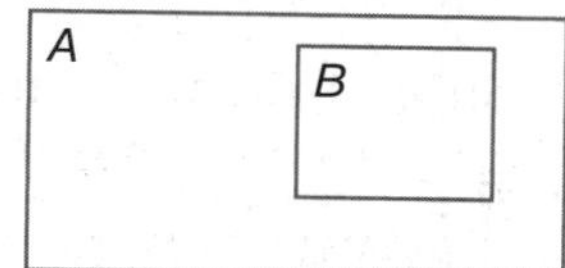


### Geometric Probability with Sectors

The area $A$ of a sector with radius $r$ and central angle $n°$ is shown below.

$$A = \frac{n}{360}\pi r^2.$$

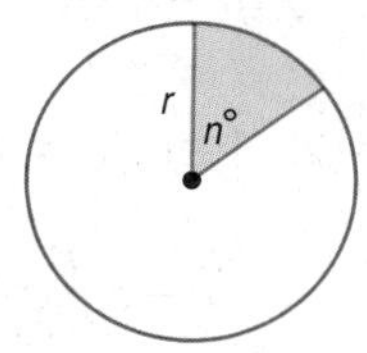


If a point in circle is chosen at random, then the $P(s)$ that the point is in the sector is shown below.

$$P(s) = \frac{\text{area of sector}}{\text{area of circle}}$$

**VOCABULARY**

**geometric probability**
probability that involves a geometric measure

**sector**
a region of a circle bounded by a central angle and its intercepted arc

You should draw a diagram for each situation when calculating geometric probability.



GO ON

## Example 1

**Find the probability that a random point within the square will be in the shaded region.**

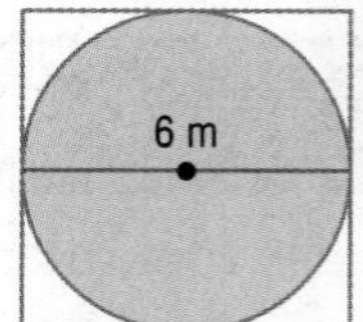


1. Let $A$ = the area of the square.

   $A = s^2 = 6^2 = 36\ m^2$

2. Let $B$ = the area of the circle.

   $B = \pi r^2 = \pi(3)^2 = 9\pi\ m^2$

3. Substitute and simplify.

   $P = \dfrac{\text{area of region } B}{\text{area of region } A} = \dfrac{9\pi\ m^2}{36\ m^2} = \dfrac{\pi}{4}$

### YOUR TURN!

**Find the probability that a random point within the rectangle will be in the shaded square.**

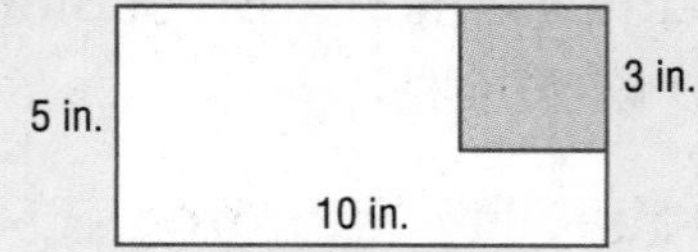


1. Let $A$ = the area of the rectangle.

   $A =$ ____________

2. Let $B$ = the area of the square.

   $B = s^2 =$ ____________

3. Substitute and simplify.

   $P = \dfrac{\text{area of region } B}{\text{area of region } A} =$ ______

## Example 2

**Find the probability that a random point within the circle will be in the shaded region.**

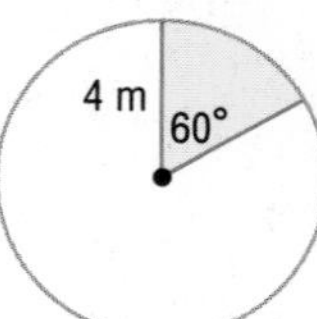


1. Let $A$ = the area of the sector.

   $A = \dfrac{n}{360}\pi r^2 = \dfrac{60}{360}\pi(4)^2 = \dfrac{8}{3}\pi\ m^2$

2. Let $B$ = the area of the circle.

   $B = \pi r^2 = \pi(4)^2 = 16\pi\ m^2$

3. Substitute and simplify.

   $P = \dfrac{\text{area of sector}}{\text{area of circle}}$

   $= \dfrac{\frac{8}{3}\cancel{\pi\ m^2}}{16\cancel{\pi\ m^2}}$

   Remember that a fraction bar is a division operation.

   $= \dfrac{8}{3} \div \dfrac{16}{1} = \dfrac{\cancel{8}}{3} \cdot \dfrac{1}{\cancel{16}_{2}} = \dfrac{1}{6}$

### YOUR TURN!

**Find the probability that a random point within the circle will be in the shaded region.**

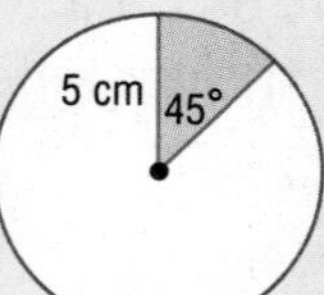


1. Let $A$ = the area of the sector.

   $A = \dfrac{n}{360}\pi r^2 = \dfrac{\square}{360}\pi(\_\_\_)^2 =$ ____ $\pi$

2. Let $B$ = the area of the circle.

   $B = \pi r^2 =$ ________ $=$ ________

3. Substitute and simplify.

   $P = \dfrac{\text{area of sector}}{\text{area of circle}}$

   $= \dfrac{\frac{25}{8}\pi\ cm^2}{\square\pi\ cm^2}$

   $=$ ________________

## Guided Practice

**Find the probability that a random point within the figure will be in the shaded region.**

**1**

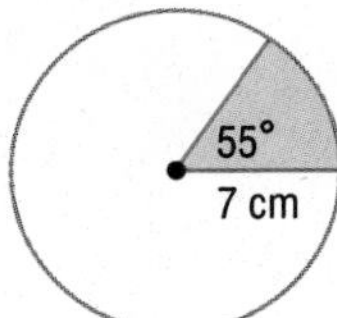


$A = \frac{n}{360}\pi r^2 = \frac{\square}{360}\pi(\_\_\_\_)^2$

$= \frac{\square}{72}(\_\_\_\_)\pi = \frac{\square}{72}\pi \text{ cm}^2$

$B = \pi r^2 = \pi(\square)^2 = \_\_\_\_$

$P = \frac{\frac{539}{72\pi}\text{ cm}^2}{\square\pi \text{ cm}^2} = \frac{\frac{539}{72}}{\square} = \frac{\square}{\square} \div \frac{\square}{\square}$

$= \frac{\square}{\square} \cdot \frac{\square}{\square} = \_\_\_\_$

**2**

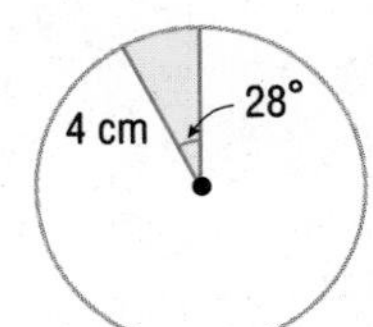


$A = \frac{n}{360}\pi r^2 = \frac{\square}{360}\pi(\_\_\_\_)^2$

$= \frac{\square}{90}(\_\_\_\_)\pi = \frac{\square}{45}\pi \text{ cm}^2$

$B = \pi r^2 = \pi(\square)^2 = \_\_\_\_$

$P = \frac{\frac{56}{45}\pi \text{ cm}^2}{\square\pi \text{ cm}^2} = \frac{\frac{56}{45}}{\square} = \frac{\square}{\square} \div \frac{\square}{\square}$

$= \frac{\square}{\square} \cdot \frac{\square}{\square} = \_\_\_\_$

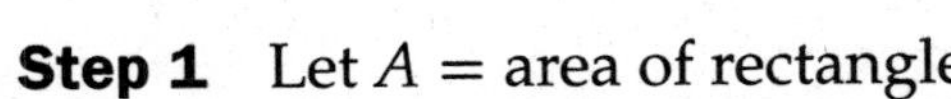

## Step by Step Practice

**3** Find the probability that a random point within the rectangle will be in the shaded region.

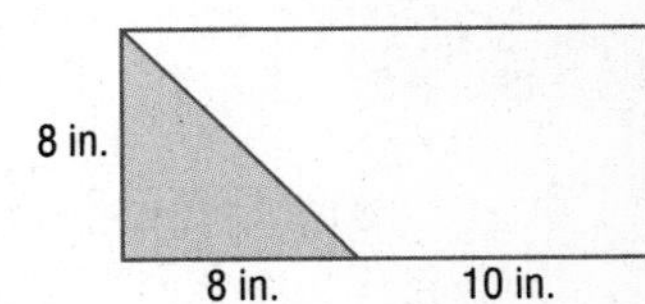


**Step 1** Let $A$ = area of rectangle.

$A = \ell w = (\_\_\_\_)(\_\_\_\_) = \_\_\_\_$

**Step 2** Let $B$ = area of triangle.

$B = \frac{1}{2}bh = \frac{1}{2}(\_\_\_\_)(\_\_\_\_) = \_\_\_\_$

**Step 3** Find the probability of a point being in the shaded region.

$\frac{\text{area of region } \square}{\text{area of region } \square} = \frac{\square}{\square} = \frac{\square}{\square}$

GO ON

**Find the probability that a random point with each figure is in the shaded region.**

4 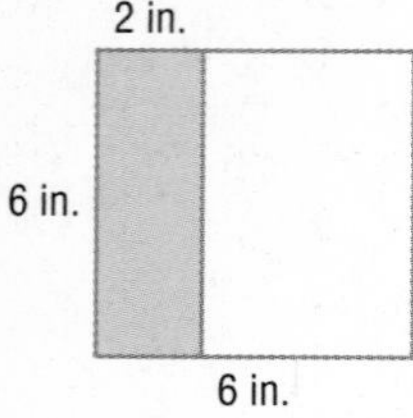


Let $A$ = the area of the square.

$A = s^2 = \square^2 = \_\_\_\_\_\_$

Let $B$ = the area of shaded rectangle.

$B = \ell w = \square \cdot \square = \square \text{ in}^2$

$P = \dfrac{\text{area of region } \square}{\text{area of region } \square} = \dfrac{\square}{\square} = \dfrac{\square}{\square}$

5 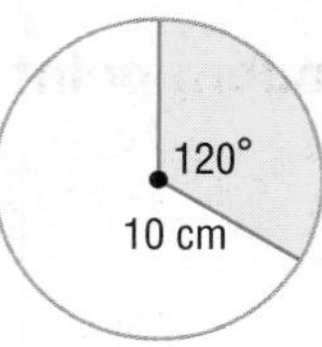


Let $A$ = the area of the sector.

$A = \dfrac{n}{360}\pi r^2 = \dfrac{\square}{360}\pi(\_\_\_\_\_\_)^2$

$= \dfrac{\square}{3}(\_\_\_\_\_\_)\ \pi = \dfrac{\square}{3}\pi \text{ cm}^2$

Let $B$ = the area of the circle.

$B = \pi r^2 = \pi(\_\_\_\_\_\_)^2$

$= \_\_\_\_\_\_\pi \text{ cm}^2$

$P = \dfrac{\text{area of sector}}{\text{area of circle}} = \dfrac{\square}{\square} = \_\_\_\_\_\_$

6 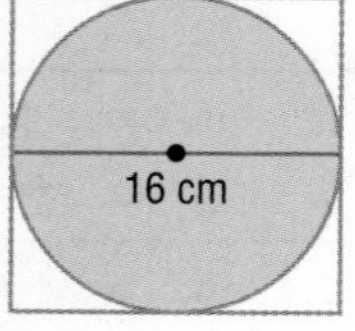


Let $A$ = the area of the square.

$A = \_\_\_\_\_\_$

Let $B$ = the area of the circle.

$B = \_\_\_\_\_\_$

$P = \dfrac{\text{area of region } \square}{\text{area of region } \square} = \_\_\_\_\_\_$

7 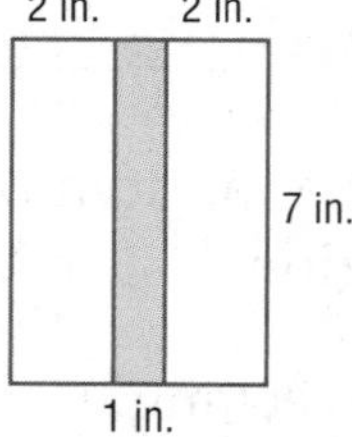


Let $A$ = the area of the rectangle.

$A = \_\_\_\_\_\_$

Let $B$ = the area of the shaded region.

$B = \_\_\_\_\_\_$

$P = \dfrac{\text{area of region } \square}{\text{area of region } \square} = \_\_\_\_\_\_$

**Find the probability that a random point within the figure is in the shaded region.**

**8**

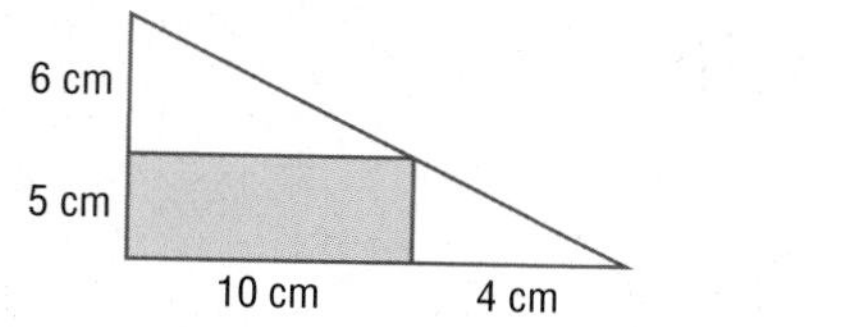


Area of traingle = ____________

Area of rectangle = ____________

$P =$ ____________

**9**

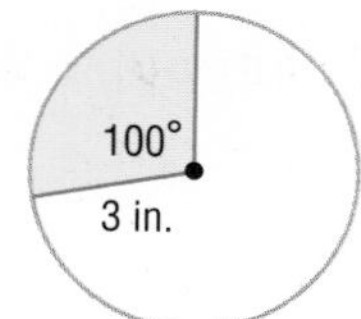


Area of circle = ____________

Area of sector = ____________

$P =$ ____________

## Step by Step Problem-Solving Practice

**Solve.**

**10** **CONSTRUCTION** The side of a house is a rectangle that is 18 feet high and 56 feet long. A window on the side of the house is a rectangle that is 3 feet long and 4 feet high. A lawn mower catches a stone and it is randomly thrown at the side of the house. What is the probability that a stone that hits the house will hit the window?

$$P = \frac{\text{area of the window}}{\text{area of the house}} = \underline{\qquad\qquad}$$

Check off each step.

_______ **Understand: I underlined key words.**

_______ **Plan: To solve the problem, I will** ____________________.

_______ **Solve: The answer is** ____________________.

_______ **Check: I checked my answer by** ____________________.



GO ON

## Skills, Concepts, and Problem Solving

**Find the probability that a random point within the figure is in the shaded region.**

11 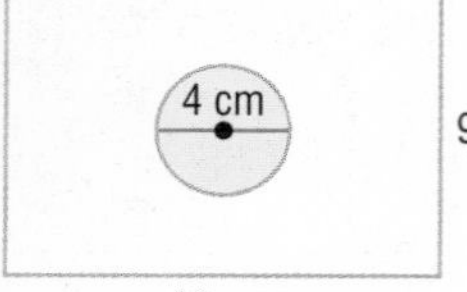


_______________

12 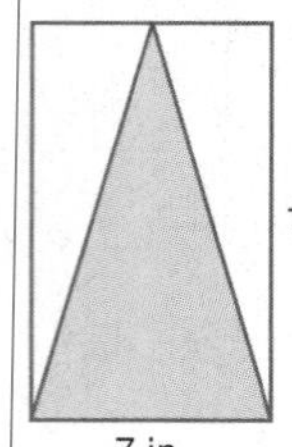


_______________

13  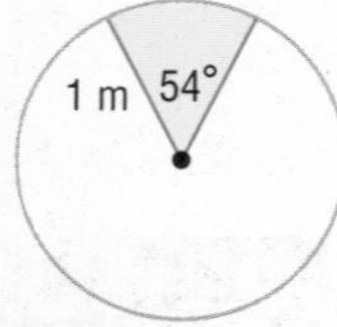


_______________

14 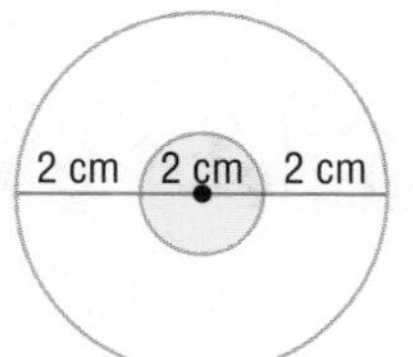


_______________

**Solve. Write the answer in simplest form.**

15 **GAMES** A carnival game has a dart board in the shape of the figure shown to the right. To win a prize, you must hit one of the identical shaded circles. What is the probability that a dart will hit a shaded circle?

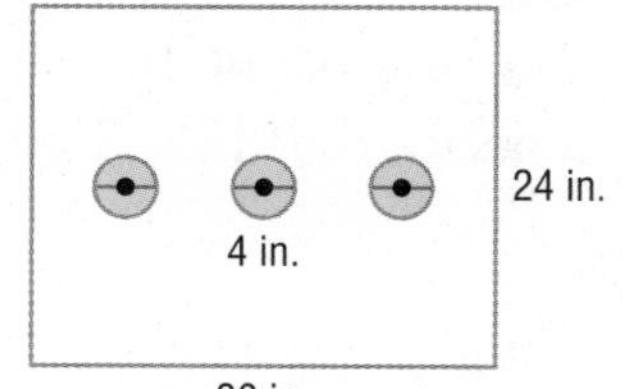


_______________

16 **GRAPHS** The pie graph to the right displays the hair colors of the sophomore class at Canterbury High School. What is the probability that a random student has brown hair?

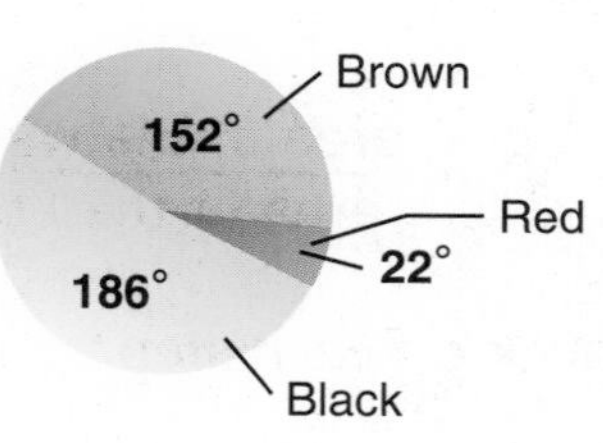


_______________

**Vocabulary Check** **Write the vocabulary word that completes each sentence.**

17 Probabilities that involve geometric measures are _______________.

18 A(n) _______________ is the region of a circle bounded by its central angle and the intercepted arc.

19 **Reflect** A diameter is drawn on a circle. Does this create a sector? Explain.

_______________

_______________

STOP

Chapter 5

# Progress Check 2 (Lessons 5-3, 5-4 and 5-5)

**A bag contains 5 red marbles, 4 green marbles, 3 blue marbles and 3 pink marbles. Find each probability.**

1 A red marble is picked, and without replacing it, a blue marble is picked.

_______________

2 A green marble is picked, and after replacing it, another green marble is picked.

_______________

3 A pink marble is picked, and after replacing it, a green marble is picked.

_______________

4 A red marble is picked, and without replacing it, a pink marble is picked.

_______________

**A group of shapes are placed in a box. Five of the objects are red balls, 4 are yellow cubes, 7 are green pyramids, 3 are red diamonds, 4 are blue octagons, and 2 are yellow diamonds. One object is picked at random. Find each probability.**

5 The object has a blue or yellow shape.

_______________

6 The object is red or a pyramid.

_______________

7 The object is a diamond or is red.

_______________

8 The object has a green or red shape.

_______________

**Find the probability that a randomly chosen point within each figure is in the shaded region.**

9 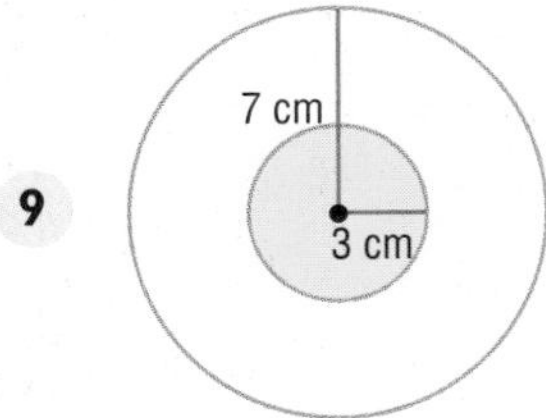


_______________

10 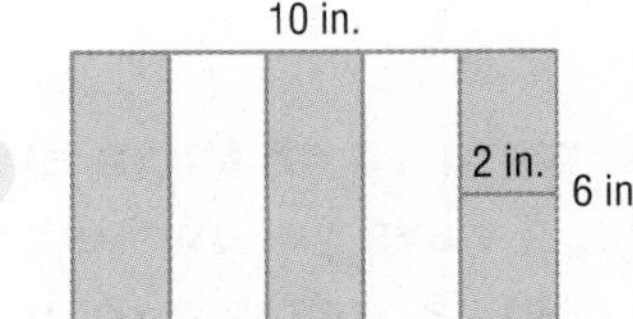


_______________

**Solve. Write the answer in simplest form.**

11 **BOARD GAMES** James needs at least a five on his next roll of the die in order to win the board game. What is the probability that he will roll a 5 or a 6 on his next roll to win the game?

_______________

Chapter 5

# Chapter Test

**Simplify each factorial.**

1 $3! =$ ________ 2 $(12 - 6)! =$ ________ 3 $5! =$ ________

**Find each permutation or combination.**

4 ${}_7C_2 =$ ________ 5 ${}_{11}C_5 =$ ________ 6 ${}_{18}C_3 =$ ________

7 ${}_3P_1 =$ ________ 8 ${}_9P_5 =$ ________ 9 ${}_6P_4 =$ ________

10 ${}_{14}C_3 =$ ________ 11 ${}_4C_1 =$ ________ 12 ${}_6C_2 =$ ________

13 ${}_{11}P_7 =$ ________ 14 ${}_8P_2 =$ ________ 15 ${}_{15}P_3 =$ ________

**A bag contains 3 red tiles, 5 pink tiles, 8 white tiles, 6 green tiles, and 2 black tiles. Find each probability.**

16 A pink tile is picked, and without replacing it, a black tile is picked.

________

17 A green tile is picked, and after replacing it, another green tile is picked.

________

18 A red tile is picked, and after replacing it, a white tile is picked.

________

19 A white tile is picked, and without replacing it, a pink tile is picked.

________

**A group of cards with pictures on them are placed in a stack. 6 of the cards have red circles, 4 have yellow squares, 5 have green squares, 7 have red triangles, 8 have yellow circles, and 5 have green triangles. One card is picked at random. Find each probability.**

20 The card has a red or yellow shape.

________

21 The card has a green shape or a square.

________

**Find the probability that a randomly chosen point within the figure is in the shaded region.**

**22** 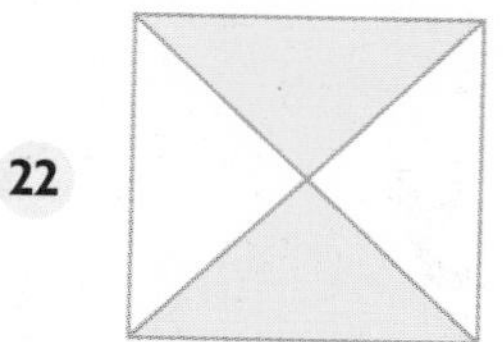 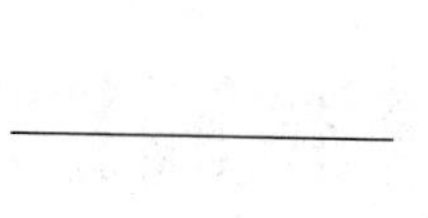 ____________

**23** 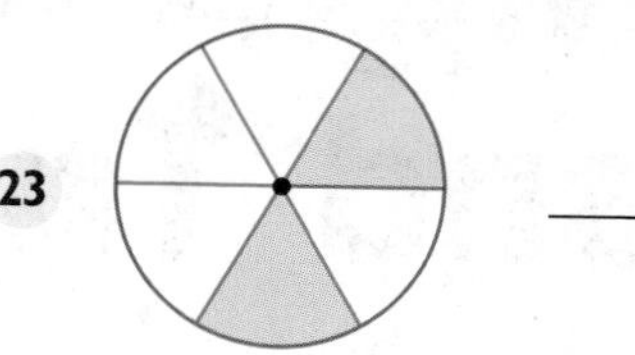 ____________

**24** 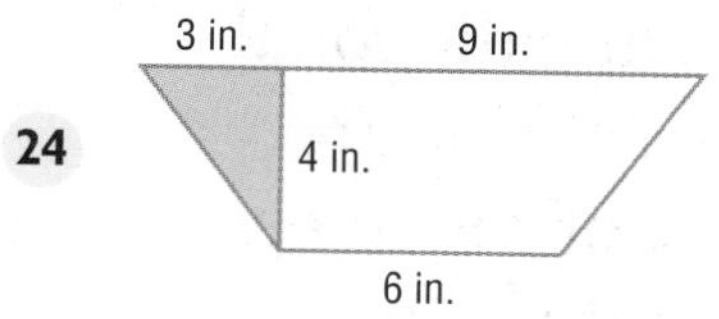

____________

**25** 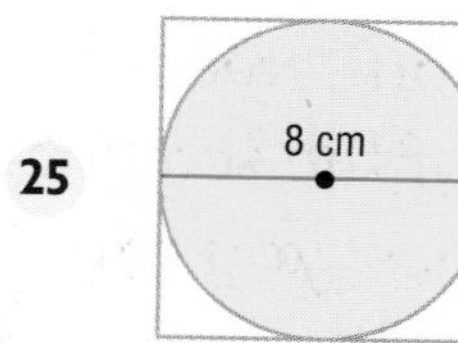

____________

**Solve.**

**26** **FOOD** The Sandwich Shoppe offers 18 items for sandwiches. How many different sandwiches can be created with 7 items on each?

________________________

**27** **BOOKS** There are 5 non-fiction books, 7 fiction books, 9 biographies, and 3 books of short stories on your bedroom bookshelf. In how many ways can you select 4 books to read in order?

________________________

**Correct the mistake.**

**28** Omar has 5 CDs: 2 heavy metal, 2 rock, and 1 jazz. He has determined that the probability of randomly selecting the jazz CD first and a heavy metal CD second without replacing the first CD is $\frac{1}{5}$. Did he calculate the probability correctly? If not, what is the correct answer?

________________________

STOP

Chapter 6

# Right Triangles and Trigonometry

## *Which TV will fit on my wall?*

Television screens are described according to the diagonal length of the screen. Because most television screens are rectangles, you can find the diagonal length $c$ of the screen by using the Pythagorean Theorem. If the length of the screen is 32 inches and the height of the screen is 24 inches, then $32^2 + 24^2 = c^2$. The diagonal length $c$ is 40 inches.

**STEP 1 Chapter Pretest** Are you ready for Chapter 6? Take the Chapter 6 Pretest to find out.

**STEP 2 Preview** Get ready for Chapter 6. Review these skills and compare them with what you will learn in this chapter.

| What You Know | What You Will Learn |
|---|---|
| You know how to find the square of a number.<br>**Example:** $10^2 = 10 \cdot 10$<br>$= 100$<br>**TRY IT!**<br>**1** $5^2 =$ ______ **2** $13^2 =$ ______<br>**3** $9^2 =$ ______ **4** $15^2 =$ ______ | *Lesson 6-1*<br>Use the Pythagorean Theorem to find the third side of a right triangle when you are given the lengths of the other two sides.<br>$a^2 + b^2 = c^2$<br>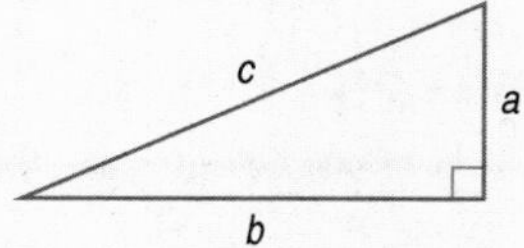 |
| You know how to find the value of $x$ in a pair of similar figures by writing and solving a proportion.<br>**Example:**<br>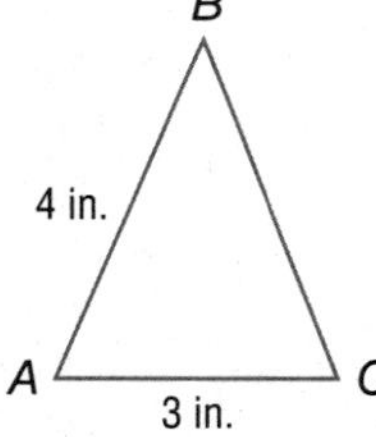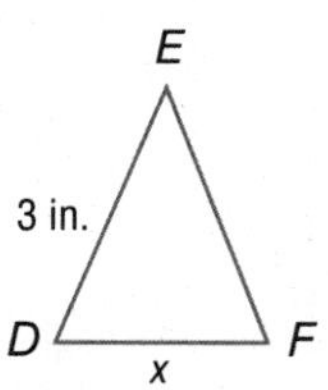<br>$\frac{AB}{DE} = \frac{AC}{DF}$ Write a proportion.<br>$\frac{4}{3} = \frac{3}{x}$ Substitute values.<br>$4x = 3(3)$ Find the cross products.<br>$4x = 9$ Simplify.<br>$x = 2\frac{1}{4}$ Divide each side by 4.<br>So, $x = 2\frac{1}{4}$ inches. | *Lesson 6-5*<br>To solve a right triangle means to use the given measures to find all of the angle measures and side lengths.<br>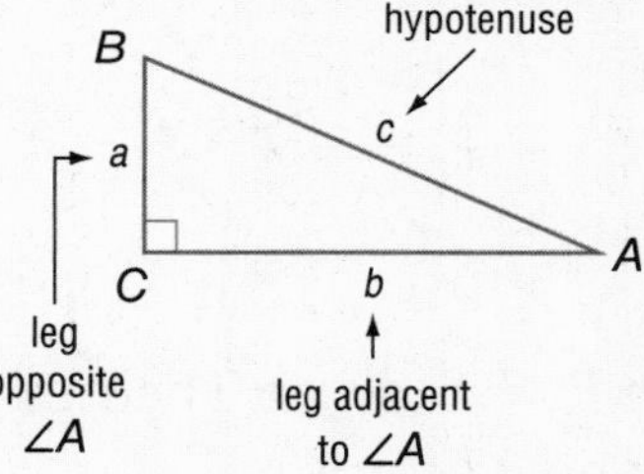<br>$\sin \angle A = \frac{\text{length of leg opposite } \angle A}{\text{length of hypotenuse}} = \frac{a}{c}$<br>$\cos \angle A = \frac{\text{length of leg adjacent } \angle A}{\text{length of hypotenuse}} = \frac{b}{c}$<br>$\tan \angle A = \frac{\text{length of leg opposite } \angle A}{\text{length of adjacent}} = \frac{a}{b}$ |

# Lesson 6-1 The Pythagorean Theorem

## KEY Concept

Use the **Pythagorean Theorem** in a right triangle when you are given any two side lengths to find the third side length.

$a^2 + b^2 = c^2$

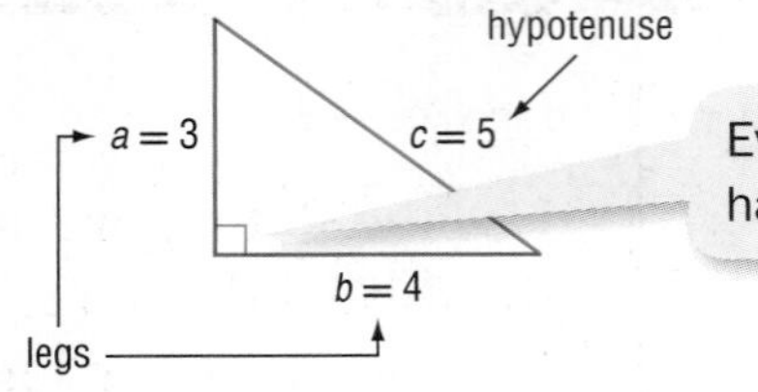


$3^2 + 4^2 = 5^2$

$9 + 16 = 25$

$25 = 25$

### VOCABULARY

**hypotenuse**
the side opposite the right angle in a right triangle; the longest side of a right triangle

**legs of a triangle**
two sides that form the right angle in a right triangle

**Pythagorean Theorem**
the sum of the squares of the lengths of the legs in a right triangle is equal to the square of the length of the hypotenuse; $a^2 + b^2 = c^2$

The Pythagorean Theorem is true for all right triangles.

## Example 1

**Find the length of the hypotenuse of the right triangle.**

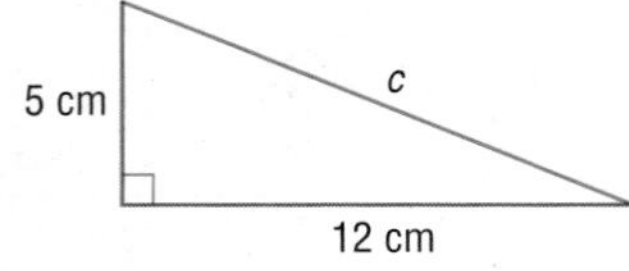


1. Use the Pythagorean Theorem. Substitute 5 for *a* and 12 for *b*.

$a^2 + b^2 = c^2$

$5^2 + 12^2 = c^2$

2. Solve for *c*.

$25 + 144 = c^2$

$169 = c^2$ Add.

$\sqrt{169} = \sqrt{c^2}$ Take the square root of each side.

$13 = c$

3. The length of the hypotenuse is 13 cm.

## YOUR TURN!

**Find the length of the hypotenuse of the right triangle.**

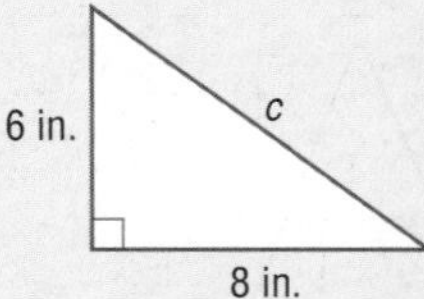


1. Use the Pythagorean Theorem. Substitute 6 for *a* and 8 for *b*.

$a^2 + b^2 = c^2$

$\square^2 + \square^2 = c^2$

2. Solve for *c*.

_____ + _____ $= c^2$

_____ $= c^2$

$\sqrt{\_\_\_} = \sqrt{\_\_\_^2}$

_____ = _____

3. The length of the hypotenuse is ______.

## Example 2

**Find the length of the missing leg of the right triangle.**

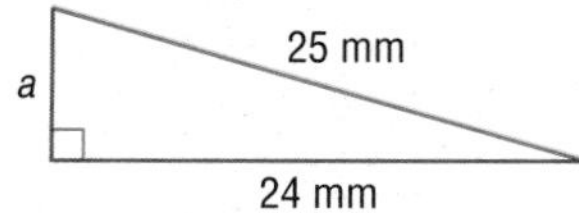


1. Use the Pythagorean Theorem. Substitute 24 for $b$ and 25 for $c$.

$a^2 + b^2 = c^2$

$a^2 + 24^2 = 25^2$

2. Solve for $a$.

$a^2 + 576 = 625$

$-576 \quad -576$ Subtract.

$a^2 = 49$

$\sqrt{a^2} = \sqrt{49}$ Take the square root of each side.

$a = 7$

3. The length of the leg is 7 mm.

**YOUR TURN!**

**Find the length of the missing leg of the right triangle.**

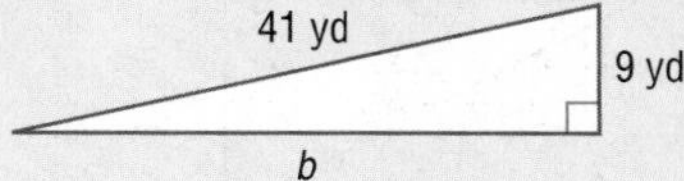


1. Use the Pythagorean Theorem. Substitute 9 for $a$ and 41 for $c$.

$a^2 + b^2 = c^2$

$9^2 + b^2 = \square^2$

2. Solve for $b$.

_______ $+ b^2 =$ _______

_______ _______ Subtract.

$b^2 =$ _______

$\sqrt{b^2} = \sqrt{\_\_\_\_\_\_}$ Take the square root.

$b =$ _______

3. The length of the leg is __________.

## Guided Practice

**Identify the hypotenuse of each right triangle.**

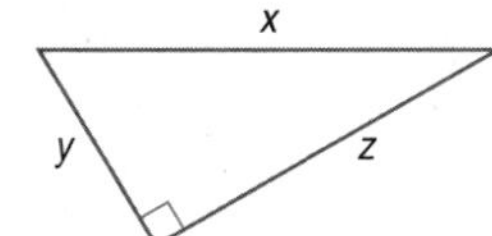


The hypotenuse of the triangle is _____.

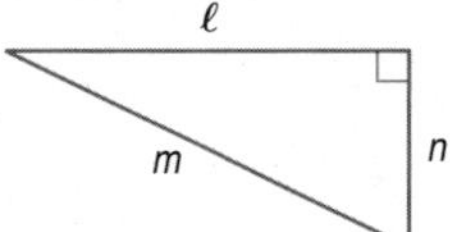


The hypotenuse of the triangle is _____.

**The chart below contains the lengths of the sides of right triangles. Complete the chart.**

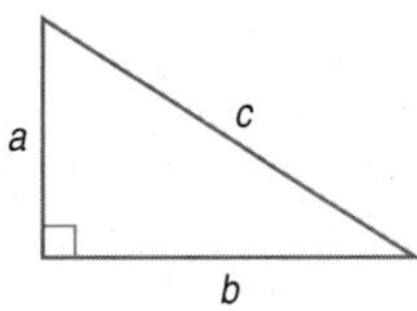


| | a | b | c | $a^2$ | $b^2$ | $a^2 + b^2$ | $c^2$ |
|---|---|---|---|---|---|---|---|
| 3 | 6 | 8 | 10 | ____ | ____ | ____ | ____ |
| 4 | 8 | 15 | 17 | ____ | ____ | ____ | ____ |
| 5 | 7 | 24 | 25 | ____ | ____ | ____ | ____ |
| 6 | 12 | 5 | 13 | ____ | ____ | ____ | ____ |

## Step by Step Practice

7 What is the length of the missing leg of the right triangle?

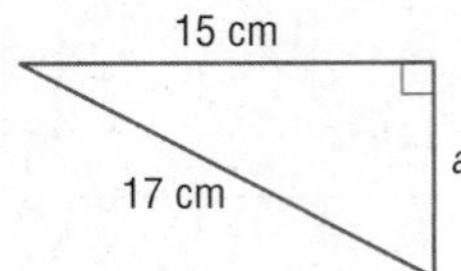


**Step 1** Use the Pythagorean Theorem. Substitute 15 for $b$ and 17 for $c$.

$$a^2 + b^2 = c^2$$

$$a^2 + \_\_\_\_\_^2 = \_\_\_\_\_^2$$

**Step 2** Solve for $a$.

$a^2 + \_\_\_\_\_ = \_\_\_\_\_$

$\_\_\_\_\_ \quad \_\_\_\_\_$ Substract.

$a^2 = \_\_\_\_\_$

$\sqrt{a^2} = \sqrt{\_\_\_\_}$ Take the square root.

$a = \_\_\_\_\_$

**Step 3** The length of the leg is _____.

**Find the missing length in each right triangle.**

8

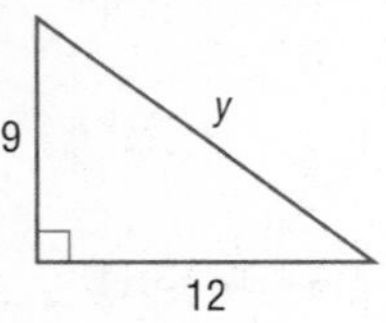


$a^2 + b^2 = c^2$

$\_\_\_\_^2 + \_\_\_\_^2 = \_\_\_\_^2$

$\_\_\_\_ + \_\_\_\_ = \_\_\_\_^2$

$\_\_\_\_ = \_\_\_\_^2$

$\sqrt{\_\_\_\_} = \sqrt{\_\_\_\_^2}$

$\_\_\_\_ = y$

9

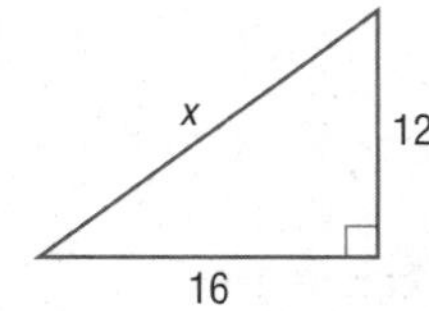


$a^2 + b^2 = c^2$

$\_\_\_\_^2 + \_\_\_\_^2 = \_\_\_\_^2$

$\_\_\_\_ + \_\_\_\_ = \_\_\_\_^2$

$\_\_\_\_ = \_\_\_\_^2$

$\sqrt{\_\_\_\_} = \sqrt{\_\_\_\_^2}$

$\_\_\_\_ = x$

**10**

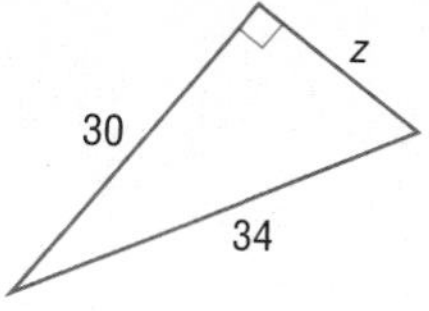


$a^2 + b^2 = c^2$

______ + ______ = ______

$\sqrt{\_\_\_\_} = \sqrt{\_\_\_\_}$

$x =$ ______

**11**

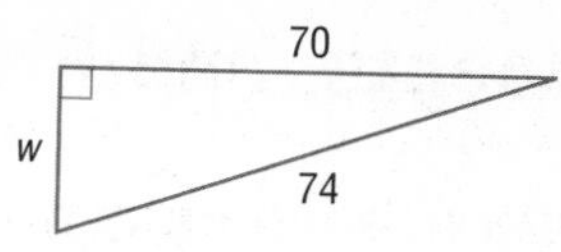


$a^2 + b^2 = c^2$

______ + ______ = ______

$\sqrt{\_\_\_\_} = \sqrt{\_\_\_\_}$

______ = ______

## Step by Step Problem-Solving Practice

**Solve.**

**12** **PAINTING** A 25-foot ladder is placed 15 feet from the base of a building. How high up on the building does the ladder reach?

Draw a picture. Then use the Pythagorean Theorem to find the missing length.

$$a^2 + b^2 = c^2$$

Check off each step.

______ **Understand: I underlined key words.**

______ **Plan: To solve the problem, I will** ________________________________.

______ **Solve: The answer is** ________________________________.

______ **Check: I checked my answer by** ________________________________.

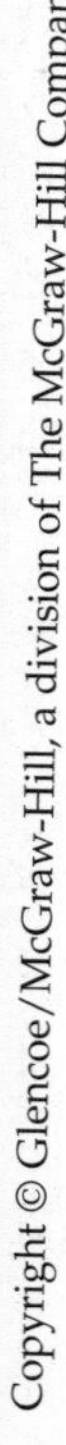



GO ON

## Skills, Concepts, and Problem Solving

**Find the missing length in each right triangle to the nearest tenth.**

**13** 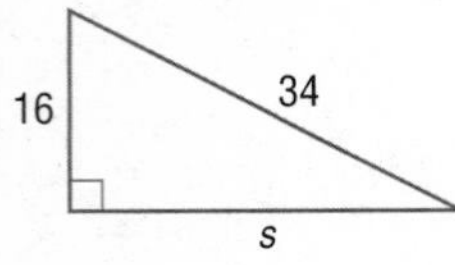


$s =$ ______________

**14** 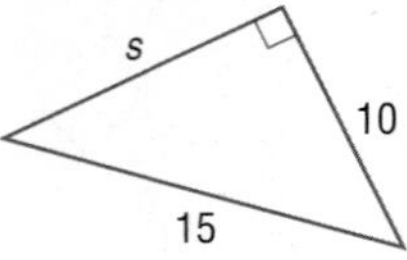


$s =$ ______________

**15** 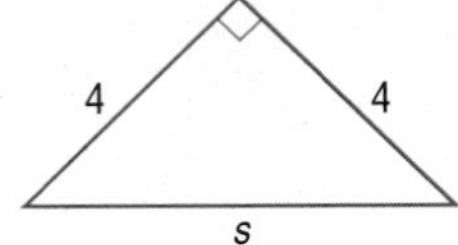


$s =$ ______________

**16** 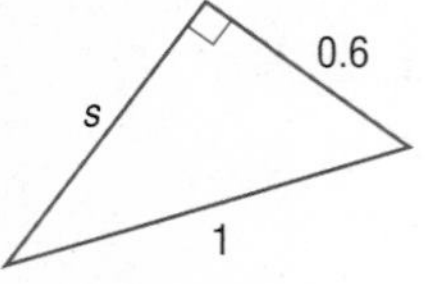


$s =$ ______________

**17** 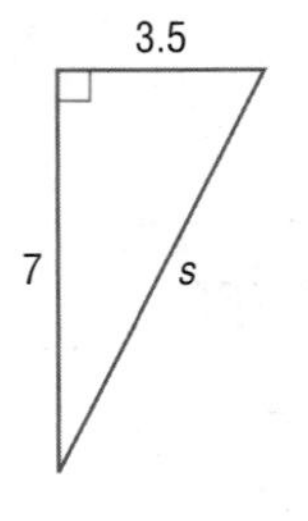


$s =$ ______________

**18** 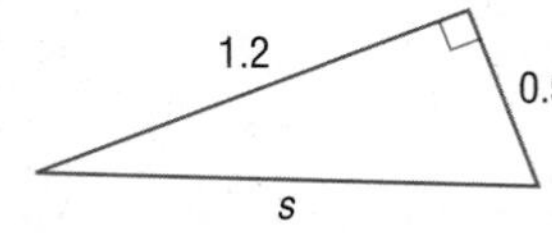


$s =$ ______________

**Solve.**

**19** **KITES** A kite is tangled in the top of a tree as shown. The kite string is 30 feet long. It is 18 feet from the base of the tree. How tall is the tree?

______________________________

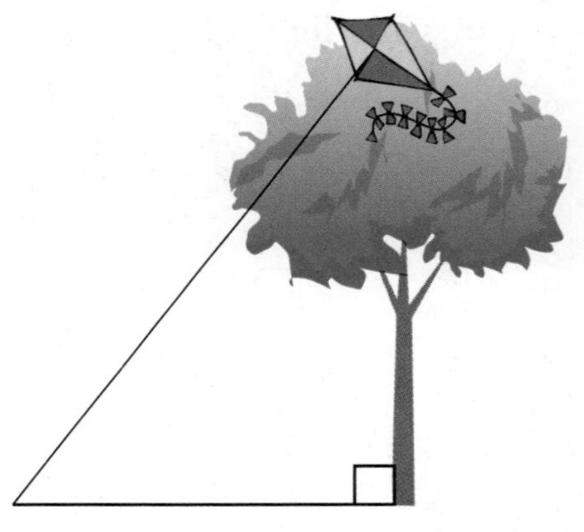

**20** **CONSTRUCTION** To reinforce a wall measuring 12 feet by 9 feet, a carpenter nails a board diagonally across the wall. How long is the board he used?

______________________________

**Vocabulary Check** **Write the vocabulary word that completes each sentence.**

**21** The side opposite the right angle in a right triangle is the ______________________.

**22** The______________________ states that the sum of the squares of the legs of a right triangle is equal to the square of the hypotenuse.

**23** **Reflect** The sides of a triangle measure 6, 9, and 14. Is the triangle a right triangle? Explain.

______________________________________________

______________________________________________

STOP

## Lesson 6-2 Sine Ratio

### KEY Concept

You can use trigonometric functions like the **sine** function to find measures in a right triangle.

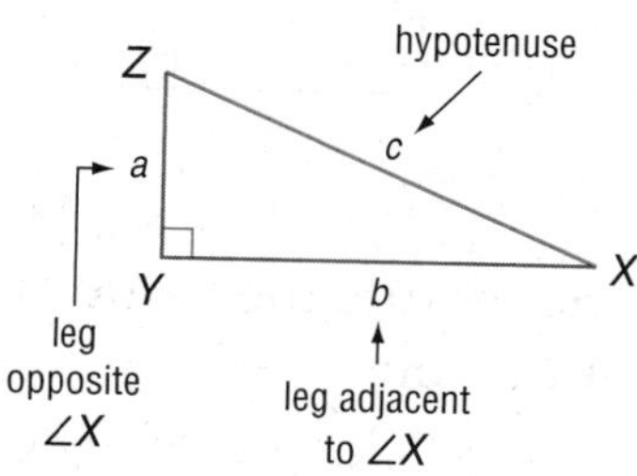


$$\text{sine } \angle X = \frac{\text{length of leg opposite } \angle X}{\text{length of hypotenuse}} = \frac{a}{c}$$

In practice, sine $\angle X$ is usually written as sin $X$.

**VOCABULARY**

**sine**
the ratio of the length of the opposite leg of an angle and the length of the hypotenuse

Notice that for the sine ratio that the denominator is the length of the hypotenuse.

### Example 1

**Find sin *R*.**

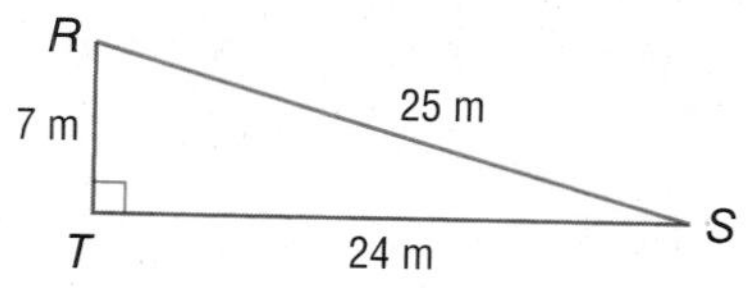


1. $\sin R = \dfrac{\text{length of leg opposite } \angle R}{\text{length of hypotenuse}}$
2. Substitute the opposite leg and hypotenuse.

$\sin R = \dfrac{24}{25}$

### YOUR TURN!

**Find sin *N*.**

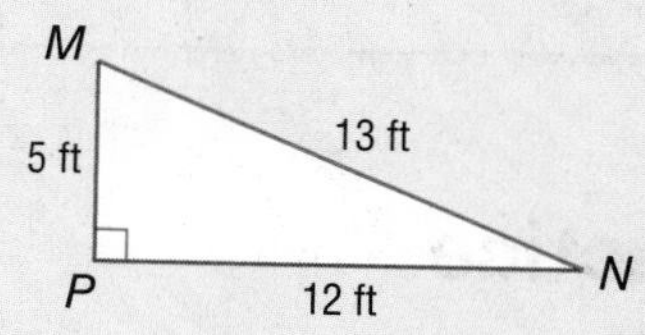


1. $\sin N = \dfrac{\text{length of leg opposite } \angle N}{\text{length of hypotenuse}}$
2. Substitute the opposite leg and hypotenuse.

$\sin N = \dfrac{\square}{\square}$

GO ON

## Example 2

**Find the missing measure to the nearest tenth.**

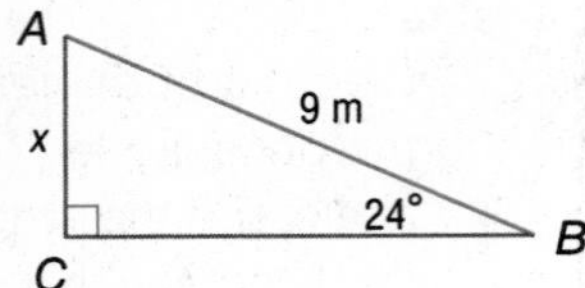


1. Substitute the values you know in the function.

$$\sin B = \frac{\text{length of leg opposite } \angle B}{\text{length of hypotenuse}}$$

$$\sin 24° = \frac{x}{9}$$

2. Solve for $x$.

Remember, $\frac{9}{1} = 9$.

$9 \cdot \sin 24° = \frac{x}{9} \cdot \frac{9}{1}$ Multiply each side by 9.

$3.7 \approx x$ Use a calculator.

3. The length of the leg is about 3.7 m.

### YOUR TURN!

**Find the missing measure to the nearest tenth.**

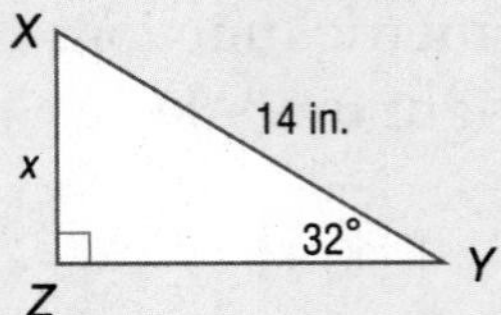


1. Substitute the values you know in the function.

$$\sin Y = \frac{\text{length of leg opposite } \angle Y}{\text{length of hypotenuse}}$$

$$\sin 32° = \frac{\square}{\square}$$

2. Solve for $x$.

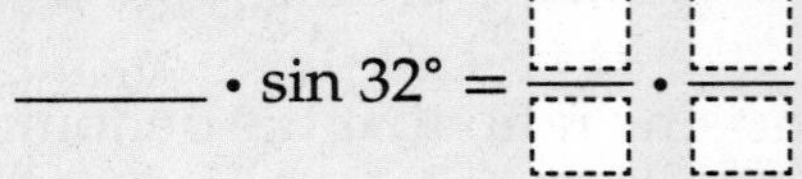

_____ $\cdot \sin 32° = \frac{\square}{\square} \cdot \frac{\square}{\square}$ Multiply each side by 14.

_____ $\approx$ _____ Use a calculator.

3. The length of the leg is about ________.

## Guided Practice

**Find the sine of ∠A.**

1

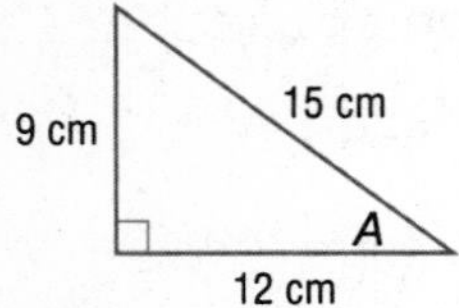


$$\sin A = \frac{\text{length of leg } \square \ \angle A}{\text{length of } \square}$$

$$= \frac{\square}{\square} = \frac{\square}{\square}$$

2

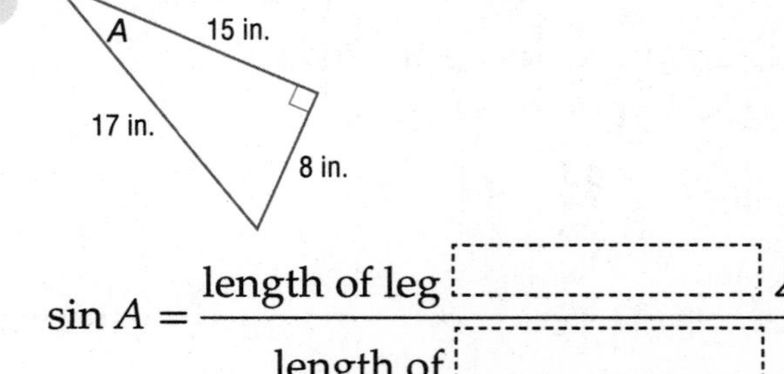


$$\sin A = \frac{\text{length of leg } \square \ \angle A}{\text{length of } \square}$$

$$= \frac{\square}{\square}$$

**Find the sine of ∠B.**

**3**

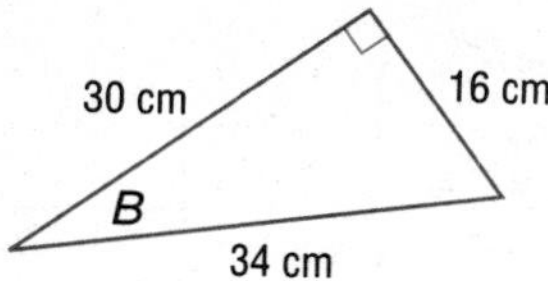


$\sin B = \dfrac{\text{opposite}}{\text{hypotenuse}} = \dfrac{\square}{\square} = \dfrac{\square}{\square}$

**4**

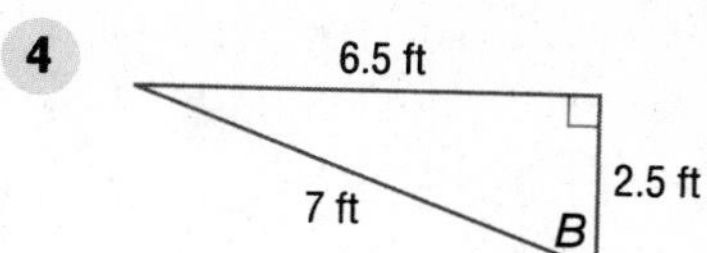


$\sin B = \dfrac{\text{opposite}}{\text{hypotenuse}} = \dfrac{\square}{\square} = \dfrac{\square}{\square}$

## Step by Step Practice

**5** Find the missing measure to the nearest tenth.

**Step 1** $\sin A = \dfrac{\text{length of leg opposite } \angle A}{\text{length of hypotenuse}}$

**Step 2** $\sin 22^\circ = \dfrac{\square}{\square}$

**Step 3** Solve for $x$.

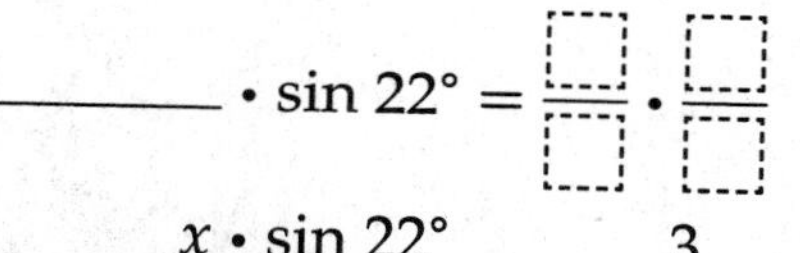

$\_\_\_\_\_ \cdot \sin 22^\circ = \dfrac{\square}{\square} \cdot \dfrac{\square}{\square}$ Multiply each side by $x$.

$\dfrac{x \cdot \sin 22^\circ}{\square} = \dfrac{3}{\square}$ Divide each side by sin 22°.

$x \approx \_\_\_\_\_$ Use a calculator.

**Step 4** The length of the leg is about _____.

**Find each missing measure to the nearest tenth.**

**6**

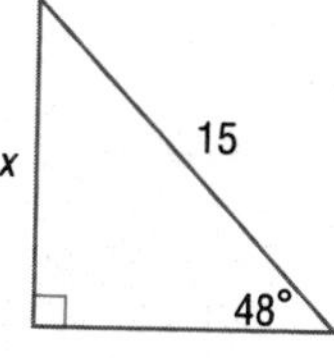


$\sin 48^\circ = \dfrac{\square}{\square}$

$\_\_\_\_\_ \cdot \sin 48^\circ = x$

$\_\_\_\_\_ \approx x$

**7**

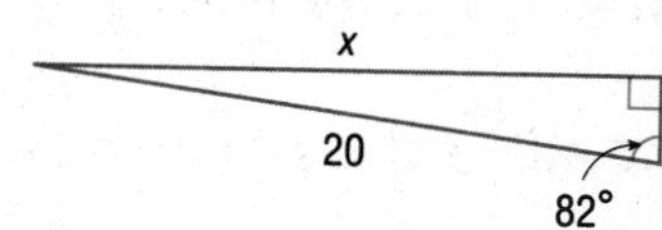


$\sin 82^\circ = \dfrac{\square}{\square}$

$\_\_\_\_\_ \cdot \sin 82^\circ = x$

$\_\_\_\_\_ \approx x$

GO ON

**Find each missing measure to the nearest tenth.**

**8**

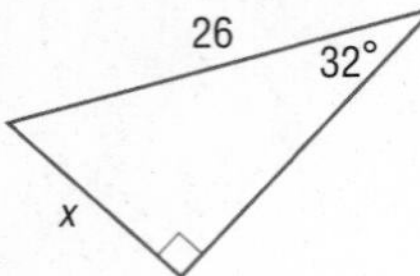


$\sin 32° = \frac{\square}{\square}$

_______ $\cdot \sin 32° = x$

_______ $\approx x$

**9**

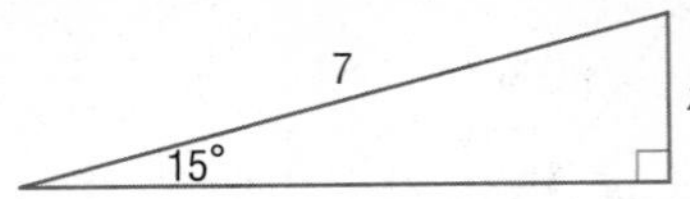


$\sin 15° = \frac{\square}{\square}$

_______ $\cdot \sin 15° = x$

_______ $\approx x$

## Step by Step Problem-Solving Practice

**Solve.**

**10** **CAMPING** The Turner family set up a tent. One side of the tent opening is 12 feet long. It creates a 38° angle with the ground. What is the height of the tent?

$$\sin \angle A = \frac{\text{length of leg opposite } \angle A}{\text{length of hypotenuse}}$$

Check off each step.

_______ **Understand: I underlined key words.**

_______ **Plan: To solve the problem, I will** _______________________________.

_______ **Solve: The answer is** _______________________________.

_______ **Check: I checked my answer by** _______________________________.

## Skills, Concepts, and Problem Solving

**Find the sine of ∠A.**

**11**

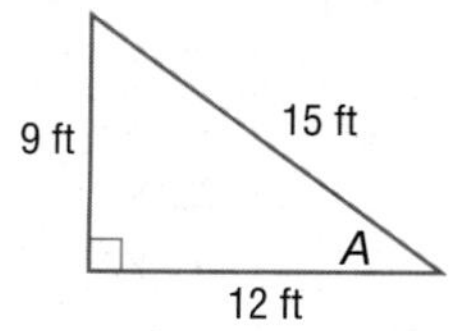


sin $A =$ ________

**12**

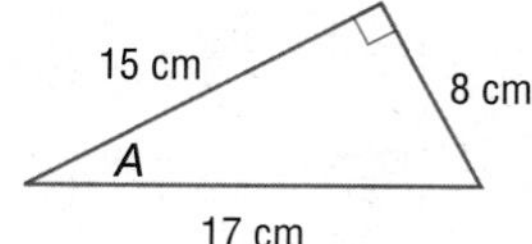


sin $A =$ ________

**13**

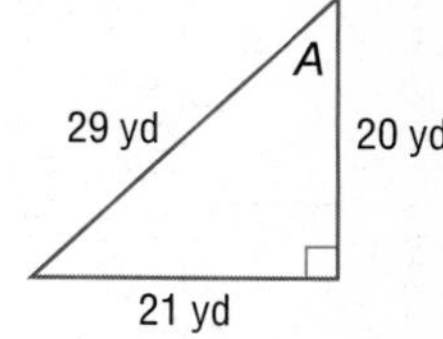


sin $A =$ ________

**Find each missing measure to the nearest tenth.**

**14**

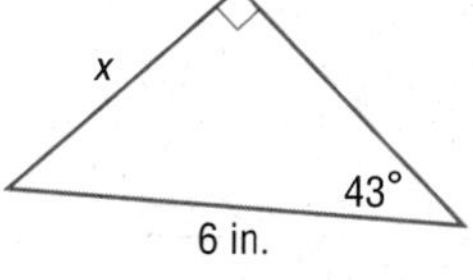


$x \approx$ ________

**15**

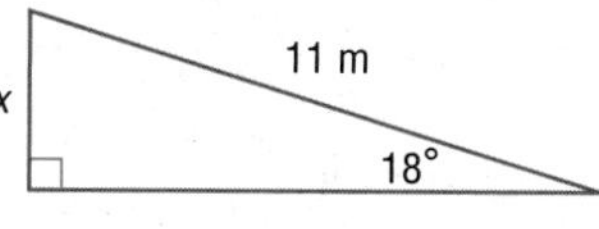


$x \approx$ ________

**16**

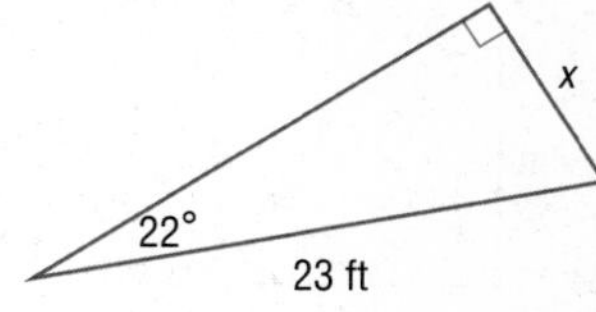


$x \approx$ ________

**Solve.**

**17** **CARPENTRY** A carpenter is cutting a rectangular piece of wood that is 5 ft long at a 45° angle as shown. To the nearest tenth, how long will the cut be?

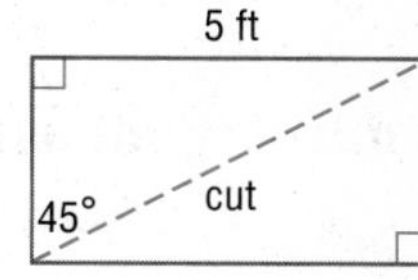


________________

**18** **SOCCER** Tara and Julie are both standing on the sideline at a soccer game. The ball is directly across from Tara, but Julie must run at an angle to get to the ball. If Julie is 32 feet from the ball at an angle of 28°, how far is Tara from the ball?

________________

**Vocabulary Check** **Write the vocabulary word that completes each sentence.**

**19** In a right triangle, given the length of the leg opposite of an angle and the length of the hypotenuse, use the ________ function to find the measure of the angle.

**20** **Reflect** Will the sine ratio ever be greater than 1? Explain.

________________

________________

STOP

Chapter 6

# Progress Check 1 (Lessons 6-1 and 6-2)

**Find each length s to the nearest tenth.**

**1**

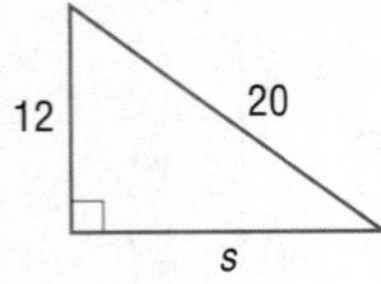


$s \approx$ ________

**2**

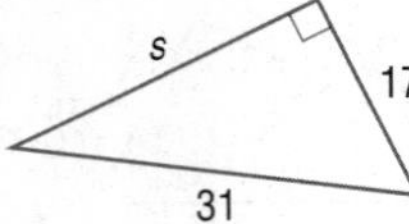


$s \approx$ ________

**3**

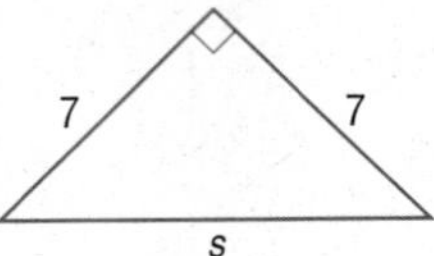


$s \approx$ ________

**Find the sine of ∠A.**

**4**

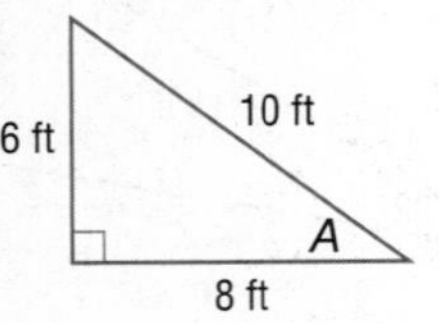


$\sin \angle A =$ ________

**5**

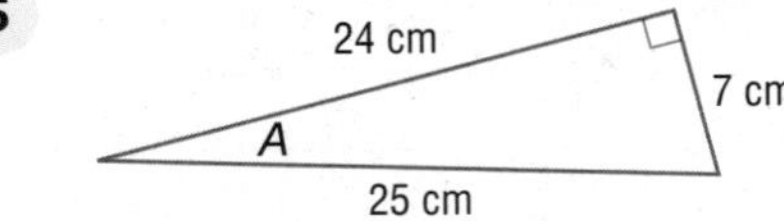


$\sin \angle A =$ ________

**6**

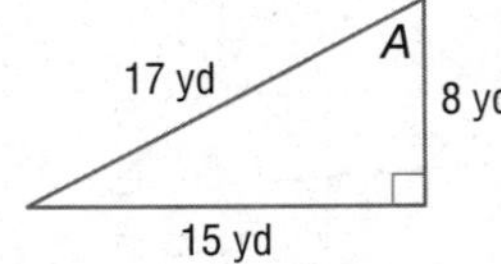


$\sin \angle A =$ ________

**Find each missing measure to the nearest tenth.**

**7**

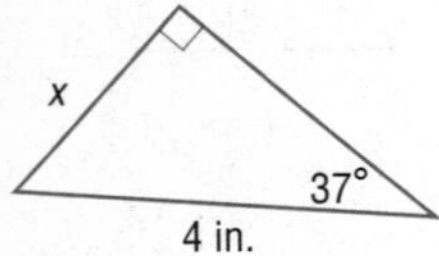


$x \approx$ ________

**8**

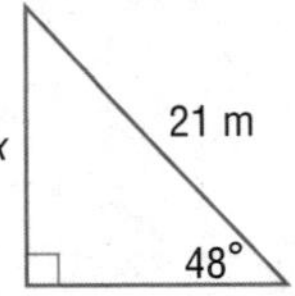


$x \approx$ ________

**9**

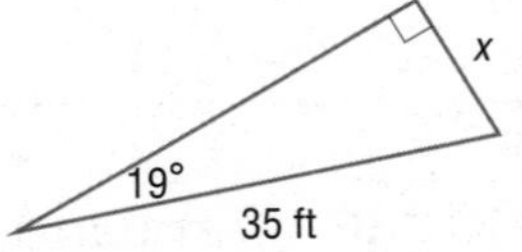


$x \approx$ ________

**Solve.**

**10** **CONSTRUCTION** Jared wants to ensure the gate opening to his backyard remains square. He reinforces the gate with a board running diagonally across the gate. The gate door is 4.5 ft tall by 6 ft wide. How long is the board he used?

________________________________

Lesson 6-3

# Cosine Ratio

## KEY Concept

Use trigonometric ratios to find unknown measures in a right triangle when you are given side lengths or angle measures.

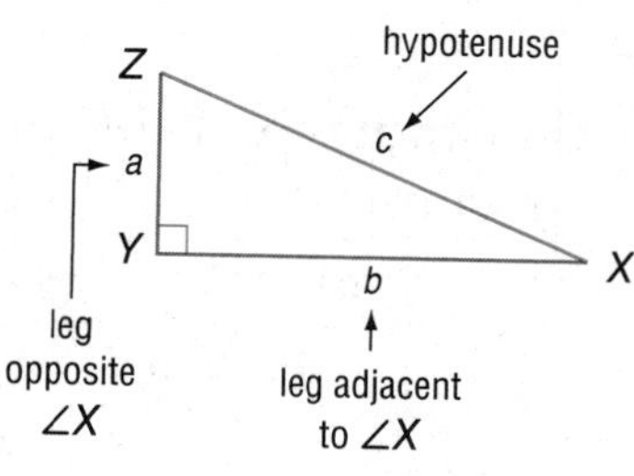


$$\text{cosine } \angle X = \frac{\text{length of leg adjacent } \angle X}{\text{length of hypotenuse}} = \frac{b}{c}$$

**VOCABULARY**

**cosine**
the ratio of the length of the adjacent leg of an angle and the length of the hypotenuse

Notice that for the cosine ratio the denominator is the length of the hypotenuse.

## Example 1

**Find cos *A*.**

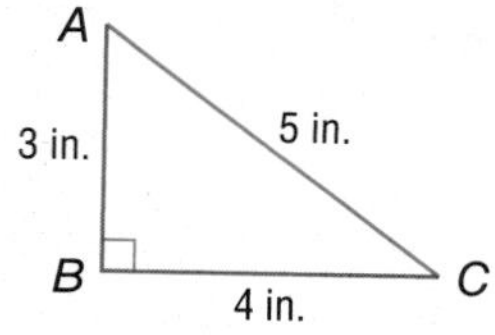


1. $\cos A = \frac{\text{length of leg adjacent } \angle A}{\text{length of hypotenuse}}$

2. Substitute the adjacent leg and hypotenuse.

$\cos A = \frac{3}{5}$

## YOUR TURN!

**Find cos *H*.**

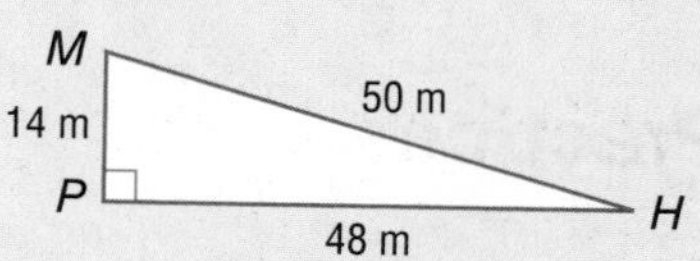


1. $\cos H = \frac{\text{length of leg adjacent } \angle H}{\text{length of hypotenuse}}$

2. Substitute the adjacent leg and hypotenuse.

$\cos H = \frac{\square}{\square} = \frac{\square}{\square}$

GO ON

## Example 2

**Find the missing measure to the nearest tenth.**

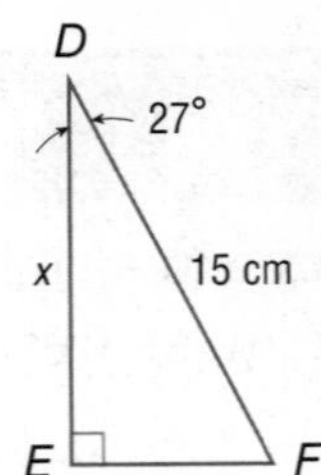


1. Substitute the values you know in the function.

$$\cos D = \frac{\text{length of leg adjacent } \angle D}{\text{length of hypotenuse}}$$

$$\cos 27° = \frac{x}{15}$$

2. Solve for $x$.

$15 \cdot \cos 27° = \frac{x}{15} \cdot \frac{15}{1}$ Multiply each side by 15.

$13.4 \approx x$ Use a calculator.

3. The length of the leg is about 13.4 cm.

**YOUR TURN!**

**Find the missing measure to the nearest tenth.**

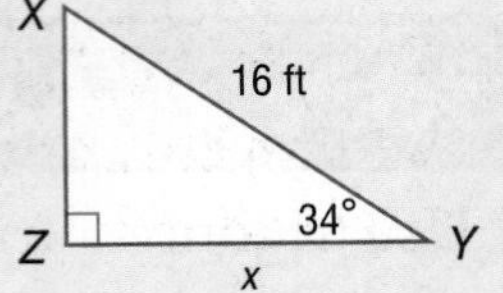


1. Substitute the values you know in the function.

$$\cos Y = \frac{\text{length of leg adjacent } \angle Y}{\text{length of hypotenuse}}$$

$$\cos 34° = \frac{\square}{\square}$$

2. Solve for $x$.

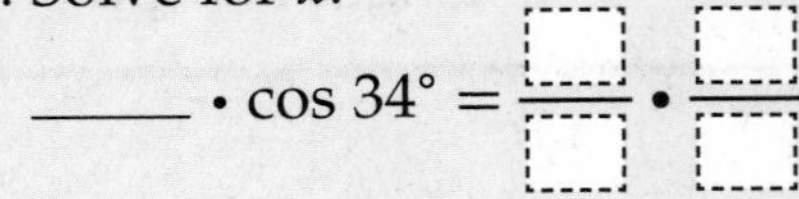

$\_\_\_\_ \cdot \cos 34° = \frac{\square}{\square} \cdot \frac{\square}{\square}$ Multiply each side by 16.

$\_\_\_\_ \approx \_\_\_\_$ Use a calculator.

3. The length of the leg is about ________.

## Guided Practice

**Find the cosine of ∠A.**

**1**

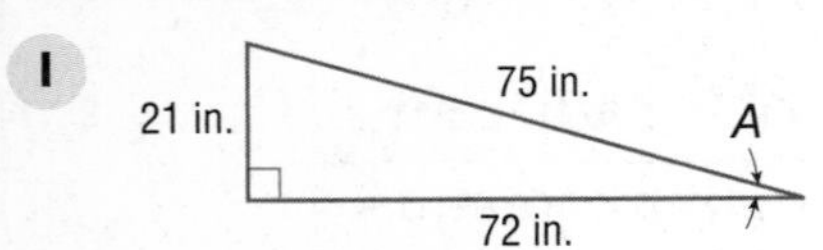


$$\cos A = \frac{\text{length of leg } \square \ \angle A}{\text{length of } \square} = \frac{\square}{\square} = \frac{\square}{\square}$$

**2**

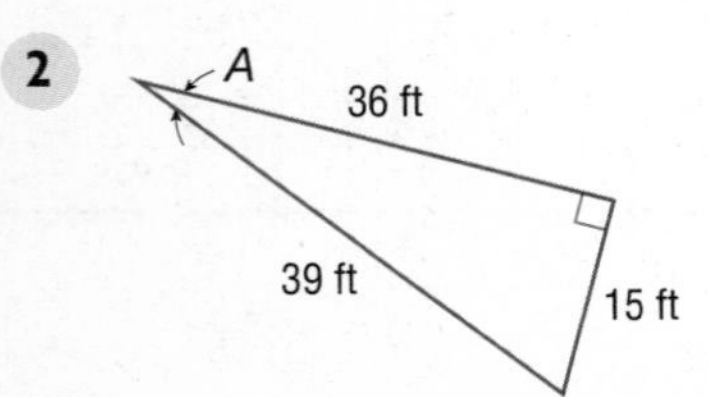


$$\cos A = \frac{\text{length of leg } \square \ \angle A}{\text{length of } \square} = \frac{\square}{\square} = \frac{\square}{\square}$$

## Step by Step Practice

**3** Find the missing measure to the nearest tenth.

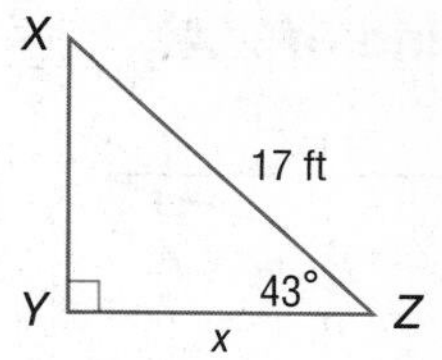


**Step 1** $\cos Z = \dfrac{\text{length of leg adjacent } \angle Z}{\text{length of hypotenuse}}$

**Step 2** $\cos 43° = \dfrac{\square}{\square}$

**Step 3** Solve for $x$. ________ $\cdot \cos 43° = \dfrac{\square}{\square} \cdot \dfrac{\square}{\square}$ Multiply both sides by 17.

________ $\approx x$ Use a calculator to solve.

**Step 4** The length of the leg is about ________.

**Find each missing measure to the nearest tenth.**

**4**

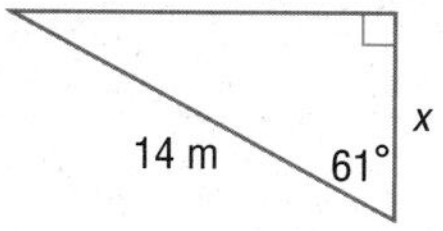


$\cos 61° = \dfrac{\square}{\square}$

________ $\cdot \cos 61° = x$

________ $\approx x$

**5**

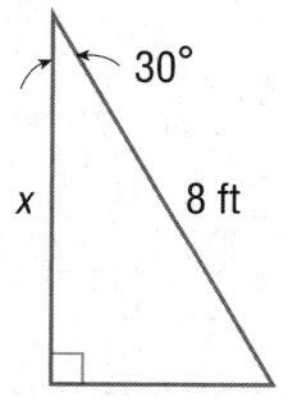


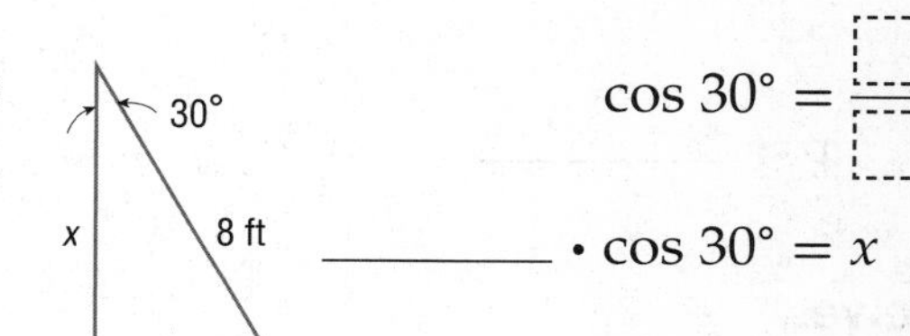

$\cos 30° = \dfrac{\square}{\square}$

________ $\cdot \cos 30° = x$

________ $\approx x$

## Step by Step Problem-Solving Practice

**Solve.**

**6** **BUILDINGS** The distance from the top of the building to the end of the building's shadow is 120 meters and creates a 54° angle. What is the height of the building?

Check off each step.

______ **Understand: I underlined key words.**

______ **Plan: To solve the problem, I will** ________________________.

______ **Solve: The answer is** ________________________.

______ **Check: I checked my answer by** ________________________.

**GO ON**

## Skills, Concepts, and Problem Solving

**Find the cosine of ∠*A*.**

**7**

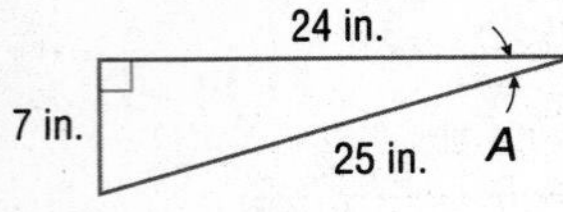


$\cos A =$ ______

**8**

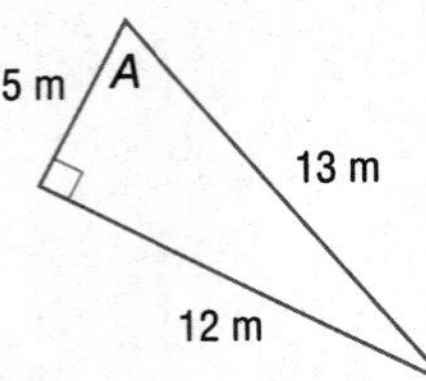


$\cos A =$ ______

**9**

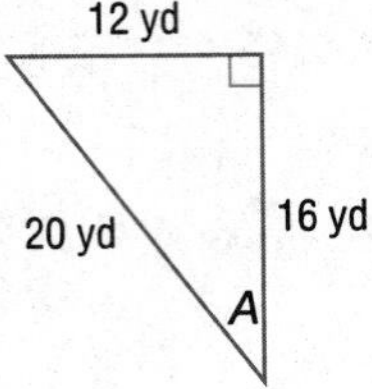


$\cos A =$ ______ $=$ ______

**Find each missing measure to the nearest tenth.**

**10**

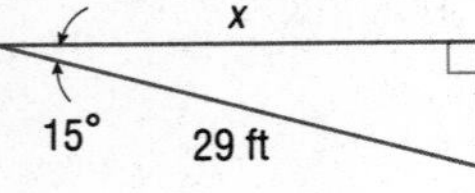


$x \approx$ ______

**11**

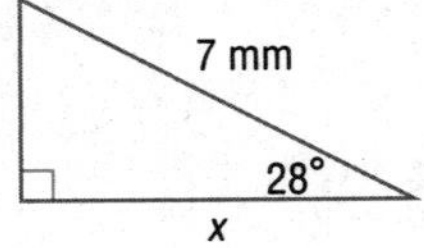


$x \approx$ ______

**12**

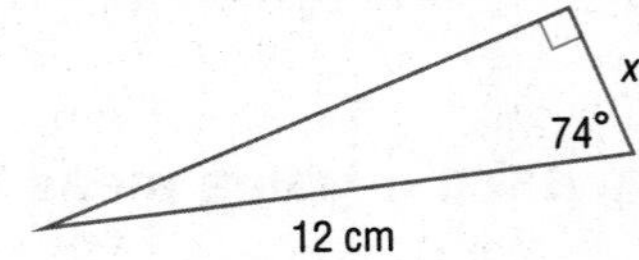


$x \approx$ ______

**Solve.**

**13** **BASEBALL** The distance between home plate and second base on a baseball diamond is about 127.28 feet. To the nearest tenth, what is the distance between home plate and third base?

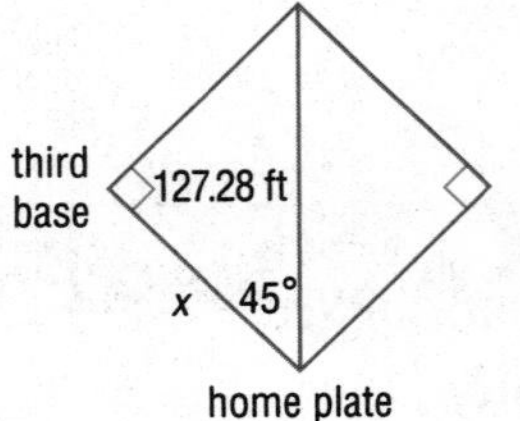


______

**14** **MOVING** A ramp for unloading a moving truck has an incline of 20°. The length of the ramp is 8 feet long. To the nearest tenth, how far from the back of the truck is the end of the ramp?

______

**Vocabulary Check** **Write the vocabulary word that completes each sentence.**

**15** In a right triangle, given the length of the leg adjacent to an angle and the length of the hypotenuse, use the ______ function to find the measure of the angle.

**16** **Reflect** Will the cosine ratio ever be greater than 1? Explain.

______

______

STOP

Lesson 6-4

# Tangent Ratio

## KEY Concept

The tangent ratio is another trigonometric ratio that you can use to find unknown measures in a right triangle when you are given side lengths and angle measures.

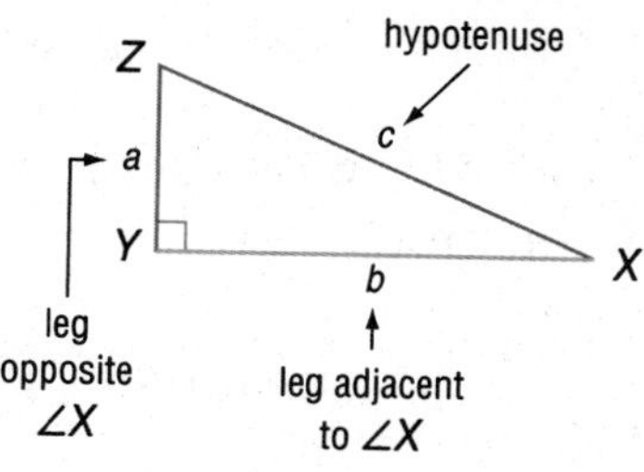


$$\tan \angle X = \frac{\text{length of leg opposite } \angle X}{\text{length of leg adjacent } \angle X} = \frac{a}{b}$$

**VOCABULARY**

**tangent**
the ratio of the length of the opposite leg of an angle and the length of the adjacent leg of the same angle

The tangent ratio uses the opposite and adjacent sides of a right triangle.

### Example 1

**Find tan *V*.**

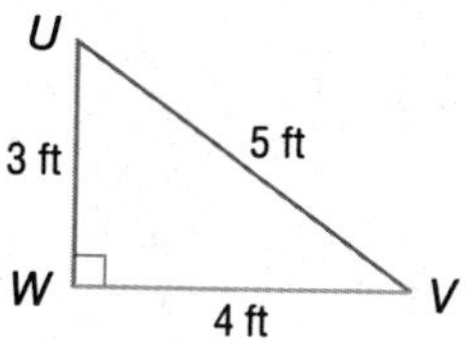


1. $\tan V = \frac{\text{length of leg opposite } \angle V}{\text{length of leg adjacent } \angle V}$

2. Substitute the opposite leg and the adjacent leg.

$\tan V = \frac{3}{4}$

**YOUR TURN!**

**Find tan *D*.**

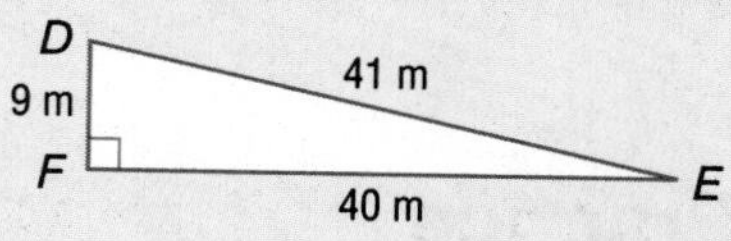


1. $\tan D = \frac{\text{length of leg opposite } \angle D}{\text{length of leg adjacent } \angle D}$

2. Substitute the opposite leg and the adjacent leg.

$\tan D = \frac{\square}{\square}$

GO ON

## Example 2

**Find the missing measure to the nearest tenth.**

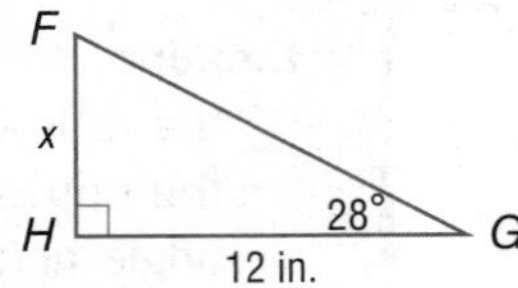


1. Substitute the values you know in the function.

$$\tan G = \frac{\text{length of leg opposite } \angle G}{\text{length of leg adjacent } \angle G}$$

$$\tan 28° = \frac{x}{12}$$

2. Solve for $x$.

$12 \cdot \tan 28° = \frac{x}{12} \cdot \frac{12}{1}$ Multiply each side by 12.

$6.4 \approx x$ Use a calculator.

3. The length of the leg is about 6.4 in.

**YOUR TURN!**

**Find the missing measure to the nearest tenth.**

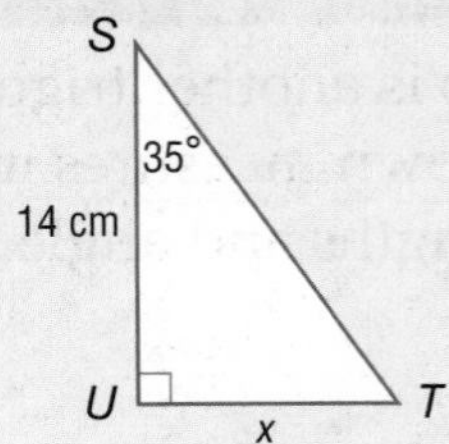


1. Substitute the values you know in the function.

$$\tan S = \frac{\text{length of leg opposite } \angle S}{\text{length of leg adjacent } \angle S}$$

$$\tan 35° = \frac{\square}{\square}$$

2. Solve for $x$.

______ $\cdot \tan 35° = \frac{\square}{\square} \cdot \frac{\square}{\square}$ Multiply each side by 14.

______ $\approx x$ Use a calculator.

3. The length of the leg is about ______.

## Guided Practice

**Find the tangent of ∠A.**

**1**

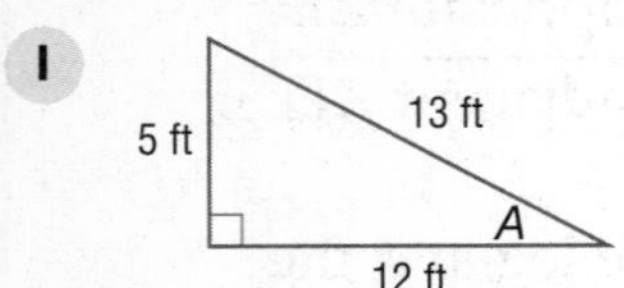


$$\tan A = \frac{\text{length of leg } \square \ \angle A}{\text{length of leg } \square \ \angle A} = \_\_\_\_\_\_$$

**2**

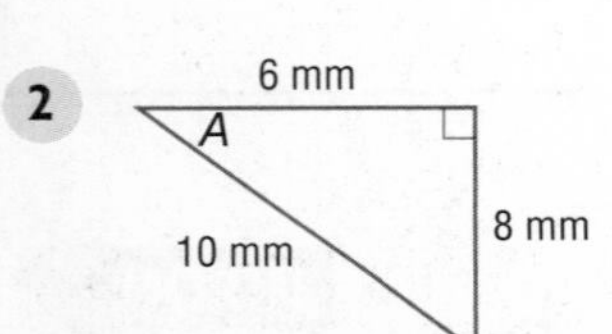


$$\tan A = \frac{\text{length of leg } \square \ \angle A}{\text{length of leg } \square \ \angle A} = \_\_\_\_\_\_$$

**Find the tangent of ∠*B*.**

**3**

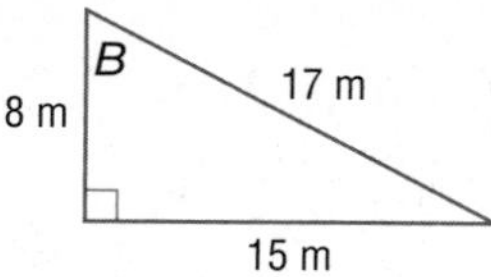


$\tan B = \dfrac{\text{opposite}}{\text{adjacent}} =$ ____________

**4**

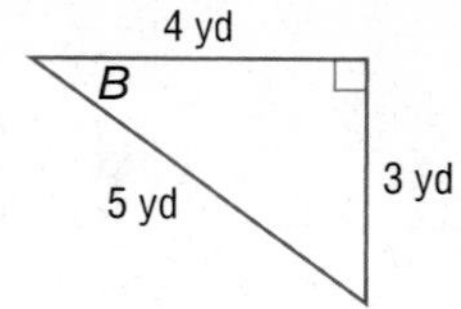


$\tan B = \dfrac{\text{opposite}}{\text{adjacent}} =$ ____________

**5**

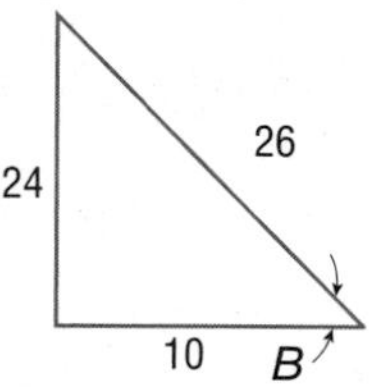


$\tan B = \dfrac{\text{opposite}}{\text{adjacent}} =$ ____________

**6**

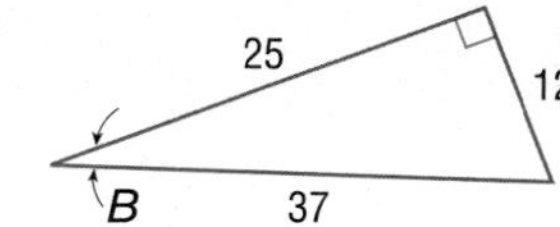


$\tan B = \dfrac{\text{opposite}}{\text{adjacent}} =$ ____________

## Step by Step Practice

**7** Find the missing measure to the nearest tenth.

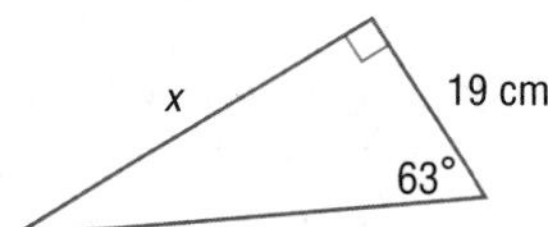


**Step 1** $\tan x = \dfrac{\text{length of leg opposite } \angle A}{\text{length of leg adjacent } \angle A}$

**Step 2** $\tan 63° = \dfrac{\square}{\square}$

**Step 3** Solve for $x$.

__________ $\cdot \tan 63° = \dfrac{\square}{\square} \cdot \dfrac{\square}{\square}$

__________ $\approx x$

**Step 4** The length of the leg is about __________.



GO ON

**Find each missing measure to the nearest tenth.**

**8** 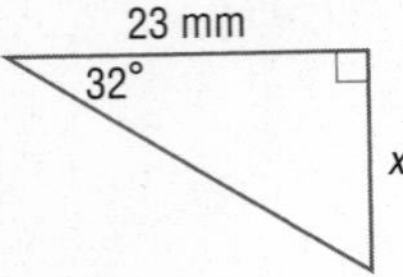


$\tan 32° = \dfrac{\square}{\square}$

______ $\cdot \tan 32° = x$

______ $\approx x$

**9** 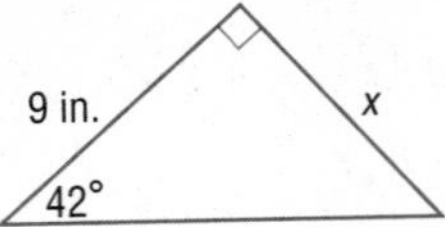


$\tan 42° = \dfrac{\square}{\square}$

______ $\cdot \tan 42° = x$

______ $\approx x$

**10** 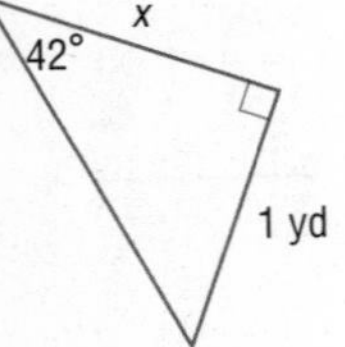


$\tan 42° = \dfrac{\square}{\square}$

______ $\cdot \tan 42° =$ ______

$x = \dfrac{\square}{\square}$

$x \approx$ ______

**11** 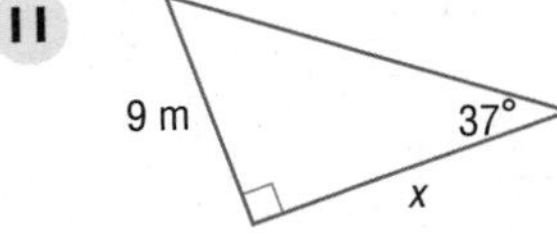


$\tan 37° = \dfrac{\square}{\square}$

______ $\cdot \tan 37° =$ ______

$x = \dfrac{\square}{\square}$

$x \approx$ ______

**12** 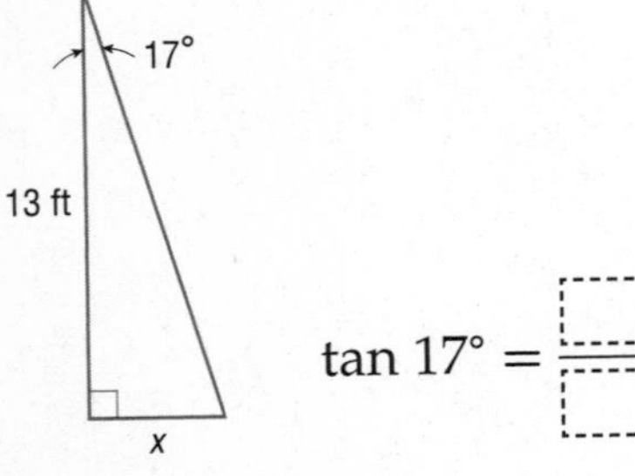


$\tan 17° = \dfrac{\square}{\square}$

______ $\cdot \tan 17° = x$

______ $\approx x$

**13** 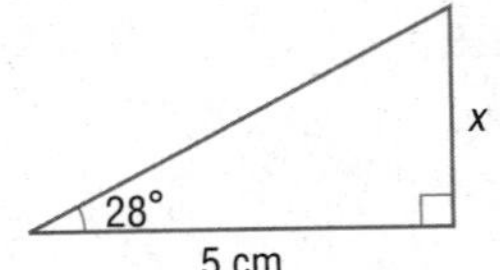


$\tan 28° = \dfrac{\square}{\square}$

______ $\cdot \tan 28° = x$

______ $\approx x$

## Step by Step Problem-Solving Practice

**Solve.**

**14** **INDIRECT MEASUREMENT** A 7 ft flag pole casts a shadow on the ground. The sun makes a 42° angle with the ground. What is the length of the shadow?

$$\tan \angle X = \frac{\text{length of leg opposite } \angle X}{\text{length of leg adjacent } \angle X}$$

Check off each step.

______ **Understand: I underlined key words.**

______ **Plan: To solve the problem, I will** ____________________.

______ **Solve: The answer is** ____________________.

______ **Check: I checked my answer by** ____________________.

## Skills, Concepts, and Problem Solving

**Find the tangent of ∠*A*.**

**15**

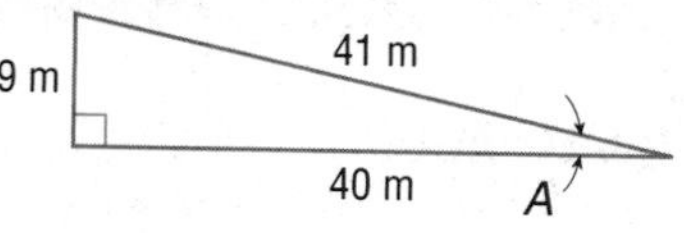


$\tan A =$ ______

**16**

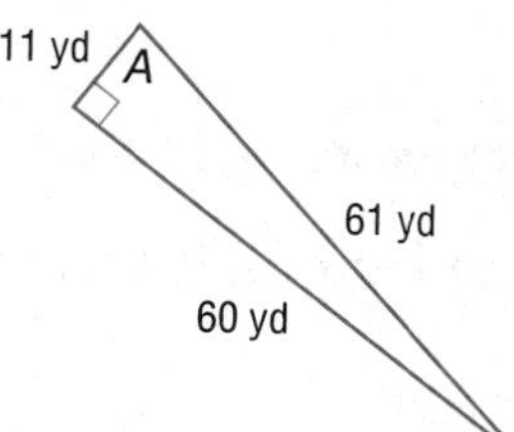


$\tan A =$ ______

**17**

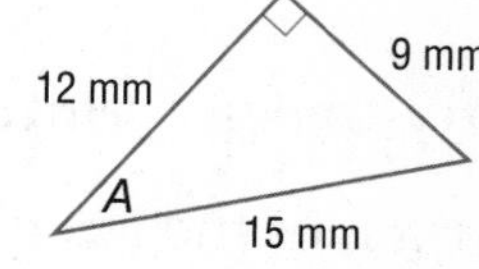


$\tan A =$ ______ $=$ ______



GO ON

**Find each missing measure to the nearest tenth.**

**18**

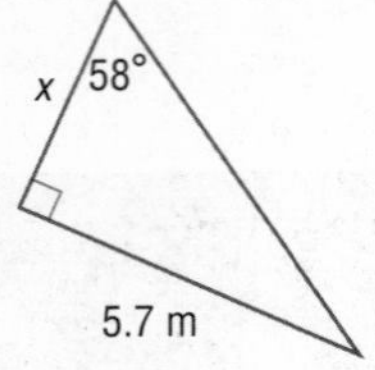


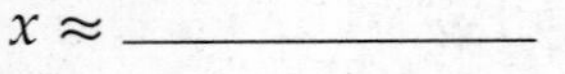

$x \approx$ ____________

**19**

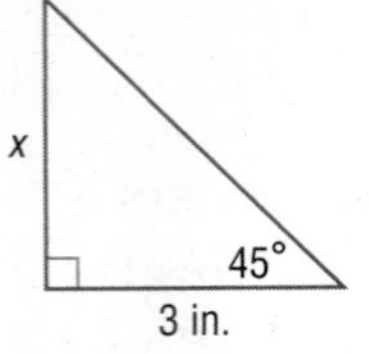


$x \approx$ ____________

**20**

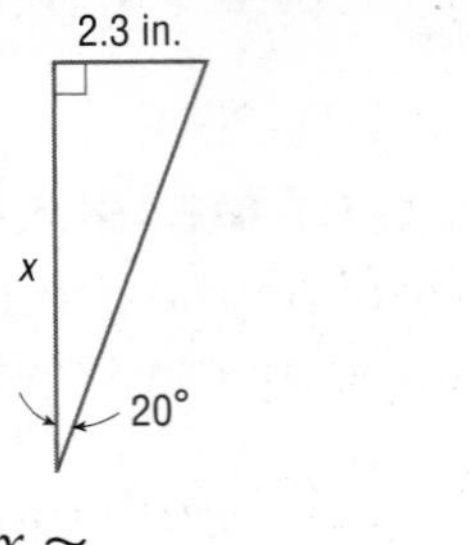


$x \approx$ ____________

**21**

x

72°

30 mm

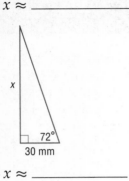

$x \approx$ ____________

**22**

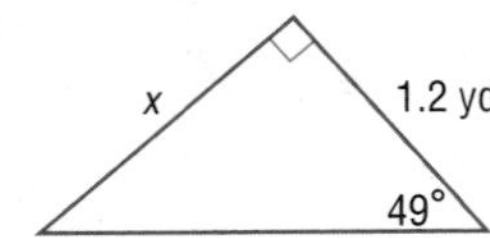


$x \approx$ ____________

**23**

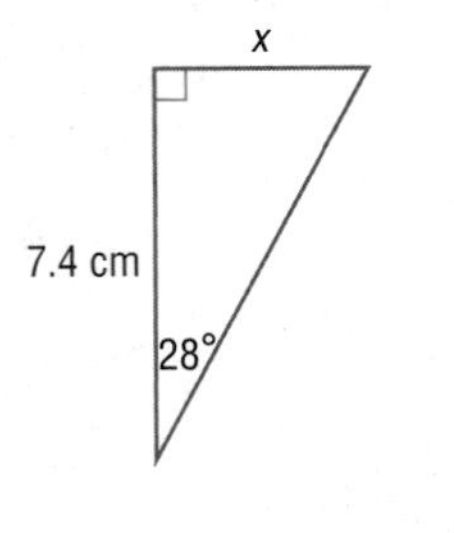


$x \approx$ ____________

**Solve.**

**24** **SKATEBOARDING** Angelo is building a skateboard ramp. It is 3.5 ft high and will rise at a 30° angle as shown. How long is the base of the ramp to the nearest tenth?

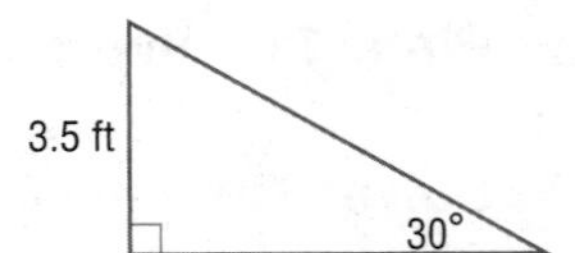


____________________________

**25** **NATURE** A birdwatcher spots a Baltimore Oriole in a tree. The line of sight of the birdwatcher is at a 48° angle with the ground. The birdwatcher is 18 ft from the base of the tree. How high in the tree is the bird?

____________________________

**Vocabulary Check** **Write the vocabulary word that completes the sentence.**

**26** The ratio of the length of the opposite leg of a given angle to the length of the adjacent leg of the same angle is the ____________ of the angle.

**27** **Reflect** The missing measurements in right triangles can be found by using the Pythagorean Theorem, the sine ratio, the cosine ratio, or the tangent ratio. What is the minimum amount of information needed to calculate a missing length?

________________________________________________

________________________________________________

STOP

Lesson 6-5

# Solve Right Triangles

## KEY Concept

To solve a right triangle means to find all of its angle measures and all of its side lengths.

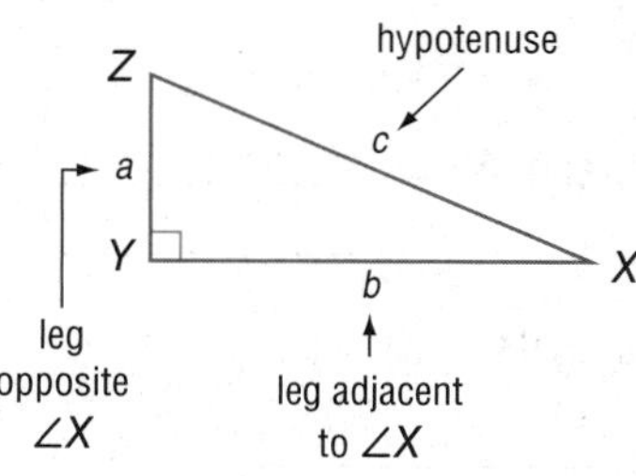


Recall that you can use trigonometric functions to find unknown measures in a right triangle.

$$\sin \angle X = \frac{\text{length of leg opposite } \angle X}{\text{length of leg hypotenuse } \angle X} = \frac{a}{c}$$

$$\cos \angle X = \frac{\text{length of leg adjacent } \angle X}{\text{length of leg hypotenuse } \angle X} = \frac{b}{c}$$

$$\tan \angle X = \frac{\text{length of leg opposite } \angle X}{\text{length of leg adjacent } \angle X} = \frac{a}{b}$$

### VOCABULARY

**cosine**
the ratio of the length of the adjacent leg of an angle and the length of the hypotenuse

**sine**
the ratio of the length of the opposite leg of an angle and the length of the hypotenuse

**tangent**
the ratio of the length of the opposite leg of an angle and the length of the adjacent leg of the same angle

Depending on the given measures, you might use all three ratios to find the angle measures and side lengths of a right triangle.



GO ON

## Example 1

**A plane is 18,000 feet above the ground. The plane begins a 3° decent to the runway. How far is the plane from the runway?**

1. Draw a triangle to model the situation.

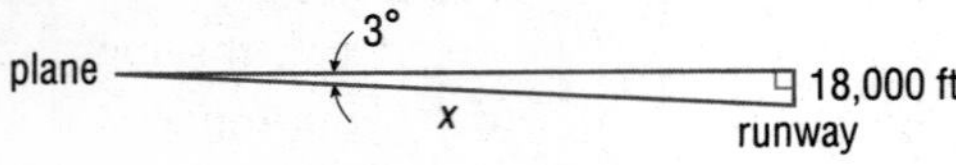


2. $\sin 3° = \frac{18,000}{x}$

Use the sine ratio because the opposite and hypotenuse are involved.

3. Solve for $x$.

$x \cdot \sin 3° = 18,000$

$x = \frac{18,000}{\sin 3°}$

$x \approx 343,932$

4. The plane is about 343,932 feet from the runway.

**YOUR TURN!**

**A plane is 300,000 feet from the runway when it begins a 4° decent. How far is the plane above the ground?**

1. Draw a triangle to model the situation.

2. $\sin 4° = \frac{\square}{\square}$

3. Solve for $x$.

______ • sin ______ = ______

______ ≈ ______

4. The plane is about ______ feet from the ground.

## Example 2

**The angle of elevation from a boat to a lighthouse is 34°. The distance from the lighthouse to the boat is 112 feet. How tall is the lighthouse?**

1. Draw a triangle to model the situation.

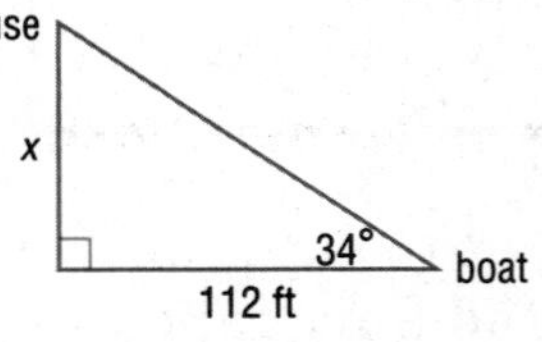


2. $\tan 34° = \frac{x}{112}$

Use the tangent ratio because the adjacent and opposite are involved.

3. Solve for $x$.

$112 \cdot \tan 34° = x$

$76 \approx x$

4. The lighthouse is about 76 ft tall.

**YOUR TURN!**

**The angle of elevation from a car to a building is 42°. The distance from the top of the building to the car is 78 ft. How far is the car from the building?**

1. Draw a triangle to model the situation.

2. $\cos 42° = \frac{\square}{\square}$

Use the cosine ratio because the hypotenuse and adjacent are given.

3. Solve for $x$.

______ • cos 42° = ______

58 ≈ ______

4. The car is about ______ feet from the building.

## Guided Practice

**Find each missing measure to the nearest tenth.**

**1**

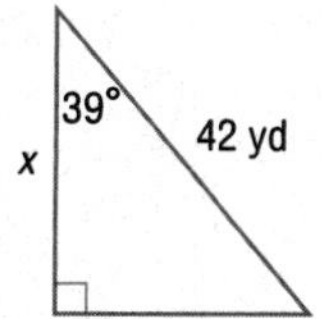


Use the ______________ ratio

because the ______________ and

the ______________ are involved.

______________ $39° = \frac{\square}{\square}$

__________ • __________ $39° =$ __________

__________ $\approx x$

**2**

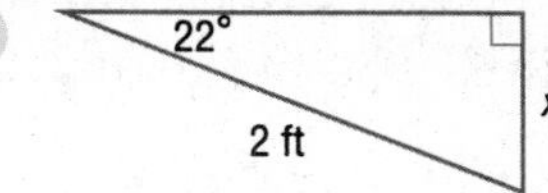


Use the ______________ ratio

because the ______________ and

the ______________ are involved.

______________ $22° = \frac{\square}{\square}$

__________ • __________ $22° =$ __________

__________ $\approx x$

**3**

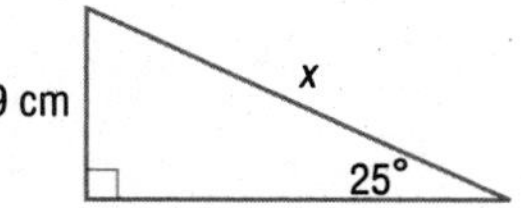


Use the ______________ ratio

because the ______________ and

the ______________ are involved.

______________ $= \frac{\square}{\square}$

$x$ • ______________ = ______________

$x = \frac{\square}{\square}$

$x \approx$ ______________

**4**

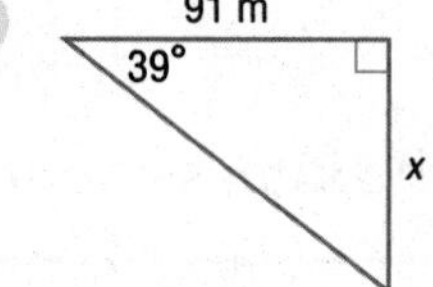


Use the ______________ ratio

because the ______________ and

the ______________ are involved.

______________ $= \frac{\square}{\square}$

______________ = ______________

______________ $\approx x$



GO ON

## Step by Step Practice

**5** **HOBBIES** The kite Dominique is flying is at a 39° angle of elevation from her hand. She has let out 55 yards of kite string. How high off the ground is the kite?

**Step 1** Draw a triangle to model the situation.

**Step 2** Use the ______________ ratio because the ______________ and the ______________ are involved.

**Step 3** ________ 39° = $\frac{\square}{\square}$

**Step 4** Solve for $x$.

________ • ________ 39° = ________

________ ≈ ________

**Step 5** The height of the kite is about ________.

**Model each situation. Find each missing measure to the nearest tenth.**

**6** From the peak of a mountain, Lynn spots an elk drinking from a creek. The angle from Lynn to the elk is 56°. If the peak Lynn is standing on is 160 meters high, how far is she from the elk?

Use the ________ ratio.

________ 56° = $\frac{\square}{\square}$

$x \approx$ ________

**7** A tree is casting a 19-foot shadow on the ground. The angle of elevation from the tip of the shadow to the top of the tree is 63°. How tall is the tree?

Use the ________ ratio.

________ 63° = $\frac{\square}{\square}$

$x \approx$ ________

## Step by Step Problem-Solving Practice

**Solve.**

**8** **SHIPPING** A ramp at a loading dock is 13 feet long. The angle of elevation of the ramp is 18°. How far is the dock above the ground?

Check off each step.

_______ **Understand: I underlined key words.**

_______ **Plan: To solve the problem, I will** _______________.

_______ **Solve: The answer is** _______________.

_______ **Check: I checked my answer by using** _______________.

## Skills, Concepts, and Problem Solving

**Model each situation. Find each missing measure to the nearest tenth.**

**9** A plane flying at 20,000 feet descends at a 2° angle to the runway. How far is the plane from the runway?

_______________

**10** A ramp has a 10° angle with the ground. If the ramp is 24 meters long, how high is the top of the ramp from the ground?

_______________

**11** A tent pole is 6 feet tall. If the angle the rope makes with the ground is 50°, how far is the support rope anchored from the pole?

_______________

**12** A diver dives at an angle of depression of 38° until she reaches a depth of 40 meters. How far has the diver swum?

_______________

**Solve. Round to the nearest tenth.**

13 **SEWING** A seamstress cuts isosceles right triangular pieces of material for a quilt. The hypotenuse of each piece is 6 inches long. What are the lengths of the sides of the triangular pieces?

_______________

14 **CARPENTRY** A carpenter frames a roof at a 32° angle with the house as shown in the diagram at the right. If the peak of the roof is in the center of the house, how high above the ground is the peak of the roof?

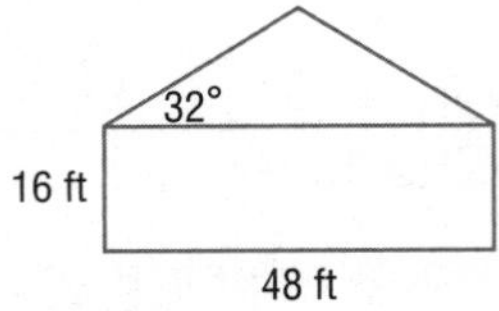


_______________

15 **NATURE** Trina is standing on a bridge and notices a boat anchored in the water. The angle of elevation from the boat to Trina is 70°. If Trina is 15 meters from the boat, how high above the water is the bridge?

_______________

**Vocabulary Check** **Write the vocabulary word that completes each sentence.**

16 Use the _______ ratio to find the ratio of the length of the opposite leg and the length of the adjacent leg in a triangle.

17 The ratio of the length of the adjacent leg of an angle to the length of the hypotenuse is the _______.

18 The _______ is the ratio of the length of the opposite leg of an angle to the length of the hypotenuse.

19 **Reflect** Carpenters often use right angles and their measurements. What other jobs involve right triangles and missing measurements? Explain.

_______________

_______________

STOP

# Progress Check 2 (Lessons 6-3, 6-4 and 6-5)

**Find the tangent of ∠A.**

**1**

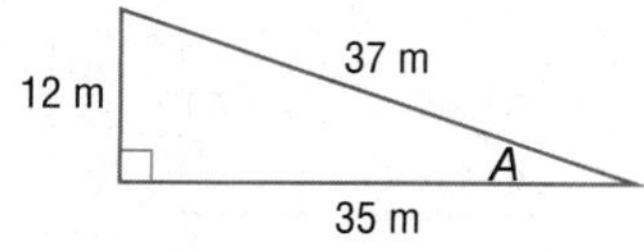


tan $A$ = ________

**2**

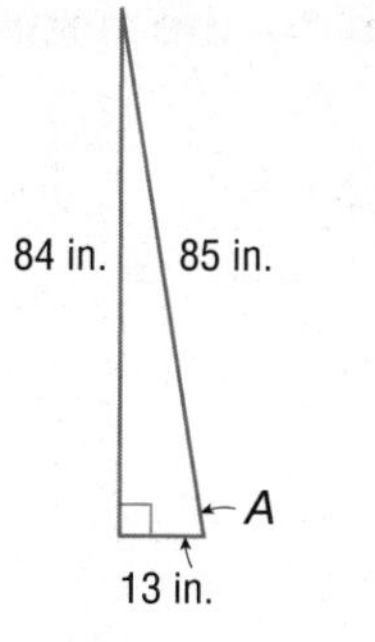


tan $A$ = ________

**3**

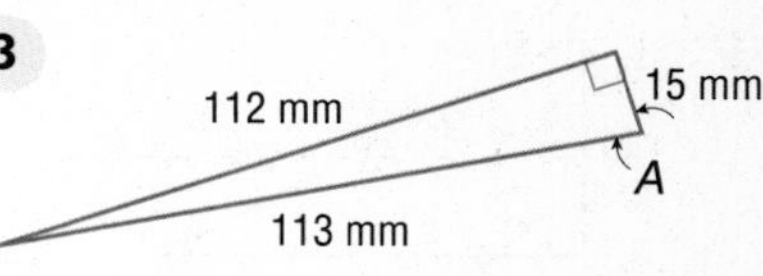


tan $A$ = ________

**Find each missing measure to the nearest tenth.**

**4**

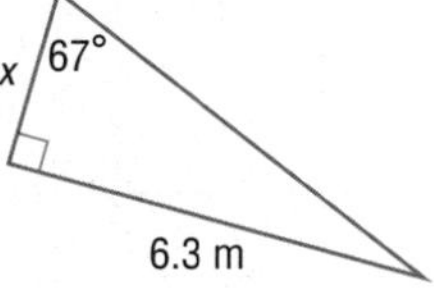


$x \approx$ ________

**5**

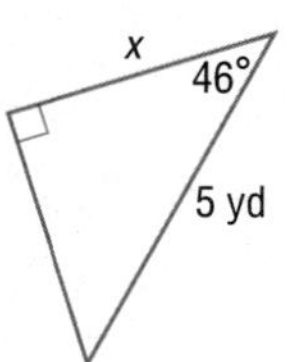


$x \approx$ ________

**6**

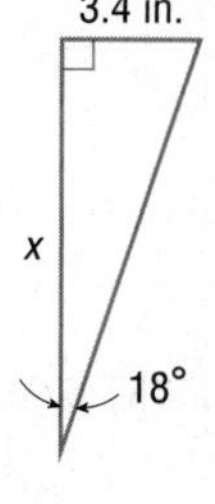


$x \approx$ ________

**Model each situation. Find each missing measure to the nearest whole number.**

**7** A plane taking off ascends at an angle of 30°. How far does it travel across land to reach an altitude of 15,000 ft?

________

**8** A skateboard ramp has a 25° angle with the ground. If the ramp is 10 ft long, how high off the ground will the skateboard be when it leaves the ramp?

________

**9** The shadow of a flag pole is 21 m long. The angle from the tip of the shadow to the tip of the flag pole is 38°. What is the height of the flag pole?

________

**10** Rohan is flying a kite. The kite has risen a vertical distance of 96 m. The angle the kite makes with the ground is 68°. What horizontal distance has the kite traveled?

________

# Chapter 6 Chapter Test

**Find the missing length in each right triangle to the nearest tenth.**

**1**
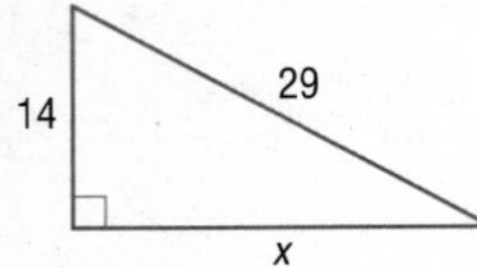


$x =$ ____________

**2**
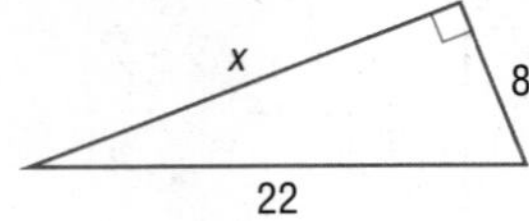


$x =$ ____________

**3**
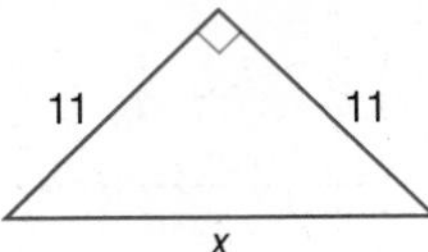


$x =$ ____________

**4**
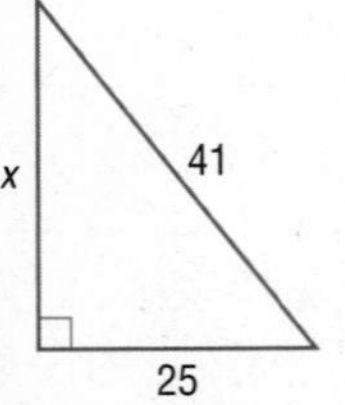


$x =$ ____________

**5**
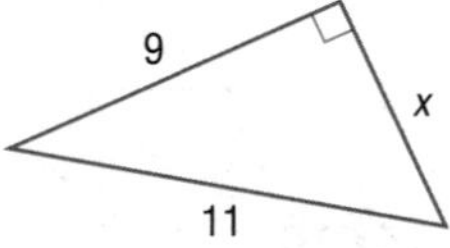


$x =$ ____________

**6**
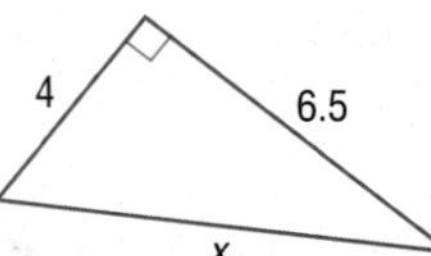


$x =$ ____________

**Find the sine of ∠A, cosine of ∠A, or tangent of ∠A.**

**7**
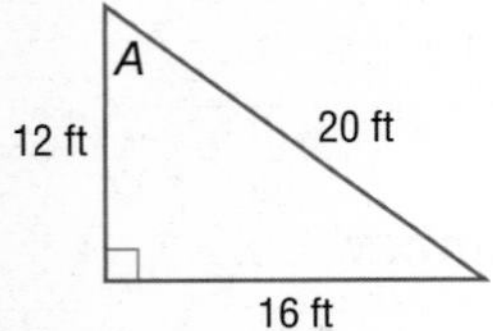


$\sin A =$ ____________

**8**
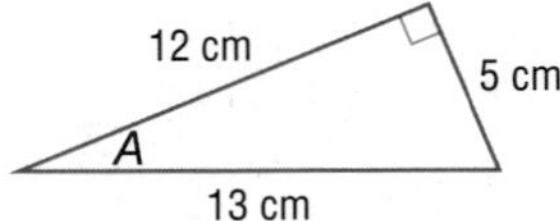


$\cos A =$ ____________

**9**
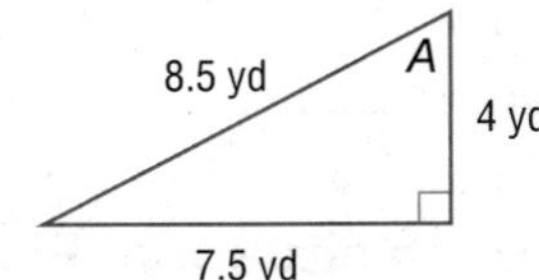


$\tan A =$ ____________

**Find each missing measure to the nearest tenth.**

**10**
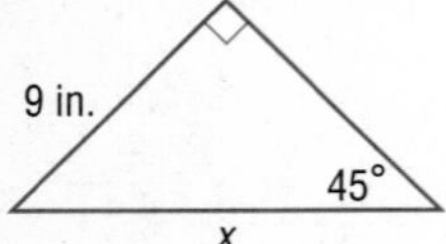


$x \approx$ ____________

**11**
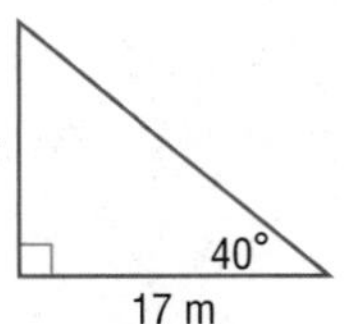


$x \approx$ ____________

**12**
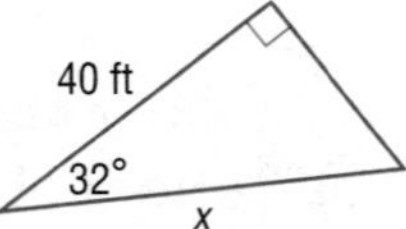


$x \approx$ ____________

**13**

30 m

22°

x

$x \approx$ ____________

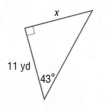

**14**
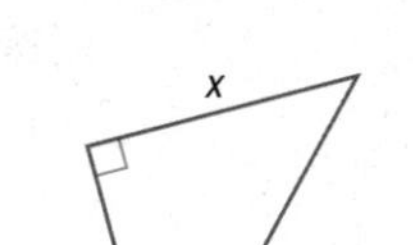


$x \approx$ ____________

**15**
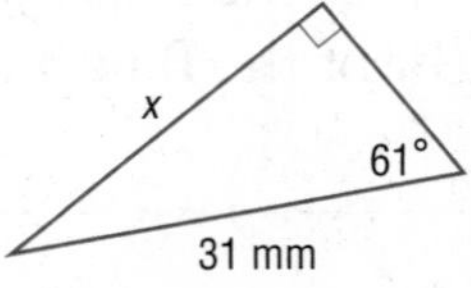


$x \approx$ ____________

**Model each situation. Find each missing measure to the whole number.**

**16** A plane travels north-northeast (45°) 70 miles to an island to deliver cargo. It then travels due west to make a second delivery. How far is the plane from its original departure?

_______________

**17** A motorcycle ramp has an angle of 23°. The top of the ramp must be 115 ft high in order for the jumper to cross the creek. How long must the ramp be?

_______________

**18** A ski lift must run parallel to the slope of the mountain. The angle of inclination for the ski lift is 47° and the distance from the base of the mountain to the peak is 9,500 ft. What is the height of the mountain?

_______________

**19** A zip line is strung from a tree at a 50° angle with the ground. The landing spot is 200 yds from the base of the tree. At what height will a person start their descent from the zip line?

_______________

**Solve. Round the answer to the nearest tenth.**

**20** **SLEDDING** Alonso is at the top (90 ft) of the local sledding hill. If Alonso's angle of descent is 72°, how far is it from to the top of the hill to the bottom of the hill?

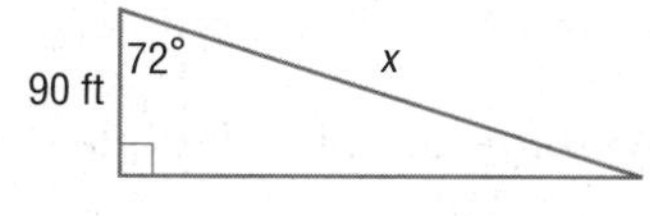

_______________________________

**Correct the mistake.**

**21** Aaron cut a square piece of wood at a 30° angle to form a triangle. The length of the cut side is 8 inches. He hopes that the side opposite the 30° angle is 5 inches long. Is he correct? If not, how long is the opposite side?

_______________________________

# Index

H

I

K

L

M

N

O

P

Q

## T



## U

## V

## Y

## Z